Lecture Notes in Physics

For information about Vols. 1–44, please contact your bookseller or Springer-Verlag.

Vol. 45: Dynamical Concepts on Scaling Violation and the New Resonances in e^+e^- Annihilation. Edited by B. Humpert. VII, 248 pages. 1976.

Vol. 46: E. J. Flaherty, Hermitian and Kählerian Geometry in Relativity. VIII, 365 pages. 1976.

Vol. 47: Padé Approximants Method and Its Applications to Mechanics. Edited by H. Cabannes. XV, 267 pages. 1976.

Vol. 48: Interplanetary Dust and Zodiacal Light. Proceedings 1975. Edited by H. Elsässer and H. Fechtig. XII, 496 pages. 1976.

Vol. 49: W. G. Harter and C. W. Patterson, A Unitary Calculus for Electronic Orbitals. XII, 144 pages. 1976.

Vol. 50: Group Theoretical Methods in Physics. 4th International Colloquium. Nijmegen 1975. Edited by A. Janner, T. Janssen, and M. Boon. XIII, 629 pages. 1976.

Vol. 51: W. Nörenberg and H. A. Weidenmüller. Introduction of the Theory of Heavy-Ion Collisions. 2nd enlarged edition. IX, 334 pages. 1980.

Vol. 52: M. Mladjenović, Development of Magnetic β-Ray Spectroscopy. X, 282 pages. 1976.

Vol. 53: D. J. Simms and N. M. J. Woodhouse, Lectures on Geometric Quantization. V, 166 pages. 1976.

Vol. 54: Critical Phenomena. Sitges International School on Statistical Mechanics, June 1976. Edited by J. Brey and R. B. Jones. XI, 383 pages. 1976.

Vol. 55: Nuclear Optical Model Potential. Proceedings 1976. Edited by S. Boffi and G. Passatore. VI, 221 pages. 1976.

Vol. 56: Current Induced Reactions. International Summer Institute, Hamburg 1975. Edited by J. G. Körner, G. Kramer, and D. Schildknecht. V, 553 pages. 1976.

Vol. 57: Physics of Highly Excited States in Solids. Proceedings 1975. Edited by M. Ueta and Y. Nishina. IX, 391 pages. 1976.

Vol. 58: Computing Methods in Applied Sciences. Proceedings 1975. Edited by R. Glowinski and J. L. Lions. VIII, 593 pages. 1976.

Vol. 59: Proceedings of the Fifth International Conference on Numerical Methods in Fluid Dynamics. 1976. Edited by A. I. van de Vooren and P. J. Zandbergen. VII, 459 pages. 1976.

Vol. 60: C. Gruber, A. Hintermann, and D. Merlini, Group Analysis of Classical Lattice Systems. XIV, 326 pages. 1977.

Vol. 61: International School on Electro and Photonuclear Reactions I. Edited by C. Schaerf. VIII, 650 pages. 1977.

Vol. 62: International School on Electro and Photonuclear Reactions II. Edited by C. Schaerf. VIII, 301 pages. 1977.

Vol. 63: V. K. Dobrev et al., Harmonic Analysis on the n-Dimensional Lorentz Group and Its Application to Conformal Quantum Field Theory. X, 280 pages. 1977.

Vol. 64: Waves on Water of Variable Depth. Edited by D. G. Provis and R. Radok. 231 pages. 1977.

Vol. 65: Organic Conductors and Semiconductors. Proceedings 1976. Edited by L. Pál, G. Grüner, A. Jánossy and J. Sólyom. 654 pages. 1977.

Vol. 66: A. H. Völkel, Fields, Particles and Currents. VI, 354 pages. 1977.

Vol. 67: W. Drechsler and M. E. Mayer, Fiber Bundle Techniques in Gauge Theories. X, 248 pages. 1977.

Vol. 68: Y. V. Venkatesh, Energy Methods in Time-Varying System Stability and Instability Analyses. XII, 256 pages. 1977.

Vol. 69: K. Rohlfs, Lectures on Density Wave Theory. VI, 184 pages. 1977.

Vol. 70: Wave Propagation and Underwater Acoustics. Edited by J. Keller and J. Papadakis. VIII. 287 pages. 1977.

Vol. 71: Problems of Stellar Convection. Proceedings 1976. Edited by E. A. Spiegel and J. P. Zahn. VIII, 363 pages. 1977.

Vol. 72: Les instabilités hydrodynamiques en convection libre forcée et mixte. Edité par J. C. Legros et J. K. Platten. X, 202 pages. 1978.

Vol. 73: Invariant Wave Equations. Proceedings 1977. Edited by G. Velo and A. S. Wightman. VI, 416 pages. 1978.

Vol. 74: P. Collet and J.-P. Eckmann, A Renormalization Group Analysis of the Hierarchical Model in Statistical Mechanics. IV, 199 pages. 1978.

Vol. 75: Structure and Mechanisms of Turbulence I. Proceedings 1977. Edited by H. Fiedler. XX, 295 pages. 1978.

Vol. 76: Structure and Mechanisms of Turbulence II. Proceedings 1977. Edited by H. Fiedler. XX, 406 pages. 1978.

Vol. 77: Topics in Quantum Field Theory and Gauge Theories. Proceedings, Salamanca 1977. Edited by J. A. de Azcárraga. X, 378 pages 1978.

Vol. 78: Böhm, The Rigged Hilbert Space and Quantum Mechanics. IX, 70 pages. 1978.

Vol. 79: Group Theoretical Methods in Physics. Proceedings, 1977. Edited by P. Kramer and A. Rieckers. XVIII, 546 pages. 1978.

Vol. 80: Mathematical Problems in Theoretical Physics. Proceedings, 1977. Edited by G. Dell'Antonio, S. Doplicher and G. Jona-Lasinio. VI, 438 pages. 1978.

Vol. 81: MacGregor, The Nature of the Elementary Particle. XXII, 482 pages. 1978.

Vol. 82: Few Body Systems and Nuclear Forces I. Proceedings, 1978. Edited by H. Zingl, M. Haftel and H. Zankel. XIX, 442 pages. 1978.

Vol. 83: Experimental Methods in Heavy Ion Physics. Edited by K. Bethge. V, 251 pages. 1978.

Vol. 84: Stochastic Processes in Nonequilibrium Systems, Proceedings, 1978. Edited by L. Garrido, P. Seglar and P. J. Shepherd. XI, 355 pages. 1978

Vol. 85: Applied Inverse Problems. Edited by P. C. Sabatier. V, 425 pages. 1978.

Vol. 86: Few Body Systems and Electromagnetic Interaction. Proceedings 1978. Edited by C. Ciofi degli Atti and E. De Sanctis. VI, 352 pages. 1978.

Vol. 87: Few Body Systems and Nuclear Forces II, Proceedings, 1978. Edited by H. Zingl, M. Haftel, and H. Zankel. X, 545 pages. 1978.

Vol. 88: K. Hutter and A. A. F. van de Ven, Field Matter Interactions in Thermoelastic Solids. VIII, 231 pages. 1978.

Vol. 89: Microscopic Optical Potentials, Proceedings, 1978. Edited by H. V. von Geramb. XI, 481 pages. 1979.

Vol. 90: Sixth International Conference on Numerical Methods in Fluid Dynamics. Proceedings, 1978. Edited by H. Cabannes, M. Holt and V. Rusanov. VIII, 620 pages. 1979.

Lecture Notes in Physics

Edited by J. Ehlers, München, K. Hepp, Zürich
R. Kippenhahn, München, H. A. Weidenmüller, Heidelberg
and J. Zittartz, Köln

Managing Editor: W. Beiglböck, Heidelberg

119

Nuclear Spectroscopy

Lecture Notes of the Workshop
Held at Gull Lake, Michigan
August 27 – September 7, 1979

Edited by G. F. Bertsch and D. Kurath

Springer-Verlag
Berlin Heidelberg GmbH 1980

Editors

George F. Bertsch
Michigan State University
East Lansing, MI 48824
USA

Dieter Kurath
Physics Division
Argonne National Laboratory
Argonne, IL 60439
USA

ISBN 978-3-540-09970-3 ISBN 978-3-540-39193-7 (eBook)
DOI 10.1007/978-3-540-39193-7

Originally published by Springer-Verlag Berlin Heidelberg New York in 1980

2153/3140-543210

PREFACE

The present status of nuclear structure physics was reviewed at a workshop held at Gull Lake, Michigan, August 27 - September 7, 1979. The main themes of current research were presented in the lectures included in the present volume.

The major tool of nuclear theory, the shell model representation, is described in the lectures of Kurath, Bertsch and Faessler. The application of group theory to simplify the description of complex spectra has developed into a high art, and is described in the lectures of Iachello. For spectral regions beyond the grasp of either the shell model or group theory, statistical considerations have an applicability that is only recently fully understood. This is described in the lectures of French. The foundations of nuclear theory in the nonrelativistic reduction of nucleon-nucleon dynamics to derive an effective Hamiltonian, is described in the lectures of Brown. The overall picture of nuclear theory is thus fairly complete, but numerous puzzles and loose ends remain. Some of these are included in an appendix to this volume by Zamick and Mekjian.

Many institutions and individuals helped make the workshop possible. We thank particularly the National Science Foundation for a grant supporting the lectures and graduate student participants, Argonne National Laboratory for supporting graduate students, and George Lauff of the Kellogg Biological Station for making the facility available as a summer school.

We finally wish to acknowledge the editorial assistance and typing by Shari Conroy for the manuscript of this volume.

G. Bertsch
D. Kurath

TABLE OF CONTENTS

I. The Nucleon-Nucleon Interaction and the Nuclear Many-Body Problem (G.E. BROWN)

1. Introduction 1
2. The Nucleon-Nucleon Interaction 1

2.1 The Long-Range Interaction 2
2.2 The Intermediate-Range Attraction 5
2.3 The Short-Range Interaction 7

3. Nuclear Matter as a Fermi Liquid 12
4. The Brueckner-Bethe Theory 14
5. A Theory of Interacting Quasiparticles and Collective Excitations 22
6. Making Contact with Nature 32

6.1 Compressibility 33
6.2 Effective Mass 33
6.3 Symmetry Energy 37
6.4 The Spin-Dependent Interactions 37
6.5 The Landau Sum Rule 42

II. The Nuclear Shell Model (D. KURATH)

1. Introduction 45
2. Basic Language of the Shell Model 48

2.1 Occupation Number Representation 48
2.2 Angular Momentum in the Spherical Basis 50
2.3 Spectroscopic Factors 51

3. Current Shell Model Programs 55

3.1 Standard Programs 55
3.2 The SU_3 Representation 58
3.3 The Glasgow Code 59

4. One Body Transition Density 60

4.1 General Formulation for One Body Operators 60
4.2 Varying Representations to Further Understanding . . 62
4.3 Stretched States in Inelastic Pion Scattering 65

III. Nuclear Vibrations (G. BERTSCH)

1. Introduction 69
2. Operators and Matrix Elements 69

2.1 Density Operator 69
2.2 Spin Density 73

3. Sum Rules 76
4. RPA 79

4.1 RPA in Coordinate Space 79
4.1.1 Reduction of RPA to Quadrature 82
4.2 Configuration Space Representations 83
4.3 Landau Theory 84
4.4 Green's Functions and the Density Responses 86
4.5 Separable Interactions and Self-Consistency 88

5. Applications 89

5.1 RPA Frequencies and Transition Densities 89
5.2 Inelastic Scattering of Nucleons and Composite Projectiles 91
5.3 The Decay of the Giant Vibrations 92

IV. Collective Description of Deformed and Transitional Nuclei (A. FAESSLER)

1. Introduction 97
2. Backbending in Deformed Nuclei 97

2.1 Formulation of the Theory 104
2.2 First Backbending in Even Mass Nuclei 106
2.3 The Second Backbending 109

3. What is the Nature of the Yrast Traps? 112

3.1 Introduction 112
3.2 Excitation and Deexcitation of High Spins in Heavy Ion Fusion Reactions 113
3.3 Yrast Traps 119

4. Description of Transitional Nuclei 122

4.1 Introduction 122
4.2 Odd-Odd Mass Nuclei 125
4.3 Even-Mass Nuclei with Zero and Two Quasi-Particle Excitations 131
4.4 Summary 135
4.5 Further Trends 136

V. Group Theory and Nuclear Spectroscopy (F. IACHELLO)

1. Introduction 140
2.1 Groups, Definitions and Examples (Lecture 1) 140

2.2 Lie Algebras 142
2.3 Isomorphism and Homomorphism 143
2.4 Cartan Classification of Lie Groups 146

3.1 Tensor Representations of the Lie Groups A_n, B_n, C_n and D_n (Lecture 2) 146

3.2 The Classification Problem. Group Chains 148
3.3 Dynamical Symmetries. Eigenvalues of the Casimir Operators 151

4.1 Group Theory of the Interacting Boson Model (Lecture 3). . 157

4.2 Algebras and Subalgebras 158
4.3 Classification of States 162

5.1 Dynamical Symmetries. Solution of the Eigenvalue Problem (Lecture 4). 166

5.2 Examples of Spectra with Dynamical Symmetries 170
5.3 Selection Rules. Matrix Elements of Operators . . . 172
5.4 Group Lattices. Broken Symmetries and Phase Transitions 174

6. Conclusions 178

VI. Statistical Spectroscopy (J.B. FRENCH)

1. Introduction and Preview 180
2. Some Simple Eigenvalue Distributions 185
3. Elementary Statistical Methods 196
4. Expectation Values, Transition Strengths and Statistical Response 203
5. Trace Evaluation; Information and Its Propagation 214
6. Fluctuations 221

6.1 Coin-Tossing Experiments 225
6.2 Two-Point Function and Some Elementary Results . . . 228

VI. Statistical Spectroscopy (continued)

7. Comments About Applications 235

7.1 Level Densities, Partition Functions, Low-Lying Spectra and Nuclear Masses 235
7.2 Expectation Values, Sum Rules and Strength Distributions 235
7.3 Symmetries 236
7.4 Fluctuations 236

Appendix - Nuclear Structure Puzzles (G. BERTSCH, L. ZAMICK and A. MEKJIAN)

7.1 The Coulomb Energy Problem 240
7.2 Spin-Orbit Interaction 242
7.3 Coriolis Interaction 243
7.4 Missing Ml Strength in Heavy Nuclei 245
7.5 Nuclear Level Densities 246
7.6 Spin Modes 248

Chapter I

THE NUCLEON-NUCLEON INTERACTION AND THE NUCLEAR MANY-BODY PROBLEM

G. E. Brown
Nordita
Copenhagen Ø

1. Introduction

In this review we wish to relate effective forces in nuclear matter and in nuclei back to the interaction between two isolated nucleons. Low-brow meson theory is used to derive the nucleon-nucleon interaction, with dispersion theoretical calculations as a guide, and a certain amount of phenomenology to pin down the parameters. A chiral picture is in the back of our mind, but since the main approach here is a semi-phenomenological one, chiral invariance is not developed in detail.

The nucleon-nucleon interaction in free space is used as a starting point to discuss effective forces in nuclear matter and in nuclei. This discussion is carried out within the framework of the Landau Fermi liquid theory, initially along the lines begun by MIGDAL [1], although our philosophy is somewhat different in that we wish to explicitly calculate, as far as possible, Fermi liquid parameters from the nucleon-nucleon interaction. We use certain crucial nuclear phenomena as a guide in this calculation, which cannot really be carried out from first principles, but must be continuously monitored by comparison of calculated results with empirical data.

Information that we gain about effective nuclear interactions makes it possible for us to discuss the problem of the binding energy of nuclear matter, although this latter quantity is not easily calculated within the framework of Fermi liquid theory.

We show, briefly, that the problem of pion-nucleus interactions can be discussed within the same framework, interaction constants being related to those in the nuclear problem through the assumption of a constituent quark structure of nucleons.

Finally, the problem of nuclear matter and neutron matter at high densities - the dense matter problem - is discussed, and implications of the parameters and mechanisms understood in the nuclear and nuclear-matter problems are indicated.

2. The Nucleon-Nucleon Interaction

The nucleon-nucleon interaction is understood, in considerable detail, as arising from the exchange of various mesons. The longest-range part results from the exchange of the lightest possible exchanged particle, the π-meson. The intermediate-range part of the interaction can be

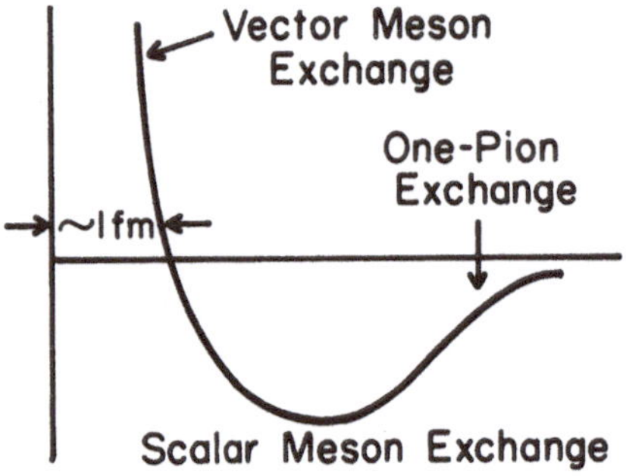

Fig. 1 Schematization of the nucleon-nucleon potential

understood as arising from the exchange of scalar mesons with various masses, really systems of two pions coupled to S = 0, T = 0, as we shall discuss in more detail later.

It is probably somewhat old-fashioned to discuss the short-range repulsion between nucleons as arising from meson exchange, since the quarkish constitution of the nucleons must manifest itself at short distances. However, the short-range repulsion has effects out to large distances $\gtrsim$ 1 fm, where it is reasonable to talk about meson exchange. Furthermore, not only qualitatively, but even semiquantitatively, preliminary results from quark calculations (within the framework of the bag model) [2] seem to provide a picture of short-range repulsions similar to that from vector-meson exchange. Speaking in a phenomenological way, one needs the wave function of relative motion of the two nucleons to become vanishingly small at short distances. This may be accomplished in many ways; as pointed out by FESHBACH [3], almost any picture involving many possible channels for the nucleon-nucleon system at short distances will leave rather little of the wave function in the incident channel; from the viewpoint of this channel, the situation looks then as if a strong short-range repulsion is operating.

We now discuss the various components of the nucleon-nucleon interaction according to their range.

2.1 The Long-Range Interaction

The longest-range part of the nucleon-nucleon interaction is mediated by the exchange of the lightest particle coupling strongly to the nucleons, the π-meson. The one-pion exchange potential can be obtained by standard procedures [4]. In momentum space it is

$$V_{OPEP}(k) = -\frac{f^2}{m_\pi^2}(\vec{\tau}_1\cdot\vec{\tau}_2)\frac{(\underset{\sim}{\sigma}_1\cdot\underset{\sim}{k})(\underset{\sim}{\sigma}_2\cdot\underset{\sim}{k})}{k^2+m_\pi^2} \tag{1-1}$$

$$= -\frac{f^2}{m_\pi^2}(\vec{\tau}_1\cdot\vec{\tau}_2)\left\{\left[\frac{(\underset{\sim}{\sigma}_1\cdot\underset{\sim}{k})(\underset{\sim}{\sigma}_2\cdot\underset{\sim}{k})-\frac{1}{3}(\underset{\sim}{\sigma}_1\cdot\underset{\sim}{\sigma}_2)k^2}{k^2+m_\pi^2}\right]\right.$$

$$\left.+\frac{1}{3}\underset{\sim}{\sigma}_1\cdot\underset{\sim}{\sigma}_2-\frac{1}{3}\frac{m_\pi^2\ \underset{\sim}{\sigma}_1\cdot\underset{\sim}{\sigma}_2}{k^2+m_\pi^2}\right\}, \qquad (1\text{-}2)$$

where in the second step we have broken V(k) up into tensor force, spin-spin δ-function and spin-spin Yukawa. Here $f^2/4\pi = .081$. This interaction can be transformed into configuration space, giving[1]:

$$V_\pi(r) = \frac{f^2}{4\pi\ m_\pi^2}(\vec{\tau}_1\cdot\vec{\tau}_2)(\underset{\sim}{\sigma}_1\cdot\underset{\sim}{\nabla})(\underset{\sim}{\sigma}_2\cdot\underset{\sim}{\nabla})\frac{e^{-m_\pi r}}{r}$$

$$= \frac{f^2}{4\pi}m_\pi(\vec{\tau}_1\cdot\vec{\tau}_2)\left\{S_{12}\left(\frac{1}{(m_\pi r)^3}+\frac{1}{(m_\pi r)^2}+\frac{1}{3\ m_\pi r}\right)e^{-m_\pi r}\right. \qquad (1\text{-}3)$$

$$\left.-\frac{4\pi(\underset{\sim}{\sigma}_1\cdot\underset{\sim}{\sigma}_2)}{3\ m_\pi^3}\delta(r)+\left(\frac{\underset{\sim}{\sigma}_1\cdot\underset{\sim}{\sigma}_2}{3}\right)\frac{e^{-m_\pi r}}{m_\pi r}\right\}$$

where r is the interparticle distance, $r = |\underset{\sim}{r}_1 - \underset{\sim}{r}_2|$, and

$$S_{12} = 3\frac{(\underset{\sim}{\sigma}_1\cdot\underset{\sim}{r})(\underset{\sim}{\sigma}_2\cdot\underset{\sim}{r})}{r^2}-(\underset{\sim}{\sigma}_1\cdot\underset{\sim}{\sigma}_2)\ . \qquad (1\text{-}4)$$

As can be seen from the above, since the coupling of the pion to nucleons is spin- and isospin-dependent, averages over either spin or isospin will make the expectation value zero in either spin-saturated or charge-conjugate nuclei or nuclear matter, averages involving equal numbers of spin-up and spin-down nucleons. Some contributions from exchange terms will be left, but these are not large. Thus, the lowest-order contributions from the tensor interaction to the binding energies of nuclei or nuclear matter are not large.

We know, however, that second-order effects of the tensor force are large. The main difference between the bound deuteron system, which has $S = 1$, and the unbound proton-proton or neutron-neutron $S = 0$ systems is that the tensor force is operative in the former. Since the tensor interaction introduces mainly intermediate states of high momentum and

---

[1] We employ relativistic units $\hbar = c = 1$.

energy when used in perturbation theory, we can choose some effective energy $\bar{E}$ for the energy of a typical intermediate state [5] and represent the higher-order effects of the tensor interaction by

$$V_{eff}(r) = -\frac{(V_{Tensor})^2}{\bar{E}} \tag{1-5}$$

$$= -\frac{1}{\bar{E}}(3 - 2\vec{\tau}_1 \cdot \vec{\tau}_2)[6 + 2\sigma_1 \cdot \sigma_2 - 2S_{12}]V_{t\pi}^2(r)$$

where $V_t(r)$ is the radial part of the pion-exchange tensor force. In the $S = 1$, $T = 0$ state,

$$V_{eff}(r) \cong -\frac{72}{E} V_{t\pi}^2(r) \tag{1-6}$$

where we drop now the S_{12} term. The factor of 72 compensates for the smallness of $(f^2/4\pi)$, so that $V_{eff}(r)$ is

$$V_{eff}(r) \cong -\frac{0.48}{\bar{E}}(m_\pi)^2 \left[\frac{1}{(m_\pi r)^3} + \frac{1}{(m_\pi r)^2} + \frac{1}{3(m_\pi r)}\right]^2 e^{-2m_\pi r} \tag{1-7}$$

Taking $m_\pi/\bar{E} \cong 1/3$ [6] and evaluating $V_{eff}(r)$ at $r = m_\pi^{-1}$ to get a rough idea of its size, we find

$$V_{eff}(\hbar/m_\pi c) \cong 0.12\ m_\pi c^2 = 17\ \text{MeV} \tag{1-8}$$

where we have here put in an $\hbar$ and c to make clear that m_π^{-1} is a length. If we estimate the strengths of Yukawa potentials of range m_π^{-1} appropriate for the ^1S and ^3S states by converting potentials which give the correct scattering ranges and effective lengths [4] to this range using

Depth x (Range) = Const.,

we find a depth of ~ 33 MeV for the ^1S potential, ~ 60 MeV for the ^3S potential[2]. Had we used a range somewhat smaller than m_π^{-1} for the effective potential, which would appear more appropriate looking at the effective potential (1-7), then we would have found that V_{eff} would easily explain the difference between triplet and singlet interactions.

We cannot take the expression (1-7), which would imply a range of $(2m_\pi)^{-1}$ for V_{eff}, literally, since the pion-exchange potential will be

---

[2]The "rule of thumb", is that in shell-model calculations the ^3S interaction should be chosen ~ (10/6) times the ^1S one, as in the Rosenfeld interaction.

modified by other interactions, especially by the tensor part of the ρ-exchange potential, at short distances.

2.2 The Intermediate-Range Attraction

The intermediate-range attraction in the nucleon-nucleon interaction is now understood as coming from the exchange of correlated pairs of pions which are in a $J=0$, $T=0$ state. This potential looks like

$$V_\sigma(r) = -\frac{1}{2\pi}\int \rho(t)\,\frac{e^{-\sqrt{t}\,r}}{r}\,d\,t \tag{1-9}$$

where $\rho(t)$ is a weighting function governing the amount of exchange of each $J=0$, $T=0$ system of mass $\sqrt{t}$. Given the amplitudes for pion-nucleon scattering, this amplitude is straightforward to compute by dispersion relations [8], although complex technology is involved since the iterated one-pion-exchange potential must be removed.

Most of the contribution to $V_\sigma(r)$ comes from processes involving virtual isobars, as shown in Fig. 2.

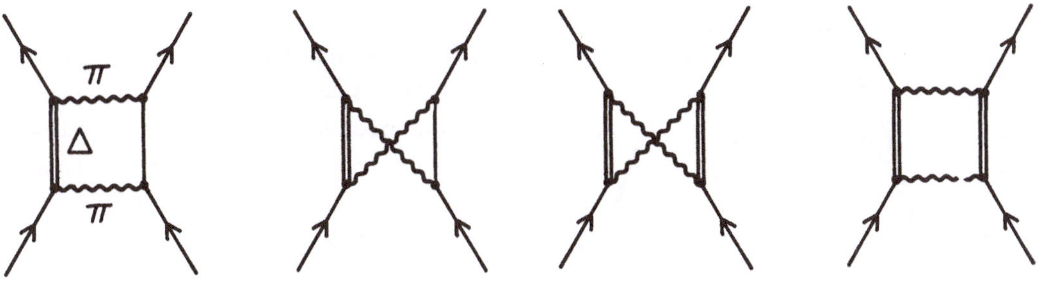

Fig. 2 Contributions to the intermediate-range potential $V_\sigma(r)$

Contributions to the isospin-dependence from crossed and uncrossed pion exchange cancel to a good accuracy [9], so that, practically speaking, the processes of Fig. 2 give an interaction of the form, (1-9), which is independent of spin and isospin. Such an interaction is of the type envisaged in the σ-model [10] as arising from the exchange of the iso-scalar $S=0$, σ-meson; therefore we label the interaction V, but the "σ" has nothing to do with spin.

Indeed, allowing for rescattering of the π-mesons, which is accomplished in the σ-model through their coupling to the σ-meson, one would more realistically have processes like those shown in Fig. 3. Even though the input σ mass in the σ-model may be large, $\gtrsim$ 1 GeV, processes such as those shown in Fig. 3 where it couples to pions will lower considerably the effective mass with which it enters into the nucleon-nucleon interaction. Although the σ-meson may be heavy, and cannot propagate far (only a distance $\sim m_\sigma^{-1}$), the two pions may form a system

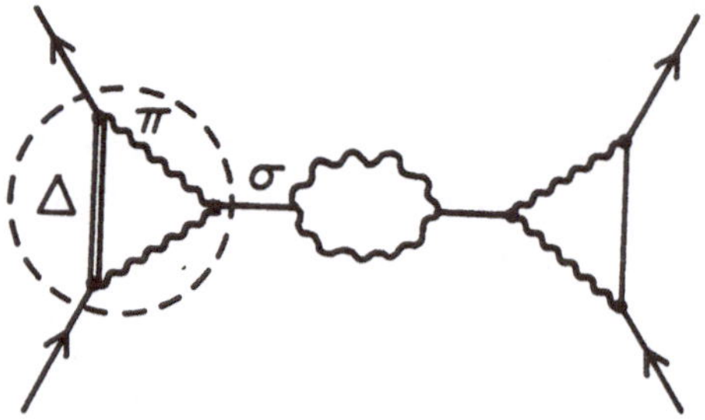

Fig. 3 The process, Fig. 2a, extended to include pion rescattering. The pion scattering is envisaged here as occurring through the coupling of the pions to a σ-meson, although a more general description may be necessary. The part of the graph encircled by the dashed line may be described as a vertex correction to the σNN coupling.

of much lower mass. Indeed, explicit calculations [11] give for processes such as shown in Fig. 3 the weighting function $\rho(t)$ shown in Fig. 4.

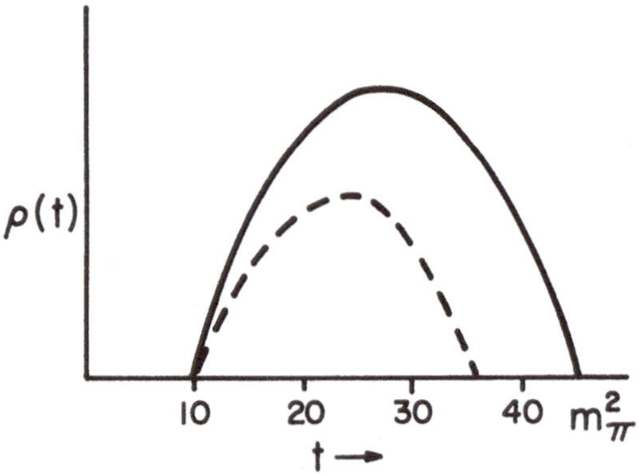

Fig. 4 The weighting function $\rho(t)$ for the exchange of two-pion systems coupled to $J = 0$, $T = 0$. The dashed line represents the weighting function after subtraction of the iterated one-pion-exchange potential and, therefore, the true weighting function for the potential corresponding to σ-exchange.

It should be noted that processes involving intermediate isobars are not the only ones contributing to $V_\sigma(r)$ of (1-9). After removal of the iterated one-pion-exchange interaction, one has processes such as shown in Fig. 5, in which two pions are "in the air at the same time"; i.e., in Fig. 5a the second pion is emitted before the first one is absorbed.

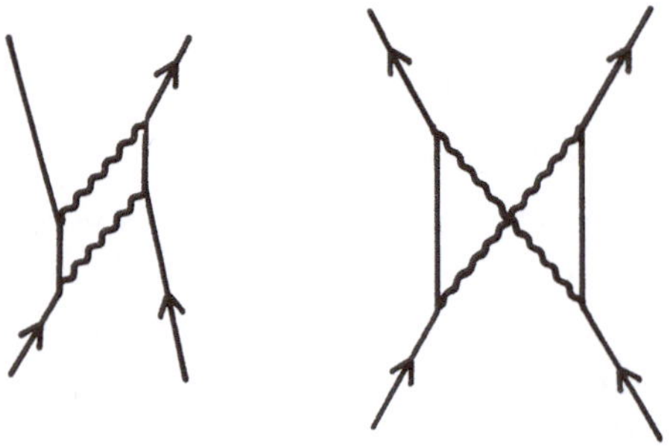

Fig. 5 Two-pion exchange processes involving intermediate nucleons which contribute to $V_\sigma(r)$.

Such processes are properly included in the dispersion-theoretical calculations [8] and in the considerations of [11].

2.3 The Short-Range Interaction

Whereas, as noted earlier, it may appear naive to discuss the short-range attraction in terms of meson exchange, the nucleon couplings to the heavy vector mesons are so strong, that the corresponding interactions reach out to well beyond the relevant Compton wavelengths m_ρ^{-1} and m_ω^{-1}; in fact, these effects are still large at ~ 1 fm, which is ~ 4 times m_ω^{-1}.

Originally, the existence of vector mesons was postulated in order to account for the spin-orbit interaction in nuclei [12]. In the hydrogen atom one has a spin-orbit force, of the Thomas form,

$$V_{s.o.} = \frac{\hbar^2}{m_n^2 c^2} \underset{\sim}{\sigma} \cdot \underset{\sim}{L} \frac{1}{r} \frac{dV_c}{dr} , \qquad (1\text{-}10)$$

where V_c is the central Coulomb potential. This spin-orbit interaction results from employing the Coulomb interaction in the relativistic Dirac equation. Vector mesons behave like heavy photons, and when the vector-meson exchange potential is used in a relativistic two-particle equation, a short-range spin-orbit interaction (of range m_ω^{-1}) results. This turns out to be what is needed in the shell model. From empirical properties of the spin-orbit force, especially as seen in nucleon-nucleon scattering, Breit predicted approximately the mass of the ω-meson, which was seen only later. The attractive feature of the relation between short-range repulsion and spin-orbit force in the boson exchange model has not been reproduced yet in the quark model.

In terms of quarks, p and n, the ω-meson can be written as

$$|\omega\rangle = \frac{1}{2} [p\bar{p} + n\bar{n}] \qquad (1\text{-}11)$$

so that the coupling of the ω to any particular hadron depends only on the number of p and n quarks in the hadron. Thus, the coupling of the ω to the Λ-particle is 2/3 of its coupling to the nucleon. Similarly, the ρ-meson can be written

$$|\rho^0\rangle = \frac{1}{2}[p\bar{p} - n\bar{n}] \qquad (1\text{-}12)$$

insofar as the isospin behavior is concerned. From the above scheme, one finds for the ratio of couplings to nucleons

$$\frac{g^2_{\omega NN}}{4\pi} = 9 \, \frac{g^2_{\rho NN}}{4\pi} \qquad (1\text{-}13)$$

From the known $g^2_{\rho NN}$ this gives

$$\left.\frac{g^2_{\omega NN}}{4\pi}\right|_{SU(3)} = 9\ \frac{g^2_{\rho NN}}{4\pi} \cong 4.5. \qquad (1\text{-}14)$$

We have used here the subscript "SU(3)" since the factor of 9 is unchanged by the introduction of strange quarks.

Aside from the spin-orbit interaction, the main effect from ω-meson exchange is the short-range repulsion

$$V_\omega(r) = \frac{g^2_{\omega NN}}{4\pi}\,\frac{e^{-m_\omega r}}{r}\ . \qquad (1\text{-}15)$$

Although the coupling constant given by (1-14) is large, analysis of empirical data, especially from nucleon-nucleon scattering, has consistently demanded a $g^2_{\omega NN}/4\pi$ much larger than given by the above argument; namely,

$$(g^2_{\omega NN}/4\pi)_{emp} \sim 10 - 20. \qquad (1\text{-}16)$$

The large effective ω-coupling can be understood by realizing [9] that combined π- and ρ-exchange, such as that shown in Fig. 6, contributes in just the ω-channel. Inclusion of the additional repulsions brings one up to the empirically required strengths.

Nucleon-nucleon interactions involving intermediate vector mesons, such as the spin-orbit one shown in (1-10), are singular for short r, going as r^{-3}.

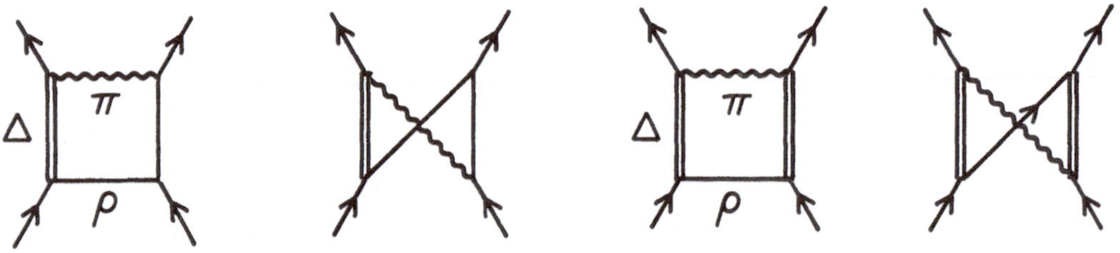

Fig. 6 Combined π- and ρ-exchange processes involving intermediate isobars which contribute in the ω-exchange channel.

This behavior can be corrected by introducing vertex functions $\Gamma_{\omega NN}$ or $\Gamma_{\rho NN}$ of finite range. A good idea of the range of these vertex functions can be obtained [13] by looking at the nucleon form factors, which in the so-called vector-meson dominance model, can be viewed as arising from the γ-ray first coupling to the vector mesons, as shown in Fig. 7. From the empirical behavior of the form factors at high momentum transfer

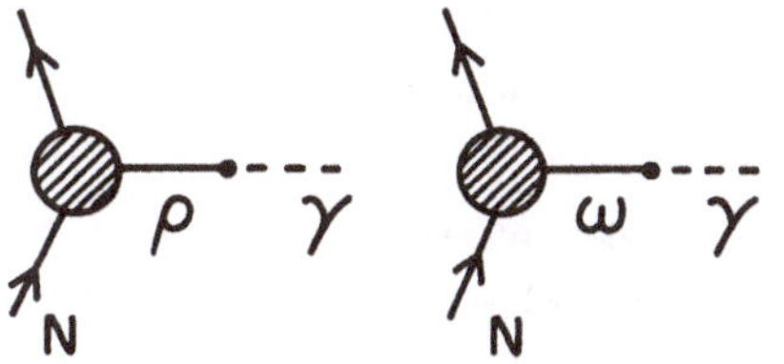

Fig. 7 Dominant contributions to the nucleon form factors

one can obtain the $\Gamma_{\omega NN}$ and $\Gamma_{\rho NN}$ as functions of momentum transfer.

In the Little Bag Model [14] the above ideas (and those of [19]) can be made explicit. In good approximation, the electromagnetic form factor of the proton can be described as

$$F^V(q^2) = \frac{e}{2}\left(\frac{m_V^2}{q^2 + m_V{}^2}\right)^2 , \qquad (1\text{-}17)$$

as regards the isovector part, with a similar expression for the iso-scalar form factor; i.e.,

$$F(q^2) = F^S + \tau_3 F^V \qquad (1\text{-}18)$$

where τ_3 is the third component of isospin. Empirically, m_V is slightly larger than m_ρ or m_ω, the vector meson masses. In the little bag picture, one power of $m_V^2/(q^2 + m_V^2)$ would come from the vector meson propagator (see Fig. 7) and the other, from the form factor of the bag. Thus, the bag form factor is

$$F_B(q^2) \cong m_V^2/(q^2 + m_V^2) . \qquad (1\text{-}19)$$

Fourier transforming this, we find $F_B(r)$ to be of extent $\sim \hbar/m_V c$, or ~ 0.3 fm, in configuration space. Of course, the Fourier transform of (1-19), taken literally, would be of Yukawa form, with a long tail, whereas in the bag model the form factor arises from quarks which are confined in a small volume with a rather abrupt surface. Thus, our above arguments should be taken only as indicative of the size of the region over which the form factor must be modified.

The ρ-meson is coupled to nucleons by both a vector and tensor coupling

$$\delta L_\rho = \frac{g_\rho}{\sqrt{4\pi}}\left\{ \bar{\psi}\, \frac{(p + p')_\mu}{2m}\, \vec{\tau}\cdot\vec{\rho}^{\,\mu}\, \psi + \bar{\psi}\, \frac{1+K_V}{2m}[\underset{\sim}{\sigma} x \underset{\sim}{k}]\cdot\, \underset{\sim}{\vec{\rho}}\cdot\vec{\tau}\, \psi \right\} \qquad (1\text{-}20)$$

where p and p' are initial and final nucleon momenta, respectively, m_n is the nucleon mass and K_V is a constant which, in the vector dominance model, would be given by the anomalous moment of the nucleon

$$K_V = 3.7 \qquad (1\text{-}21)$$

Effects from the vector coupling of the ρ tend to be swamped by the much larger ones from the ω, although the ρ-exchange interaction has a τ-dependence which should be distinguishable, in principle, from that of ω exchange. Effects from the tensor coupling of the ρ tend to dominate those from the tensor coupling of the ω, which we have not discussed. The latter would involve a $1+K_S$, which is very small compared with $(1+K_V)$, if one uses the vector dominance model as a guide.

One can rewrite the tensor ρ coupling as

$$\delta L_T = \frac{f_\rho}{\sqrt{4\pi}m_\rho} \bar{\psi}\, [\underset{\sim}{\sigma} \times \underset{\sim}{k}] \cdot \vec{\underset{\sim}{\rho}} \cdot \vec{\tau}\psi \tag{1-22}$$

where

$$f_\rho = g_{\rho NN}(1+K_V)\; m_\rho/2m \tag{1-23}$$

From the accepted value [15],

$$\frac{g^2_{\rho NN}}{4\pi} = 0.5, \tag{1-24}$$

one would find

$$\frac{f^2_\rho}{4\pi} = 1.86 \quad \text{(vector dominance)}. \tag{1-25}$$

In fact, determination of f_ρ from the nucleon form factor [16] demands a larger f_ρ,

$$\frac{f^2_\rho}{4\pi} \sim 4.86, \tag{1-26}$$

and nuclear phenomena are easier to understand [17,18] with this larger value of $f^2_\rho/4\pi$. This larger f_ρ would seem to violate the vector dominance assumption; it can be understood in a simple way by assuming that there is a direct vector coupling of γ-rays to nucleons, rather than all of the coupling going through the ρ-meson [19].

The nucleon-nucleon interaction arising from ρ-exchange with tensor couplings is

$$\begin{aligned} V_\rho(k) &= -\frac{f^2_\rho}{m^2_\rho}(\vec{\tau}_1\cdot\vec{\tau}_2)\frac{[\underset{\sim}{\sigma}_1 \times \underset{\sim}{k}][\underset{\sim}{\sigma}_2 \times \underset{\sim}{k}]}{k^2+m^2_\rho} \\ &= -\frac{f^2_\rho}{m^2_\rho}(\vec{\tau}_1\cdot\vec{\tau}_2)\left\{\frac{\underset{\sim}{\sigma}_1\cdot\underset{\sim}{\sigma}_2\, k^2 - \underset{\sim}{\sigma}_1\cdot\underset{\sim}{k}\;\underset{\sim}{\sigma}_2\cdot\underset{\sim}{k}}{k^2+m^2_\rho}\right\} \end{aligned} \tag{1-27}$$

$$= -\frac{f_\rho^2}{m_\rho^2}(\vec{\tau}_1\cdot\vec{\tau}_2)\left\{-\frac{[\underset{\sim}{\sigma}_1\cdot\underset{\sim}{k}\;\underset{\sim}{\sigma}_2\cdot\underset{\sim}{k}-\frac{1}{3}\underset{\sim}{\sigma}_1\cdot\underset{\sim}{\sigma}_2\,k^2]}{k^2+m_\rho^2}+\frac{2}{3}\underset{\sim}{\sigma}_1\cdot\underset{\sim}{\sigma}_2-\frac{2}{3}\frac{\underset{\sim}{\sigma}_1\cdot\underset{\sim}{\sigma}_2\,m_\rho^2}{k^2+m_\rho^2}\right\} \tag{1-27}$$

where, in the last step, decomposition into irreducible tensors has been carried out. Writing V_ρ in configuration space,

$$V_\rho(r) = \frac{f_\rho^2}{4\pi}m_\rho\,\vec{\tau}_1\cdot\vec{\tau}_2\left\{-S_{12}\left(\frac{1}{(m_\rho r)^3}+\frac{1}{(m_\rho r)^2}+\frac{1}{3(m_\rho r)}\right)e^{-m_\rho r}\right.$$
$$\left.-\frac{8\pi}{3}\frac{(\underset{\sim}{\sigma}_1\cdot\underset{\sim}{\sigma}_2)}{m_\rho^3}\delta(\underset{\sim}{r})+\frac{2}{3}(\underset{\sim}{\sigma}_1\cdot\underset{\sim}{\sigma}_2)\frac{e^{-m_\rho r}}{m_\rho r}\right\} \tag{1-28}$$

Comparison of the above with the pion-exchange potential, (1-2), shows the ρ-exchange tensor interaction to have opposite sign. This means that the radial part of the $\vec{\tau}_1\cdot\vec{\tau}_2$ tensor potential will be of the form shown in Fig. 8. The interaction at short distances ($r \lesssim 0.6$ fm) will be modified by the highly repulsive central interaction coming from ω-exchange. The fact that the ρ-exchange tensor potential cuts off the singular π-exchange tensor potential at short distances is very important for a number of phenomena which we shall discuss in the next chapters.

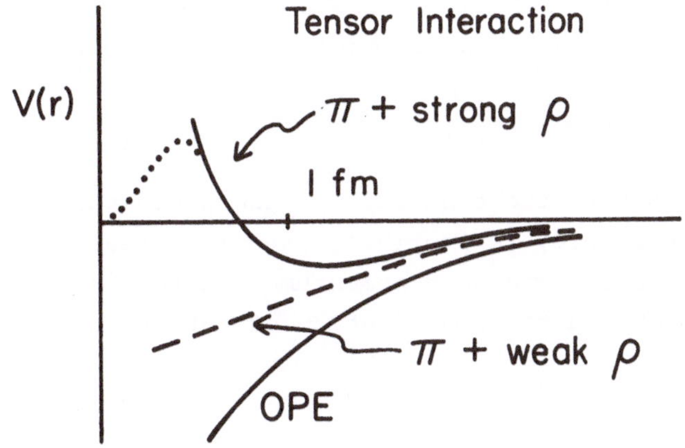

Fig. 8 Radial behavior of the $\vec{\tau}_1\cdot\vec{\tau}_2$ tensor potential from strong ρ ($f_\rho^2 = 4.86$) and weak ($f_\rho^2 = 1.86$) exchange. The dotted line indicates the sort of modification the strong short-ranged ω-exchange will have on the effective potential from strong ρ exchange

3. Nuclear Matter as a Fermi Liquid

We shall formulate the problems of nuclear matter within the framework of the Landau Fermi liquid theory. Our aim is to make a microscopic theory, in which we calculate the Landau Fermi liquid parameters, beginning from the nucleon-nucleon force. Of course, such a calculation cannot be made rigorously at the present time, but by keeping one hand on empirical quantities, we can make connections.

Our approach here is different from that of MIGDAL [1], whose theory of finite Fermi systems is a derivative of the Landau theory, with certain (major) added assumptions. The Migdal theory is strictly phenomenological, fitting the parameters of a quasiparticle interaction of assumed form, so as to reproduce empirical phenomena, much in the spirit of Landau's discussion of liquid He^3, although matters are more complicated in nuclei because of the finite nature of the system.

We are admittedly not on firm ground in calculating Landau parameters from a microscopic model in the case of nuclear matter, because there is not really any small parameter to expand in. We shall try to make plausible a picture of interacting quasiparticles and collective excitations, showing that both single-particle and collective features must be treated on an equal footing. We shall appeal to the richness of experimental data which has accumulated in recent years, for confirmation of this approach.

There are many excellent reviews of the Landau Fermi liquid theory [20,21], in addition to the elegant original articles of LANDAU [22], so we shall not try to rereview this subject, but shall indicate operationally how it can be applied to give a description of nuclear matter.

Landau started from a kinetic equation for quasiparticles

$$\frac{\partial n_p}{\partial t} + \frac{\partial n_p}{\partial \underset{\sim}{x}} \frac{\partial \varepsilon}{\partial \underset{\sim}{p}} - \frac{\partial n_p}{\partial \underset{\sim}{p}} \frac{\partial \varepsilon}{\partial \underset{\sim}{x}} = I(n), \tag{1-29}$$

where n_p is the quasiparticle number[3], ε and p are the energy and momentum of the quasiparticle, respectively, and I is the collision term. In the process of writing down a momentum flux tensor π_{ik} for quasiparticles, so as to obtain the conservation of total momentum

$$\frac{\partial}{\partial t}\int d^3x \int p_i \; n \; d\tau = 0 \tag{1-30}$$

---

[3] The above n_p can be generalized [21] to $n_p \to n_p(q) = a^+_{p+q}\, a_p$, where a^+_{p+q} and a_p are Fermi creation and annihilation operators, respectively. In this way, the above equation is extended to describe particle-hole excitations.

where

$$d\tau = g\,\frac{d^3p}{2\pi^3} \qquad (1\text{-}31)$$

g being the degeneracy, Landau was led to the assumption: (See the discussion in § 2 of [23].)

$$\frac{\delta E}{\delta n(p)} = \varepsilon(p)\,. \qquad (1\text{-}32)$$

It is clear that this assumption is true in a Hartree-Fock theory; however, the import is much more general, although the idea of quasiparticles moving in average fields due to other particles is back of the Landau theory.

We reiterate that Landau was led to the central assumption (1-32) by conservation laws; in this case, conservation of energy and momentum. This shows what a powerful tool conservation laws can be.

Landau considered the total energy of the system to be a functional of occupation number n(p) of all of the quasiparticles

$$E = \{n(p_1),\ n(p_2),\ \ldots\}\ . \qquad (1\text{-}33)$$

Thus, the $\varepsilon(p)$ in (1-32) are a functional of the $\delta n(p)$'s. To describe collective phenomena, where many of the δn's enter, one needs to carry out the variation of E to second order,

$$\delta E = E'-E = \Sigma\, \varepsilon^{(0)}(\underset{\sim}{p})\ \delta n(\underset{\sim}{p}) + \frac{1}{2}\sum_{\underset{\sim}{p},\underset{\sim}{p}'} f(\underset{\sim}{p},\underset{\sim}{p}')\,\delta n(\underset{\sim}{p})\ \delta n(\underset{\sim}{p}') \qquad (1\text{-}34)$$

<u>defining</u> $f(\underset{\sim}{p},\underset{\sim}{p}')$. Here we have indicated vectors explicitly, since the dependence of f on angle between $\underset{\sim}{p}$ and $\underset{\sim}{p}'$ is crucial. Spin and isospin variables have been suppressed here, as elsewhere, for simplicity. It follows from (1-34) that

$$\varepsilon(\underset{\sim}{p}) = \frac{\delta E}{\delta n(\underset{\sim}{p})} = \varepsilon^{(0)}(\underset{\sim}{p}) + \sum_{p'} f(\underset{\sim}{p},\underset{\sim}{p}')\ \delta n(\underset{\sim}{p}')\ . \qquad (1\text{-}35)$$

The quantity $f(\underset{\sim}{p},\underset{\sim}{p}')$ is clearly the interaction between quasiparticles, since it is the change in energy with removal of quasiparticles in states $\underset{\sim}{p}$ and $\underset{\sim}{p}'$. It is crucial that in this removal, the occupation numbers of the other quasiparticles, $n_{p''}$, $\underset{\sim}{p}'' \neq \underset{\sim}{p}$ or $\underset{\sim}{p}'$, be kept constant. It turns out then, that $f(\underset{\sim}{p},\underset{\sim}{p}')$ is the quasiparticle-quasihole interaction. This will be clear from examples worked later, if the reader is not already familiar with this point.

The reader may have difficulty with the following conceptual point: what does one mean by functional differentiation with respect to quasiparticle number? Landau made the crucial assumption that there is a one-to-one correspondence between quasiparticle excitations in the interacting system and particle excitations in the noninteracting system. Consequently, in order to make functional differentiations in the interacting system, one can simply remove bare particles in the noninteracting system and see how the resulting changes propagate through the interacting system.

Explicit models are easily worked out when the energy is expressed in Rayleigh-Schrödinger perturbation theory, because in this theory the connection with bare-particle occupation number is clear.

We shall here work out the connection between the Brueckner-Bethe theory, which begins from Rayleigh-Schrödinger perturbation theory and makes partial summations, and the Landau theory. If one wishes to make a microscopic description of nuclear matter or nuclei, one must, in some way, get from the bare forces with extremely strong repulsive cores, which cannot be handled in perturbation theory, to effective interactions (pseudopotentials) which are well behaved at short distances. The Brueckner-Bethe theory shows how to combine the strong short-range repulsion with correlation in the wave function which keep the particles apart over the range of this repulsive interaction, so as to give a well-believed effective interaction.

The Brueckner-Bethe theory is only a starting point, and we must then go on to include collective effects. These turn out to be very important. The questionable part of our formalism is probably whether we include these to sufficient accuracy.

4. The Brueckner-Bethe Theory

In the Brueckner theory, it is realized that, because of the strong short-ranged interactions between nucleons, ordinary perturbation theory cannot be used, and the pair interactions must be summed to all orders. One defines a G-matrix by

$$(\underset{\sim}{k}_1\underset{\sim}{k}_2|G|\underset{\sim}{k}_1\underset{\sim}{k}_2) = (\underset{\sim}{k}_1\underset{\sim}{k}_2|V|\underset{\sim}{k}_1\underset{\sim}{k}_2) - \frac{1}{2}\sum_{k_3,k_4>k_F} \frac{(\underset{\sim}{k}_1\underset{\sim}{k}_2|V|\underset{\sim}{k}_3\underset{\sim}{k}_4)(\underset{\sim}{k}_3\underset{\sim}{k}_4|G|\underset{\sim}{k}_1\underset{\sim}{k}_2)}{k_3{}^2/2m + k_4{}^2/2m - \varepsilon_1 - \varepsilon_2} \tag{1-36}$$

where the Pauli principle has been put in explicitly. Exchange is assumed to be included, i.e., the matrix elements of V are direct minus exchange terms. Here ε_1 and ε_2 are hole energies, inclusive of self-energy insertions,

$$\varepsilon(k) = \frac{k^2}{2m} + \sum_{k_2<k_F} (k\,k_2|G|k\,k_2). \tag{1-37}$$

It is usually convenient not to put self-energy insertions in particle lines, but rather to group them with other many-body clusters and to evaluate the entire sum [6]. There is then an energy gap between hole and particle states, which turns out to be awkward for making connection with the Landau theory. This difficulty can be overcome by introduction of a model space [23].

In order to show the connection to Landau theory, we write the Brueckner expression for the energy

$$E = \sum_{k_1} (k_1{}^2/2m)\,n(k_1) + \frac{1}{2}\sum_{k_1,k_2} (k_1k_2|G|k_1k_2)\,n(k_1)\,n(k_2) \tag{1-38}$$

in which we explicitly write in the dependence on particle occupation number, and then carry out the variations to obtain $\varepsilon(p)$ and $f(p,p')$. Equations (1-36) and (1-37) are then rewritten

$$(k_1k_2|G|k_1k_2) = (k_1k_2|V|k_1k_2) - \sum \frac{(k_1k_2|V|k_3k_4)(1-n(k_3))(1-n(k_4))(k_3k_4|G|k_1k_2)}{(k_3{}^2/2m) + (k_4{}^2/2m) - \varepsilon_1 - \varepsilon_2}, \tag{1-39}$$

$$\varepsilon(k) = \delta E/\delta n(k) = (k^2/2m) + \sum_{k_2} (kk_2|G|kk_2)\,n(k_2) + \frac{1}{2}\sum_{k_1k_2} (k_1k_2|\delta G/\delta n(k)|k_1k_2)\,n(k_1)\,n(k_2) \tag{1-40}$$

The $\varepsilon(k)$ above in (1-40) is distinguished from that in (1-37) and from those to be used in the denominator of (1-39) by the last term involving $\delta G/\delta n(k)$. This is often called a rearrangement term, and enters into the Landau $\varepsilon(k)$, so that the Landau $\varepsilon(k_F)$ corresponds to the actual removal energy of the quasiparticle, whereas the Brueckner $\varepsilon(k_F)$ in (1-37) does not.

Now the variations can be carried out to obtain $f(\tilde{k},\tilde{k}')$. Whereas the $n(k)$ refer to be occupation numbers in the noninteracting, reference state (think of Rayleigh-Schrödinger perturbation theory), removal of a particle in this state leads to the removal of a quasiparticle, since particles are assumed to transform smoothly into quasiparticles as the interaction is included.

Note from (1-40) that the variation

$$\delta\varepsilon(k)/\delta n(k')$$

has to be carried out with respect to

(i) the $n(k_2)$,

(ii) the $n(k_3)$ and $n(k_4)$ in (1-39) for G,

(iii) the $n(k)$ appearing in ε_1 and ε_2 in the denominator of (1-39).

Terms arising from (ii) and (iii), which are respectively, corrections to the Pauli operator, and self-energy coming from removal of a particle in state k', are often called rearrangement terms.

Let us carry out these variations, replacing, for simplicity, G by V on the right-hand side of (1-39) and (1-40). This gives a theory which is essentially second-order Rayleigh-Schrödinger perturbation theory, but with the Hartree-Fock single-particle energies for the holes, i.e., we simplify (1-40) to

$$\varepsilon(k) = (k^2/2m) + \sum_{k_2} (\underset{\sim}{k}\underset{\sim}{k}_2|G|\underset{\sim}{k}\underset{\sim}{k}_2)\, n(k_2) \tag{1-41}$$

We obtain the following contributions to f(k,k') from the variations listed as (i), (ii), (iii):

$$f_{(i)}(\underset{\sim}{k},\underset{\sim}{k}') = (\underset{\sim}{k}\underset{\sim}{k}'|V|\underset{\sim}{k}\underset{\sim}{k}') + \frac{1}{2}\sum_{k_3,k_4} \frac{(\underset{\sim}{k}\underset{\sim}{k}'|V|\underset{\sim}{k}_3\underset{\sim}{k}_4)(\underset{\sim}{k}_3\underset{\sim}{k}_4|V|\underset{\sim}{k}\underset{\sim}{k}')}{\varepsilon_k + \varepsilon_{k'} - (k_3{}^2/2m) - (k_4{}^2/2m)} (1-n_3)(1-n_4),$$

$$f_{(ii)}(\underset{\sim}{k},\underset{\sim}{k}') = -\sum_{k_2,k_4} \frac{(\underset{\sim}{k}\underset{\sim}{k}_2|V|\underset{\sim}{k}\underset{\sim}{k}_4)(\underset{\sim}{k}'\underset{\sim}{k}_4|V|\underset{\sim}{k}\underset{\sim}{k}_2)}{\varepsilon_k + \varepsilon_2 - (k'^2/2m) - (k_4{}^2/2m)} n_2(1-n_4)$$

$$-\sum_{k_2,k_4} \frac{(\underset{\sim}{k}_2\underset{\sim}{k}'|V|\underset{\sim}{k}\underset{\sim}{k}_4)(\underset{\sim}{k}\underset{\sim}{k}_4|V|\underset{\sim}{k}_2\underset{\sim}{k}')}{\varepsilon_2 + \varepsilon_{k'} - (k^2/2m) - (k_4{}^2/2m)} n_2(1-n_4) + \tag{1-42}$$

$$\frac{1}{2}\sum_{k_1,k_2} \frac{(\underset{\sim}{k}_1\underset{\sim}{k}_2|V|\underset{\sim}{k}\underset{\sim}{k}')(\underset{\sim}{k}\underset{\sim}{k}'|V|\underset{\sim}{k}_1\underset{\sim}{k}_2)}{\varepsilon_1 + \varepsilon_2 - (k^2/2m) - (k'^2/2m)} n_1 n_2,$$

$$f_{(iii)}(\underset{\sim}{k},\underset{\sim}{k}') = -(\underset{\sim}{k}\underset{\sim}{k}'|V|\underset{\sim}{k}\underset{\sim}{k}') \sum_{k_2,k_3,k_4} \frac{(\underset{\sim}{k}\underset{\sim}{k}_2|V|\underset{\sim}{k}_3\underset{\sim}{k}_4)(\underset{\sim}{k}_3\underset{\sim}{k}_4|V|\underset{\sim}{k}\underset{\sim}{k}_2)}{[\varepsilon_k + \varepsilon_2 - (k_3^2/2m) - (k_4^2/2m)]^2} n_2(1-n_3)(1-n_4)$$

$$-\sum_{k_2,k_3,k_4}{}' \frac{(k_2 k'|V|k_2 k')(kk_2|V|k_3 k_4)(k_3 k_4|V|kk_2)}{[\varepsilon_k + \varepsilon_2 - (k_3{}^2/2m) - (k_4{}^2/2m)]^2} n_2(1-n_3)(1-n_4). \tag{1-42}$$

We draw these terms graphically as shown in Fig. 10. Terms with two internal hole lines, except for Fig. 10(ii)c, have been consistently dropped. The graphs, Figs. 10(i), would both be included in the G matrix $(\underset{\sim}{k}\underset{\sim}{k}'|G|\underset{\sim}{k}\underset{\sim}{k}'')$, but otherwise one can go directly to the G-matrix case with the full G in the right-hand side of (1-36), simply by changing all matrix elements of V into those of G.

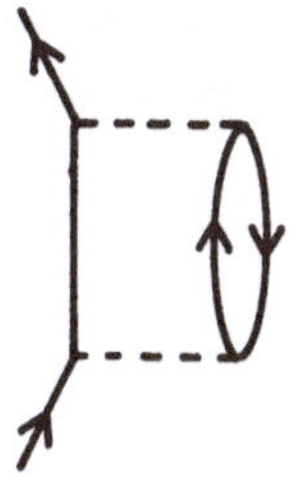

Fig. 9 Second-order self-energy insertion for a single particle. The dashed lines here represent the interactions V.

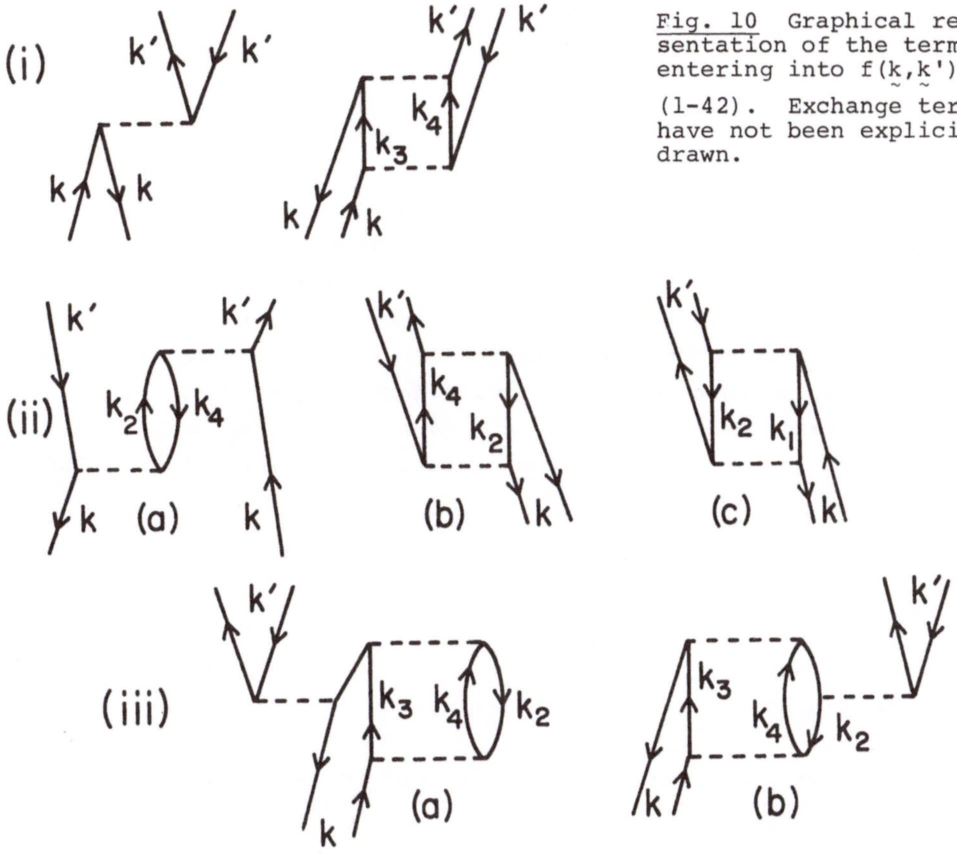

Fig. 10 Graphical representation of the terms entering into $f(\underset{\sim}{k},\underset{\sim}{k}')$, (1-42). Exchange terms have not been explicitly drawn.

Since $f(\underset{\sim}{k},\underset{\sim}{k}')$ is the particle-hole interaction in the long-wavelength limit, we have drawn it that way in Fig. 10, both the particle and hole having momentum $\underset{\sim}{k}$ (or $\underset{\sim}{k}'$), so that the total momentum is zero. The magnitude of $\underset{\sim}{k}$ (or $\underset{\sim}{k}'$) must be k_F, since only on the Fermi surface can a particle and hole have equal momentum. One can think of f as the limit of a particle-hole interaction between excitations of momentum $\underset{\sim}{q}$, as $|\underset{\sim}{q}| \to 0$.

The fact that $f(\underset{\sim}{k},\underset{\sim}{k}')$ is the particle-hole interaction in the long-wavelength limit can be made clearer by the following argument in which we think of a problem involving a weak interaction V which need be handled only to first order. Consider first the particle-hole interaction connecting particle-hole states of total momentum $\underset{\sim}{q}$ as shown in Fig. 11. One can think of this interaction as

$$\Gamma = \sum_{k'k} (k'+q,k|V|k',k+q) \; (a_{k'}{}^{\dagger}a_{k'+q})^{\dagger}(a_k{}^{\dagger}a_{k+q}), \qquad (1\text{-}43)$$

with the $(a_k{}^{\dagger}a_{k+q})$ acting to the right, the $(a_{k'}{}^{\dagger}a_{k'+q})^{\dagger}$ acting to the left. That is, the matrix element of Γ between particle-hole states $a^{\dagger}_{\underset{\sim}{k}_1+\underset{\sim}{q}}\, a_{\underset{\sim}{k}_1}|0)$ and $a^{\dagger}_{\underset{\sim}{k}_2+\underset{\sim}{q}}\, a_{\underset{\sim}{k}_2}|0)$, where $|0)$ is the vacuum, will give the interaction

$$(k_2+q,k_1|V|k_2,k_1+q)$$

shown in Fig. 11.

Such matrix elements enter into the description of collective excitations, the collective excitation being a linear combination of particle-hole states with coherent phases. For example, plasma oscillations are described in the random-phase approximation in such a way, where the interparticle Coulomb interaction is V. The momentum of the excitation is $\underset{\sim}{q}$, so that in the long-wavelength limit, $\underset{\sim}{q} \to 0$. In this limit, the operator pair $(a_k{}^{\dagger}a_{k+q})$ in (1-43) becomes $(a_k{}^{\dagger}a_k) = n_k$; similarly, $(a_{k'}{}^{\dagger}a_{k'+q})$ becomes $n_{k'}$. Consequently, the expectation value of Γ becomes

$$(0|\Gamma|0) = \sum_{k_1k_2} (k_2,k_1|V|k_2,k_1)\,n(k_2)\,n(k_1) \qquad (1\text{-}44)$$

i.e., it has the same form as the potential energy term in (1-39). Thus, removing the n's by functional differentiation gives the matrix element $(\underset{\sim}{k}_2,\underset{\sim}{k}_1|V|\underset{\sim}{k}_2,\underset{\sim}{k}_1)$.

To this order, we could have carried out the same considerations for the particle-particle interaction with the same conclusions. However the second-order particle-particle interaction will have terms like those in Fig. 12. Whereas the matrix element corresponding to the process Fig.

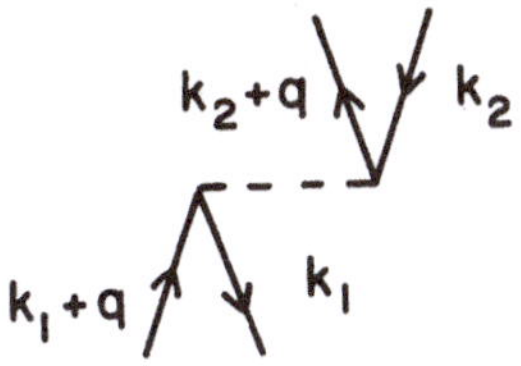

Fig. 11 Particle-hole interaction connecting two states of total momentum q.

12(b) is just that of Fig. 10(ii)a, with the left-hand line pointing up instead of down, the matrix element corresponding to the process Fig. 12(a) is missing in (1-42). The reason is simple. If we redraw

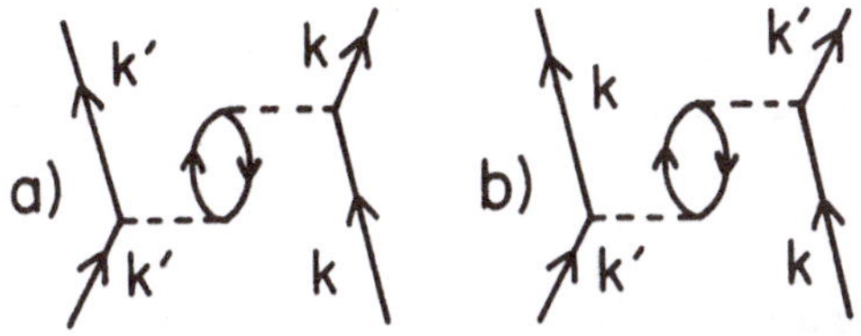

Fig. 12 Typical terms in the second-order particle-particle interactions.

Fig. 12(a) as a particle-hole interaction as is done in Fig. 13, it is reducible; that is, it can be obtained by putting together two first-order interactions of the type shown in Fig. 11. But when the interaction, Fig. 11, is used as the kernel in an integral equation such as occurs in the random-phase approximation, it is automatically iterated to all orders. Thus, reducible diagrams should not (and do not) occur in f.

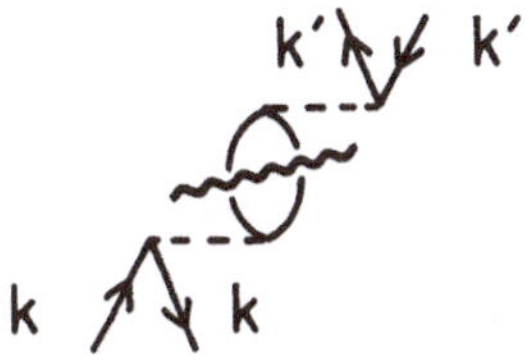

Fig. 13 Figure 12(a) redrawn as a particle-hole interaction

In nuclear matter, the expression for $f(\underset{\sim}{k},\underset{\sim}{k}')$ must be generalized to a matrix in spin and isospin space, because of the two spins and two kinds of particles. It can be written

$$f(\underset{\sim}{k},\underset{\sim}{k}') = (\pi^2/2m^* k_F)\ \{F + F'\ \vec{\tau}_1 \cdot \vec{\tau}_2 + G\ \underset{\sim}{\sigma}_1 \cdot \underset{\sim}{\sigma}_2 + G'\ \vec{\tau}_1 \cdot \vec{\tau}_2\ \underset{\sim}{\sigma}_1 \cdot \underset{\sim}{\sigma}_2\} \qquad (1\text{-}45)$$

where F,F',G and G' are dimensionless functions of the angle between $\underset{\sim}{k}$ and $\underset{\sim}{k}'$, and $2m\ k_F/\pi^2$ is the density of states on the Fermi surface[4].

[4]MIGDAL [1] uses the symbols f,g for the dimensionless quantities we call F,G. Our notation is more usual in other applications of the Landau theory. Migdal employs the density of states $m^* k_F/\pi^2$ appropriate to liquid He^3. Thus F = 2f etc.

In general, tensor invariants must also be introduced [18,24]. However, these do not seem to be quantitatively important for the applications we consider here and, for simplicity, we suppress them.

Each of the above coefficients is expanded in a Legendre series

$$F = \Sigma F_\ell P_\ell(\cos \theta_L) \tag{1-46}$$

where we label the Landau angle by θ_L. The angle θ_L has nothing to do with the scattering angle; thus far we have considered only forward scattering. It has to do with the velocity dependence of the interaction, as we make clear in the following example.

Let us take a simple schematic interaction

$$f(\underset{\sim}{k},\underset{\sim}{k}') = f_0 + \hat{k}\cdot\hat{k}'\, f_1 \tag{1-47}$$

where f_0 and f_1 are numbers. This interaction is then a special example of the interaction, (1-46), where only the first two terms are nonzero. The term in f_1 describes the full momentum dependence of this simplified force. Let us now compute the effective mass of a quasiparticle at the Fermi surface interacting with other quasiparticles*. To obtain this, we calculate the difference in self energies, Σ,

$$\Sigma(k_f + \delta k) - \Sigma(k_f) = \frac{(k_f + \delta k)^2 - k_f^{\,2}}{2}\left(\frac{1}{m^*} - \frac{1}{m}\right) \tag{1-48}$$

of a quasiparticle slightly displaced off the Fermi surface and a quasiparticle on it. (See Fig. 14.)

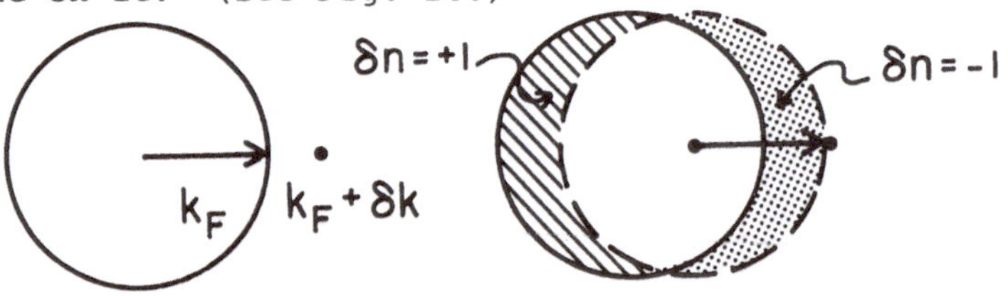

Fig. 14 Calculation of the self energy of a particle of momentum $k_f+\delta k$ by shifting the Fermi sea, as in b) so the particle is at k_f.

In calculating $\Sigma(k_F + \delta k)$, we can change to a calculation of $\Sigma(k_f)$, shifting the Fermi sea δk to the left, as shown in Fig. 14. Such a shift will not change the interaction between single quasiparticle and quasiparticles in the Fermi sea, since it is Galilean invariant. In calculating the interaction now one must take into account that $\delta n = \pm 1$ in the regions shown. We find

*This calculation has been carried out very elegantly by Landau using Galilean invariance.

$$\sum(k_f + k) - \sum(k_f) = -\frac{4}{(2\pi)^3} f_1 \int_{-1}^{1} \frac{\underset{\sim}{k}\cdot\underset{\sim}{k}'}{k_f^2} 2\pi k_f^2 \delta k \cos\theta \, d(\cos\theta) \quad (1\text{-}49)$$

$$= -\frac{2k_f^2}{3\pi^2} f_1 \quad \delta k = k_f \, \delta k \left(\frac{1}{m^*} - \frac{1}{m}\right)$$

where $\hat{k}\cdot\hat{k}' = \cos\theta$.

From (1-49) we find

$$\frac{m^*}{m} = 1 + \frac{F_1}{3} \quad (1\text{-}50)$$

where $F_1 = \frac{2k_f m^* f_1}{\pi^2}$. (1-51)

The expansion (1-46) is assumed to converge rapidly. In liquid He^3, the first two terms are found to be sufficient for most applications, and we shall assume the same to be true for nuclear matter.

Our strategy in finite nuclei will be somewhat different. In §1 it was shown that the long-range part of the nucleon-nucleon force came from pion exchange. As will be shown in §5, exchange terms from this interaction give the main part of F_1, although the $V_\rho(r)$, (1-28), also makes contributions. It seems to be a good approximation to handle the interaction from all other exchanges in zero-range approximation, treating it as a $\delta(\underset{\sim}{r}_{12})$ function; e.g., note that the weighting function for σ-exchange, Fig. 4, peaks for a mass of $\sim 5m_\pi$, corresponding to a range of $\hbar/(5m_\pi c)$, small compared with the average interparticle spacing. Those interactions treated as δ-functions will contribute only to the $\ell = 0$ multipoles: F_0, F_0', G_0, G_0'. Rather than introducing the F_1 which would arise from the V_π of (1-3) and the V_ρ of (1-28), we shall handle these two interactions explicitly, in a shell-model basis.

The coefficient F_0 can be related to the compressibility K, F_0' to the symmetry energy β. These relations are standard, and we refer the reader to [1]. We summarize them below

Table 1 Relations between Landau parameters and observables

Quantity	Expression
Compression modulus	$K = 6 \frac{k_f^2}{2m^*}(1 + F_0)$
Effective mass m^*	$\frac{m^*}{m} = 1 + \frac{F_1}{3}$
Symmetry energy β	$\beta = \frac{1}{3}\frac{k_f^2}{2m^*}(1 + F_0')$

The compression modulus

$$K = r_0^{\,2} \frac{d^2E}{d\,r_0^{\,2}} \tag{1-52}$$

is that used in Brueckner theory; it is a factor of 9 larger than in Migdal's formula, which corresponds to defining K as $\rho_0^{\,2}(d^2E/d\rho_0^{\,2})$, ρ_0 being the equilibrium density. The r_0 in the above formula is given by $(4\pi/3)r_0^{\,3} = \rho_0^{\,-1}$.

The symmetry energy is defined in the following way: If one has an excess of neutrons, then a term appears in the energy per particle

$$\delta\, E_\tau = \beta\left(\frac{N-Z}{A}\right)^2 . \tag{1-53}$$

Empirical values of K, m^* and β are somewhat subject to interpretation; we shall discuss them in detail later.

5. A Theory of Interacting Quasiparticles and Collective Excitations

Earlier work [25] clarified the connection between Brueckner theory and Landau theory, but was not quantitatively successful, because it did not properly include effects of collective excitations. We can think of the quasiparticle energies being composed of two types of contributions, such as shown in Fig. 15.

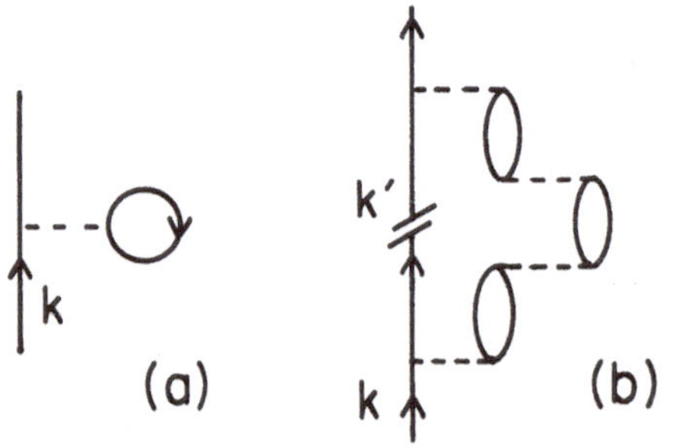

Fig. 15 The two types of contributions to the particle self energy. Here the wavy line represents the G-matrix. There is some double counting between a) and b), because the G-matrix sums ladder contributions to all orders, so that the second-order term must be removed from b).

It is clear that clothing the particle as in Fig. 15b, with collective excitations will change observable quantities in important ways. The quasiparticle builds up correlations in the medium around it, so that the clothed quasiparticle is substantially larger than the bare particle, and harder to compress; thus, the compression modulus will be increased. Inclusion of the processes, Fig. 15b, also changes the effective mass, since it represents processes in which the quasiparticle carries its collective-excitation clothing along with it. In the case of liquid He^3, such processes increase the effective mass to $m^* \simeq 3m$

[26,27]. We shall indicate later that the effective mass in finite nuclei is also substantially changed by inclusion of collective effects.

From functional differentiation of the processes Fig. 15b we will obtain new contributions to $f(\underset{\sim}{k},\underset{\sim}{k}')$. To begin with, we consider the changes we get by differentiating with respect to $n(\underset{\sim}{k}')$ where $\underset{\sim}{k}'$ refers to the intermediate particle line as shown in Fig. 15b. We shall return to a discussion of other contributions later. Functionally differentiating with respect to $n(\underset{\sim}{k}')$ gives the contribution, Fig. 16, to $f(\underset{\sim}{k},\underset{\sim}{k}')$.

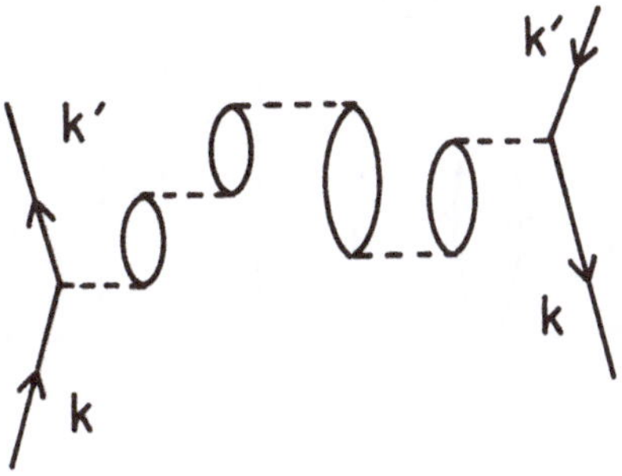

Fig. 16 Contribution of the collective effect, Fig. 14b, to the particle-hole interaction $f(\underset{\sim}{k},\underset{\sim}{k}')$.

We shall call this contribution the "Induced Interaction".

Now, in a complete theory, f is the complete particle-hole interaction, so that it is clear that the G-matrix in the particle-hole interactions should be replaced by f to include higher-order effects. It has also been proved in the limit of $\underset{\sim}{k}' \to \underset{\sim}{k}$, that the vertex functions on the left- and right-hand sides should be replaced by f [28].

Thus, more generally, the induced interaction is assumed to have the form, Fig. 17. The particle-hole phase space between interactions

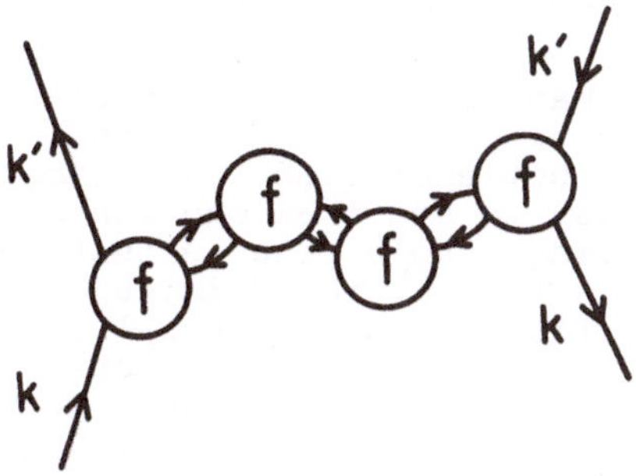

Fig. 17 Assumed general form of the induced interaction.

f is assumed to be described by Lindhard functions. It should be emphasized that assumption of the form for the Induced Interaction, Fig. 17, away from zero Landau angle $\underset{\sim}{k}' = \underset{\sim}{k}$ involves an extrapolation. We shall see that this form has certain very pleasant properties, especially with respect to preserving antisymmetry in any microscopic calculation.

We arrive then at the integral equation [28], shown graphically in Fig. 18.* In this equation, we have a kernel, encircled, consisting of processes such as one would evaluate in Brueckner theory. These

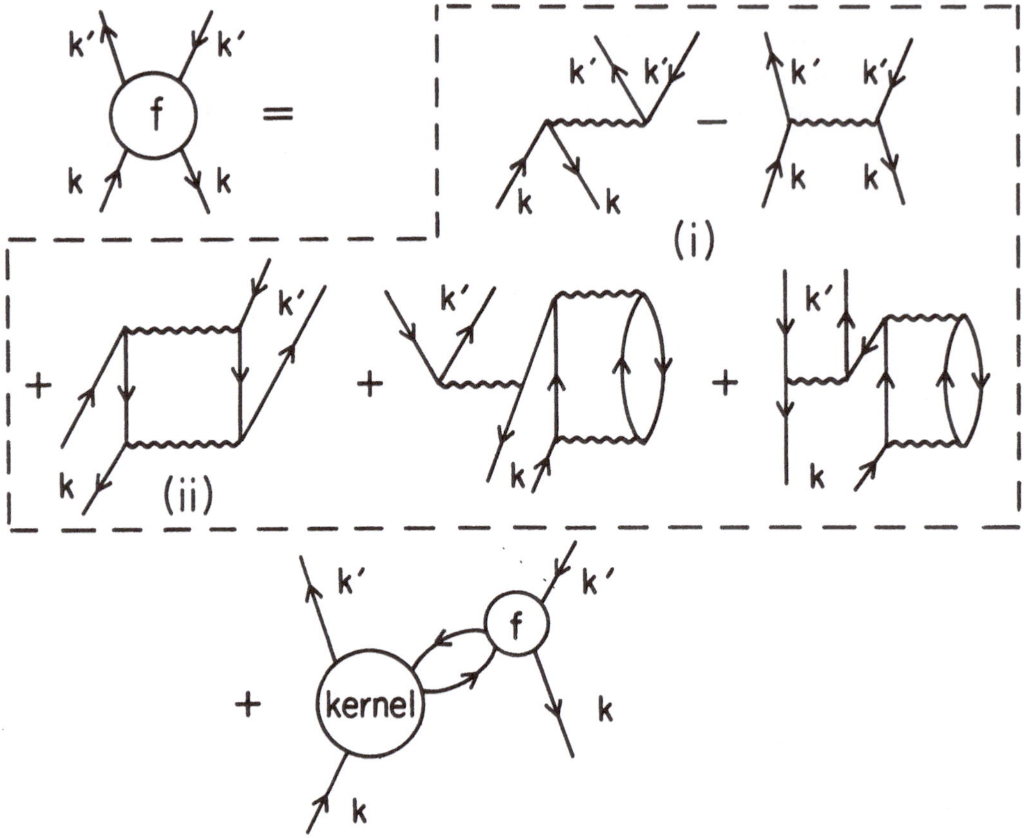

Fig. 18 Graphical representation of our integral equation. The wavy lines here denote G-matrix elements. The encircled processes will be called the kernel.

should all be irreducible in the crossed channel; i.e., one should not be able to make a cut which goes through a single particle and hole line in the crossed channel. The term which is not encircled, on the bottom, contains all reducible diagrams. It is easy to see that one can cut vertically through a single particle and single hole line in this term.

By including all irreducible processes to all orders in the kernel, we would include all processes in the solution of this integral equation, so we are essentially just rearranging the order of summation. Yet this is of importance, because the resulting equation has a much greater stability than a straightforward summation of graphs; i.e., the solution is rather insensitive to the order one goes to in the kernel.

Much of the convenience of the integral equation depicted in Fig. 18 stems from the fact that calculations to any order in the kernel will give results for the particle-particle interaction which automa-

---

*This equation, with first-order kernel, was used by M.W. Kirson (Ann. Phys. (N.Y.) 66 (1971) 624. A discussion of this equation within the framework of the schematic model was carried out by D.W.L. Sprung and A.M. Jopko, Can. Journ. Phys. 50 (1972) 2768.

tically satisfy the Landau sum rule, whereas a straightforward calculation of f to a given order in the G-matrix does not [30]. We've not yet talked about the particle-particle interaction, so a few words of explanation would be in place here.

The particle-particle interaction between quasiparticles on the Fermi surface can be expressed as [20,21,22]

$$A(\underset{\sim}{k},\underset{\sim}{k}') = \sum_{\ell} \left\{ \frac{F_\ell P_\ell(\hat{k}\cdot\hat{k}')}{1+F_\ell/(2\ell+1)} + \vec{\tau}_1\cdot\vec{\tau}_2 \frac{F_\ell' P_\ell(\hat{k}\cdot\hat{k}')}{1+F_\ell'/(2\ell+1)} \right. \tag{1-54}$$
$$\left. + \underset{\sim}{\sigma}_1\cdot\underset{\sim}{\sigma}_2 \frac{G_\ell P_\ell(\hat{k}\cdot\hat{k}')}{1+G_\ell/(2\ell+1)} + (\vec{\tau}_1\cdot\vec{\tau}_2)(\underset{\sim}{\sigma}_1\cdot\underset{\sim}{\sigma}_2) \frac{G_\ell' P_\ell(\hat{k}\cdot\hat{k}')}{1+G_\ell'/(2\ell+1)} \right\}$$

The graphical interpretation of (1-54) is straightforward, and is shown in Fig. 19. The $[1+F_\ell/(2\ell+1)]^{-1}$ in (1-54) just represent the summation of bubbles, to all orders, in the crossed channel. Note that the particle lines going in and out of the interaction f are now drawn

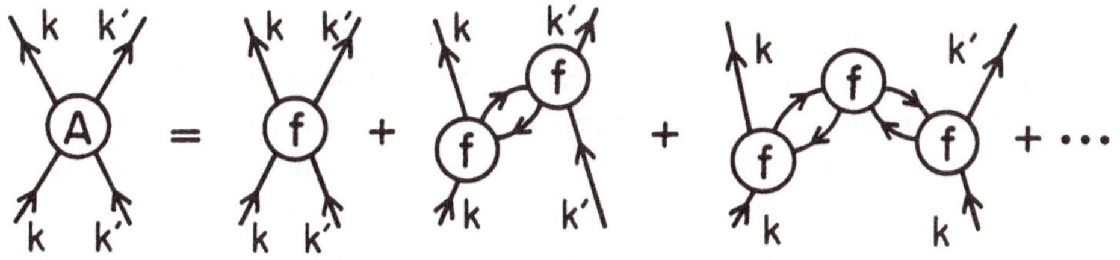

a) b) c) d)

Fig. 19 Graphical interpretation of the relationship between A and f.

appropriate for particle-particle scattering. Since $\underset{\sim}{k}$ and $\underset{\sim}{k}'$ are both on the Fermi sea, it is somewhat a matter of convention as to how they are to be drawn.

The final process on the right-hand side of Fig. 19 is topologically like the Induced Interaction, Fig. 17, except that one of the $(\underset{\sim}{k},\underset{\sim}{k}')$ pair has been interchanged. In fact, the Induced Interaction is just the exchange term corresponding to the process on the right-hand side of Fig. 19. Putting this latter process and the Induced Interaction together, we have, then, both the direct and exchange terms of this interaction, so that antisymmetry of the total particle-particle interaction $A(\underset{\sim}{k},\underset{\sim}{k}')$ is guaranteed, provided that antisymmetric kernels are used in our integral equation, Fig. 18.

Antisymmetry gives us the Landau sum rule. Landau pointed out that the forward scattering of identical particles (same spin and isospin) must vanish, because this amplitude must change sign under interchange of the particles (antisymmetry); on the other hand, the particles are identical and nothing is changed. Thus,

$$\sum_{\ell}\left\{\frac{F_\ell}{1+F_\ell/(2\ell+1)}+\frac{F_\ell'}{1+F_\ell'/(2\ell+1)}+\frac{G_\ell}{1+G_\ell/2\ell+1)}+\frac{G_\ell'}{1+G_\ell'/(2\ell+1)}\right\}=0 \tag{1-55}$$

This sum rule relates, in an important way, the various parameters of the interaction. It is badly violated in [1] and in most of the applications of the Landau theory to nuclei; in fact, in [1] all of the coefficients, F_0, F_0' etc. are positive, so there's no chance of (1-55) being satisfied.

Of course, the complete quasiparticle interaction A must satisfy (1-55). In general, A's that are constructed from calculations of f to some finite order, with subsequent use of (1-54) will not satisfy (1-55). This is clear, because the polarization bubbles are included to all orders in the direct term of the phonon-induced interaction, Fig. 19d, whereas if one calculates f to any finite order, the corresponding exchange part of the induced interaction will have polarization bubbles only to finite order. We could give references to any number of calculations in which (1-55) is violated. Our construction of the integral equation, Fig. 18, ensures the satisfaction of the sum rule (1-55) so long as an antisymmetric kernel is employed in the integral equation, regardless of the order one goes to in the G-matrix in the kernel. This is because the Induced Interaction just gives the proper exchange term for the phonon induced interaction in A. We shall see later that preservation of the Landau sum rule puts useful and important constraints on the calculation, and that solution of the integral equation, Fig. 18, is much more stable with respect to input than calculations are of f to some finite order.

We first derive the form of the Induced Interaction for the simplified case where it is sufficient to keep only the $\ell = 0$ terms in the expansions, such as (1-46), for the Landau parameters. It is straightforward to construct the term of second order in f in the Induced Interaction, and then generalize it by making the complete expression into a geometrical series. One has (see eqs. (3.12) of [29]).

$$\begin{aligned}4F_i &= \frac{F_0^{\,2}U(q,0)}{1+F_0U(q,0)}+\frac{3F_0'^{\,2}U(q,0)}{1+F_0'U(q,0)}+\frac{3G_0^{\,2}U(q,0)}{1+G_0U(q,0)}+\frac{9G_0'^{\,2}U(q,0)}{1+G_0'U(q,0)}\\ 4F_i' &= \frac{F_0^{\,2}U(q,0)}{1+F_0U(q,0)}-\frac{F_0'^{\,2}U(q,0)}{1+F_0'U(q,0)}+\frac{3G_0^{\,2}U(q,0)}{1+G_0U(q,0)}-\frac{3G_0'^{\,2}U(q,0)}{1+G_0'U(q,0)}\end{aligned} \tag{1-56}$$

$$4G_i = \frac{F_0{}^2 U(q,0)}{1+F_0 U(q,0)} + \frac{3F_0'^2 U(q,0)}{1+F_0' U(q,0)} - \frac{G_0{}^2 U(q,0)}{1+G_0 U(q,0)} - \frac{3G_0'^2 U(q,0)}{1+G_0' U(q,0)}$$

$$4G_i' = \frac{F_0{}^2 U(q,0)}{1+F_0 U(q,0)} - \frac{F_0'^2 U(q,0)}{1+F_0' U(q,0)} - \frac{G_0{}^2 U(q,0)}{1+G_0 U(q,0)} + \frac{G_0'^2 U(q,0)}{1+G_0' U(q,0)} \tag{1-56}$$

Here $U(q,0)$ is the Lindhard function $U(q,\omega)$ for $\omega = 0$,

$$U(q,0) = \left[\frac{1}{2} + \frac{1}{2}\left(\frac{q}{4p_f} - \frac{p_f}{q}\right) \ell n \left(\frac{p_f - \frac{q}{2}}{p_f + \frac{q}{2}}\right)\right] \tag{1-57}$$

with the normalization

$$U(0,0) = 1. \tag{1-58}$$

Note that the various excitations contribute, with coefficients corresponding to their statistical weight, to the spin- and isospin-independent parameter F. Note, also that although the input F's, etc., may have only $\ell = 0$ terms, the Induced Interaction will develop terms of higher order in ℓ.

The right-hand side of (1-56) can be understood in the following way. The F_0's in the numerator describe the coupling of the quasiparticle and quasihole to the density fluctuations of the exchanged mode; similarly the F_0' to the isospin fluctuation of the exchanged mode, etc. The quantity

$$\chi(q,\omega) = - \frac{U(q,\omega)}{1 + F_0 U(q,\omega)} , \tag{1-59}$$

evaluated at $\omega = 0$, since we are calculating Landau parameters, is the density-density correlation function in our approximation. This turns out to be identical with the function of Pines and Nozieres [21].

This interpretation tells us how to generalize the expression (1-56) to include $\ell = 1$ terms. In this case, the F_1 will describe coupling of the quasiparticle or quasihole to the fluctuation in current associated with the exchanged excitation, and we shall then construct the current-current correlation function $\chi_{jj}(\tilde{q},0)$ to describe the propagation of the excitations.

This can be made explicit by considering the vertex Fig. 20, in the limit $q \to 0$. Although we shall extrapolate our description to $\tilde{k}'=\tilde{k}-\tilde{q}$ with $0 \lesssim |\tilde{q}| \lesssim 2k_F$, we can only make rigorous statements for $q \to 0$; i.e., for the long-wavelength limit in the crossed channel.

As noted earlier, it was proved in the appendix to [28] that the

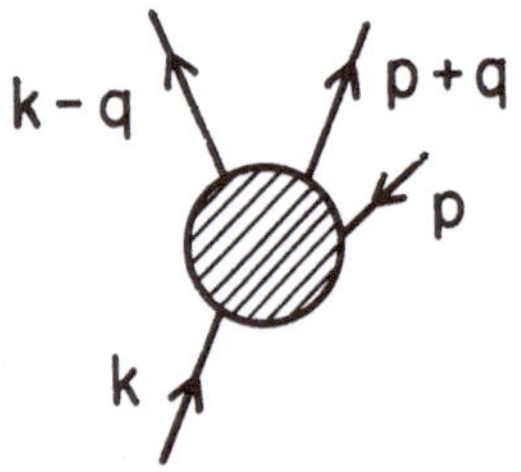

Fig. 20 Vertex function describing quasiparticle coupling to a particle-hole excitation of momentum q.

vertex function, Fig. 10, in the limit of $q \to 0$ is just $f(\hat{k},\hat{p})$, the magnitudes of both $\underset{\sim}{k}$ and $\underset{\sim}{p}$ being equal to k_F. Expanding f up to terms of $\ell = 1$,

$$f(\hat{k}\cdot\hat{p}) = f_0 + f_1(\hat{p}\cdot\hat{k}) = f_0 + f_1\{(\hat{k}\cdot\hat{q})(\hat{p}\cdot\hat{q}) + \sqrt{1-(\hat{k}\cdot\hat{q})^2}\sqrt{1-(\hat{p}\cdot\hat{q})^2}\cos\phi\} \tag{1-60}$$

where we have used the addition theorem for Legendre polynomials in the last step. The term in cos ϕ vanishes upon integration over angles.

The interaction is then

$$\Gamma = \int f(\hat{k}\cdot\hat{p})\ \delta n_p\ 4\ \frac{d^3p}{(2\pi)^3} = f_0\ \delta n_0 + \frac{f_1}{3}\ (\hat{k}\cdot\hat{q})\ \delta n_1 \tag{1-61}$$

where $$\delta n_p = \delta n_0 + \delta n_1 \cos(\hat{p}\cdot\hat{q}), \tag{1-62}$$

with

$$\delta n_0 = \frac{1}{2}\int_{-1}^{1} \delta n_p\ d\cos(\hat{p}\cdot\hat{q})$$

$$\delta n_1 = \frac{3}{2}\int_{-1}^{1} \delta n_p \cos(\hat{p}\cdot\hat{q})\ d\cos(\hat{p}\cdot\hat{q}). \tag{1-63}$$

Note that $\delta n_0 = \delta\rho$. The δn_ℓ's are multipoles in the expansion of the δn_p's associated with the particle-hole excitation of (vanishing) momentum q.

The current associated with the particle-hole excitation is

$$\underset{\sim}{j} = \rho\ \underset{\sim}{v}, \tag{1-64}$$

with ρ the density, and $\underset{\sim}{v}$ the velocity field of the excitation, given by

$$m\ \rho\ \underset{\sim}{v}\cdot\hat{q} = \sum_p \underset{\sim}{p}\cdot\hat{q}\ \delta n_p = \frac{p_f}{3}\ \delta n_1 \tag{1-65}$$

where we have used the fact* that each quasiparticle on the Fermi surface carries momentum p_f. From (1-65) we find

$$\delta n_1 = \frac{3m}{p_f} j, \tag{1-66}$$

and, since $\underset{\sim}{j}$ is in the direction of $\hat{q}$, the interaction is

$$\Gamma = f_0 \, \delta n_0 + \frac{mf_1}{p_f} \hat{k} \cdot \underset{\sim}{j} \ . \tag{1-67}$$

In order to preserve a symmetrical treatment between $\underset{\sim}{k}$ and $\underset{\sim}{k}'$ as we move away from $\underset{\sim}{q} = 0$, the $\hat{k}$ above should be interpreted [28] as $(\hat{k}+\hat{k}')/2$.

The current carried by a quasiparticle of momentum $\underset{\sim}{k}$ is proportional to $\underset{\sim}{k}$, so the final term in (1-67) represents the coupling between the current carried by the quasiparticle and that carried by the particle-hole excitation. In the $\underset{\sim}{q} = 0$ limit, the coupling can be put into the familiar form [31]

$$\frac{mf_1}{p_f} \hat{k} \cdot \underset{\sim}{j} = \frac{mf_1}{p_f} \rho \, \hat{\underset{\sim}{k}} \cdot \underset{\sim}{v} = \frac{m^* - m}{m^*} \underset{\sim}{k} \cdot \underset{\sim}{v} \tag{1-68}$$

which can be derived simply using Landau's application of Galilean invariance.

In this example, we have worked out the coupling to the longitudinal particle-hole mode, the velocity field of which will be in the direction of $\underset{\sim}{q}$. But $\frac{1}{2}(\hat{k} + \hat{k}') \cdot \underset{\sim}{q} = 0$. However, the current-current correlation function is diagonal in the long wavelength limit, so that exactly the same contribution as calculated above will be picked up from transverse modes.

The induced interaction, (1-56), can now be generalized to include the current-current couplings, giving:

$$4F_i = \frac{F_0^{\,2}\, U(q,0)}{1 + F_0 U(q,0)} + \left(1 - \frac{q^2}{4p_F^{\,2}}\right) \frac{F_1^{\,2}\, \alpha_1(q,0)}{1 + F_1\, \alpha_1(q,0)}$$

$$+ \frac{3F_0'^{\,2}\, U(q,0)}{1 + F_0'\, U(q,0)} + 3\left(1 - \frac{q^2}{4p_f^{\,2}}\right) \frac{F_1'^{\,2}\, \alpha_1(q,0)}{1 + F_1'\, \alpha_1(q,0)} \tag{1-69}$$

*The current carried by a quasiparticle in the interacting system must be equal to the current carried by a bare particle in the noninteracting system, since the interactions are translationally invariant, and therefore preserve particle current.

$$+ \frac{3G_0^2 \, U(q,0)}{1+G_0 \, U(q,0)} + 3\left(1 - \frac{q^2}{4p_f^2}\right) \frac{G_1^2 \, \alpha_1(q,0)}{1+G_1 \, \alpha_1(q,0)}$$

$$+ \frac{9G_0'^2 \, U(q,0)}{1+G_0' \, U(q,0)} + 9\left(1 - \frac{q^2}{4p_f^2}\right) \frac{G_1'^2 \, \alpha_1(q,0)}{1+G_1' \, \alpha_1(q,0)} \qquad (1\text{-}69)$$

where α_1 is essentially the current-current correlation function and obvious generalizations involving the $\alpha_1(q,0)$ of (1-56) for the other parameters.

The factor $(1 - q^2/4p_f^2)$ arises from our symmetrization between $\underset{\sim}{k}'$ and $\underset{\sim}{k}$ in describing the quasiparticle current. For $\underset{\sim}{k}' = -\underset{\sim}{k}$, the average current carried by the quasiparticle is zero, as the factor $(1 - q^2/4p_f^2)$ correctly reproduces. Here $\alpha_1(q,0)$ is given by [29]

$$\alpha_1(q,0) = \frac{1}{2}\left[\frac{3}{8} - \frac{k_F^2}{2q^2} + \left(\frac{k_F^3}{2q^3} + \frac{k_F}{4q} - \frac{3q}{32k_F}\right)\ell n\left(\frac{k_F + \frac{q}{2}}{k_F - \frac{q}{2}}\right)\right] \qquad (1\text{-}70)$$

Our integral equation (1-45) takes, finally, the deceptively simple form

$$f(\underset{\sim}{k},\underset{\sim}{k}') = f_{d(i)}(\underset{\sim}{k},\underset{\sim}{k}') + f_{d(ii)}(\underset{\sim}{k},\underset{\sim}{k}') + f_{d(iii)}(\underset{\sim}{k},\underset{\sim}{k}') \qquad (1\text{-}71)$$
$$+ f_i(f(\underset{\sim}{k},\underset{\sim}{k}')), \qquad |\underset{\sim}{k}| = |\underset{\sim}{k}'| = k_f$$

the induced interaction f_i being a function of the Landau f. The "direct interaction", shown as the encircled part of Fig. 18, contains all terms irreducible in the crossed channel and can, in principle, be extended to any order. We have used "order" here in the sense of the number of G-matrices, but one might well use some other scheme for constructing the direct interaction.

The above (1-71) can be solved by iteration. For the F_0 and F_1, it tends to be immensely stable; i.e., one can alter the input, in terms of the f_d's considerably, with little alteration in the final F_0 and F_1. We show, as example, in Fig. 21 how little final results change with inclusion of $f_{d(ii)} + f_{d(iii)}$. It can be seen that an ~20% change in the f_d-input is translated into an ~8% change in F_0.

The reasons for this stability are two-fold: (i) The important F_ℓ', G_ℓ and G_ℓ' are all positive, in nuclear matter; thus, we know the important F_ℓ must be negative, so that the Landau sum rule (1-55) be satisfied. In practice, this implies that F_0 be negative. (ii) The

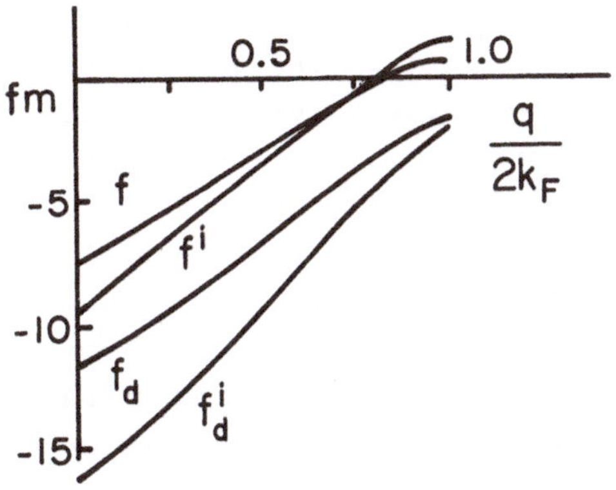

Fig. 21 The variation of the spin- and isospin-independent particle-hole interaction caused by a change in the direct interaction is illustrated. The f_d is the interaction $f_{(i)} + f_{(ii)} + f_{(iii)}$, the kernel of the integral equation illustrated in Fig. 18, and f is the solution to the integral equation. The f_d^i and f^i are the corresponding terms when only $f_{(i)}$ is used in the kernel.

Induced Interaction acts in the opposite direction to the direct interaction. To elaborate on point (ii), let us drop the spin- and isospin-dependent terms, for the moment, and consider the equation

$$F = F_d + \frac{F_0^{\,2}\, U(q,0)}{1+F_0\; U(q,0)} + \left(1 - \frac{q^2}{4p_f^{\,2}}\right)\frac{F_1^{\,2}\,\alpha_1(q,0)}{1+F_1\;\alpha_1(q,0)} \tag{1-72}$$

Suppose that we put in for F_d an amplitude which is too attractive. (An example is the $F_d{}^I$ of Fig. 21.) Then the terms involving $F_0^{\,2}$ and $F_1^{\,2}$ on the right-hand side of (1-72) will be too large and, being repulsive, will subtract off too much from F_d, giving a final F which does not differ too much from the correct one.

One can see that all of the terms in the induced Interaction, (1-69), are positive so that they all will increase the compressibility. In terms of our discussion at the beginning of the chapter, we can understand this in terms of clothing of the quasiparticles; through interaction with collective modes in the environment about them, the quasiparticles build correlations into that environment and become effectively larger in size and thus harder to compress.

Sjöberg [29,32] arrived at a value of the compression modulus $K \simeq 200$ MeV in all of his approximations. The work of [32] should be preferable to that of [29], since a model space was used there, and single-particle energies, etc. must be continuous across the Fermi surface in order to make any sense out of a microscopic calculation within the framework of the Landau theory. The work of [32] is, however, marred by a spin sum being incorrectly carried out in the calculation of G'. Thus, G' turned out to be much too small. This reflects back on the other parameters by F_0 being too small, F_0' too large, and G_0 too large, as can be seen from (1-56).

6. Making Contact with Nature

We shall now discuss the nature of the interactions in the varous channels, keeping in mind the underlying nucleon-nucleon interaction outlined in §1. It seems natural to begin with the spin- and isospin-independent interaction F, although in many ways this is the most complicated. The basic reason for the complication is that the forces in this channel arise mostly from second- and higher-order effects, of the type shown in Fig. 2 involving intermediate isobars, or of the form, (1-5), involving the second-order tensor force, or of the form, Fig. 5, involving higher-order processes with intermediate nucleons. None of the force comes from direct terms in lowest-order meson exchange, there being no elementary scalar, isoscalar meson. Once intermediate states involving virtual nucleons and/or virtual isobars are involved, the many-body problem becomes much more complicated, because we must describe the interactions of these virtual particles in the many-body system.

We can outline the type of complexity introduced by virtual intermediate states by considering interactions of virtual particles encountered in evaluation of the second-order tensor interaction, as shown in Fig. 22. This self-energy insertion would be

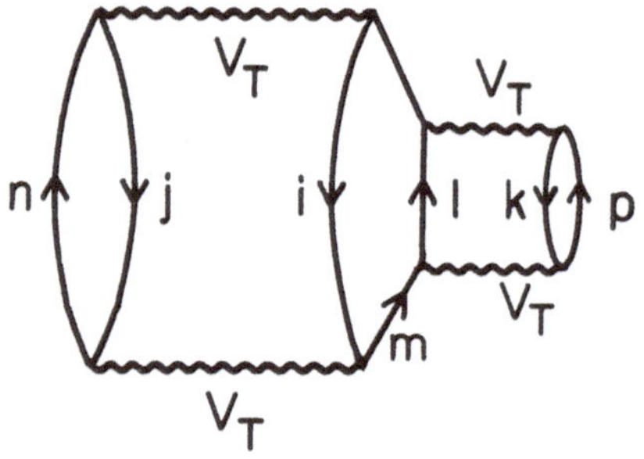

Fig. 22 Self-energy insertion for the virtual particle state m.

$$\Sigma^{(2)} = \sum_{k\ell p} \frac{|(p\ell|V_T|km)|^2}{\varepsilon_i+\varepsilon_j+\varepsilon_k-\varepsilon_\ell-\varepsilon_p-\varepsilon_n} . \qquad (1\text{-}73)$$

It is to be noted that energies ε_i, ε_j and ε_n of holes and particles not actively involved in the self-energy interaction enter into the expression for $\Sigma^{(2)}$. This is the so-called off-energy shell effect [6]. The validity of the usual treatment of such off-energy shell effects has been called into question by PANDHARIPANDE and WIRINGA [33], but similar effects arise in their work from lack of commutation of the various interaction and correlation operators, at least in the case of spin-dependent operators such as encountered with tensor forces. Typical energies for intermediate states are ~400 MeV [5,6], so off-shell self-energy insertions of highly excited virtual states come into play.

It may seem that we are simply collecting complications to impress the reader with the difficulty of the problem. But all of these complications come into play when we discuss how the nuclear system behaves under compression, and try to determine the equilibrium position of the system. It seems to us likely that the questions discussed above will only be finally settled when exact calculations; e.g., of the Monte Carlo type, can be carried out for the Fermion many-body system. In the case of the simpler Boson systems, such calculations have already been performed. In the present half-way situation, the more phenomenological Fermi liquid theory is a practical vehicle.

6.1 Compressibility

The compression modulus K defined by (1-52) is given in terms of Fermi liquid parameters by

$$K = 6\,\frac{k_F^{\,2}}{2m^*}\,(1 + F_0) \tag{1-74}$$

Calculations by SJÖBERG [29,32] have given $K \simeq 200$ MeV. These calculations suffer from an error in spin sums in evaluating G_0', which then turned out to be much too small. Instead of the values of 0.3 and -0.1 found in these papers, Sjöberg now obtains $G_0' \sim 1.30$ for the Reid soft-core potential. As can be seen from (1-56), this larger value of G_0' will increase F_0, and therefore the compressibility. Both the numerical behavior of the integral equation discussed earlier and the constraint imposed by the Landau sum rule (1-55) guarantee that F_0 will not be increased very much.

Empirically, the situation with K has been greatly clarified by the identification of the breathing mode in Pb^{208}[35,36]. By composing the compression modulus in finite nuclei out of a volume term, a surface term and symmetry-energy and Coulomb components, BLAIZOT and GRAMMATICOS [37, 38] can perform the extrapolation to infinite nuclear matter, and obtain $K \sim 220$ MeV, a value which would seem to fit in well with Sjöberg's calculations.

6.2 Effective Mass

The history of the effective-mass calculations in nuclei and nuclear matter is interesting and instructive, in that only recently have the theoretical and empirical determinations come at all together. Brueckner theory always produced effective masses $m^*/m \sim 0.6 - 0.7$, which would imply that the level density, at the Fermi surface, which is proportional to m^*, is substantially lowered by the particle interactions. On the other hand, empirical descriptions using the Nilsson model, etc., began from velocity-independent potential wells for the shell model, corresponding to $m^*/m = 1$, and then usually had to compress the levels further to achieve agreement with experiment.

The empirical behavior of the effective mass was clarified in 1964 by BROWN, GOULD and GUNN [39] and Fig. 23, which we reproduce from this paper, shows this behavior.

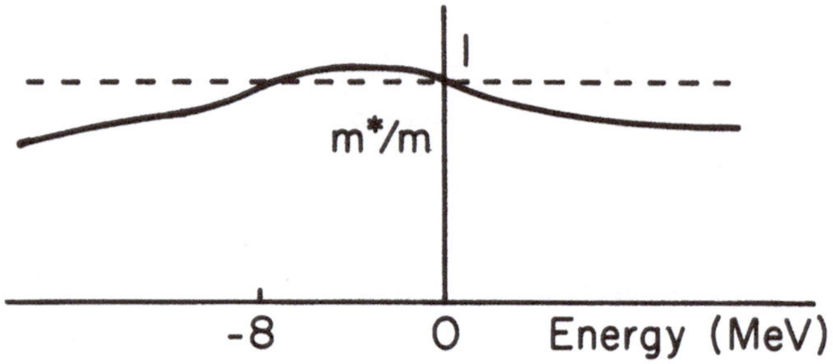

Fig. 23 Behavior of the effective mass m* as a function of distance from the Fermi surface [39]

The theoretical reasons underlying the compression of levels at the Fermi surface were given by BERTSCH and KUO [40] and elucidated further in the work of JEUKENNE, LEJEUNE and MAHAUX [41]. In these works it is pointed out that the ω-dependence of m^* is essential for the explanation of level compression. The effective mass can be expressed as

$$\frac{m}{m^*} = \left. \frac{1 + \frac{\partial}{\partial T_k} \Sigma(k,\omega)}{1 - \frac{\partial \Sigma}{\partial \omega}(k,\omega)} \right|_{\omega=\varepsilon_k} \tag{1-75}$$

where T_k is the kinetic energy

$$T_k = \frac{k^2}{2m} \tag{1-76}$$

and in the denominator $\partial\Sigma/\partial\omega$ is to be evaluated at the quasiparticle pole ε_k. As pointed out in [41] and [42], Brueckner theory calculations treated the ω-dependence in a rough way, so that the denominator in (1-75) was close to unity; actually, it is the decrease in the denominator near $\varepsilon_k = \varepsilon_F$ which causes the behavior of the effective mass shown in Fig. 23.

In the theory outlined in the previous section, the ω-dependence is included in the induced interaction, and can be handled carefully because the whole treatment is focused on quasiparticle behavior near the Fermi surface. This treatment would be the same as that in [41], were we to set the denominators in the induced interaction equal to unity; i.e., to neglect the rescattering. (In [41] terms of higher order are discussed, but not evaluated in detail.) Looking at the behavior of the induced interaction as function of Landau angle, we see that keeping the interaction in the denominator $1 + G_0' U(q,0)$, for example, will increase the large Landau angle contribution, since $U(q,0)$ becomes smaller there, and on the whole, will keep the induced interaction from dropping off sharply with angle, thus decreasing the contribution of the induced interaction to F_1, the coefficient of the $\cos\theta_L$ term. Thus, in the calculations with the complete Induced Interaction, m^*/m is increased only slightly over the value calculated from the direct interaction.

The values of $m^*/m \sim 0.7 - 0.8$ calculated by Sjöberg [29,32] may be appropriate for nuclear matter. If so, the larger m^*/m found empirically in finite nuclei may be explained by the role of surface vibrations here [42]. It is important to understand the origin of the effective mass, since in situations like that in neutron-star matter, phenomena such as superfluidity depend sensitively on m^*/m, and we have no direct way at getting at this quantity.

A simple model for calculating the effective mass can be made in terms of π- and ρ-meson exchange potentials [43]; at nuclear-matter density this model gives results close to Sjöberg's [29]. It is, of course, natural that at low densities pion-exchange gives the most rapid variation of f with Landau angle, because it is of longest range in configuration space, and therefore has a Fourier transform which varies most rapidly. It is less clear why one should take only the ρ-exchange potential, out of all the short-range interactions, into account.

In a Fermi sea, the tensor parts of the π- and ρ-exchange potentials (1-2) and (1-28) do not contribute at all, and only the exchange terms from the $\underset{\sim}{\sigma}_1 \cdot \underset{\sim}{\sigma}_2 \; \vec{\tau}_1 \cdot \vec{\tau}_2$ potentials are nonzero, the direct parts averaging out with either spin or isospin summation. In evaluating exchange terms, we consider the two processes shown in Fig. 24. From the matrix elements in Fig. 24 we can make up the elements

$$< \left(\frac{p\bar{p} \pm n\bar{n}}{\sqrt{2}} \right) \left| \vec{\tau}_1 \cdot \vec{\tau}_2 \right| \left(\frac{p\bar{p} \pm n\bar{n}}{\sqrt{2}} \right) > \; = \frac{2 \pm 4}{2} = \begin{pmatrix} 3 \\ -1 \end{pmatrix} \qquad (1\text{-}77)$$

appropriate for the exchange terms of, respectively, isoscalar (+ sign) and isovector (- sign) particle-hole matrix elements. Thus, the OPEP interaction

$$\frac{f^2}{3} \; \frac{(\underset{\sim}{\sigma}_1 \cdot \underset{\sim}{\sigma}_2)(\vec{\tau}_1 \cdot \vec{\tau}_2)}{k^2 + m_\pi^{\;2}}$$

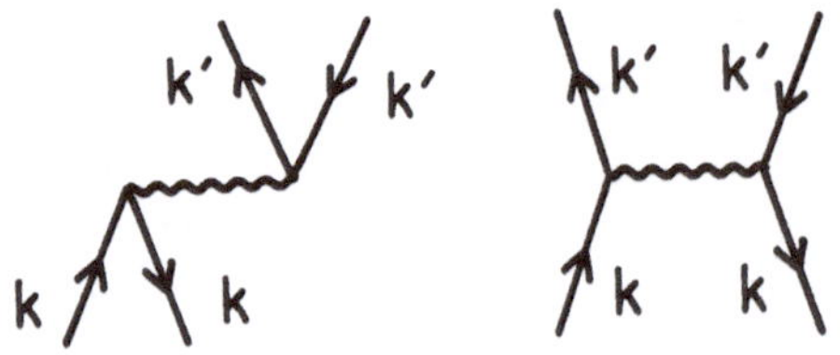

Fig. 24 Isospin matrix elements for the exchange term in the particle-hole interaction.

leads to particle-hole matrix elements

$$-3\,\frac{f^2}{k^2+m_\pi^{\,2}}\,,$$

an additional minus sign entering in because this is the exchange term.

We easily find

$$\delta F = -\frac{N_0}{4}\;\frac{3f^2}{k^2+m_\pi^{\,2}} \qquad (1\text{-}78)$$

where

$$N_0 = \frac{2\,k_f\,m^*}{\pi^2} \qquad (1\text{-}79)$$

and k^2 is now related to the Landau angle by

$$k^2 = 2\,k_f^{\,2}\,(1-\cos\theta_L). \qquad (1\text{-}80)$$

Straightforward calculation gives

$$F_{1\pi} = -\frac{9}{8}\,N_0\,\frac{f^2}{2k_f^{\,2}}\left\{\frac{2\,k_f^{\,2}+m_\pi^{\,2}}{2\,k_f^{\,2}}\,\ell n\,\frac{m_\pi^{\,2}+4\,k_f^2}{m_\pi^{\,2}} - 2\right\} \qquad (1\text{-}81)$$

which is easily evaluated at nuclear-matter densities where $k_f \cong 2\,m_\pi$ to give

$$F_{1\pi} = -0.46\ m^*/m. \qquad (1\text{-}82)$$

A similar calculation for the $\underset{\sim}{\sigma}_1\cdot\underset{\sim}{\sigma}_2\ \vec{\tau}_1\cdot\vec{\tau}_2$ part of the ρ-exchange potential gives,

$$F_{1\rho} = -\frac{9}{8}\,N_0\,\frac{\hat{f}_\rho^{\,2}}{k_f^{\,2}}\left\{\frac{2k_f^{\,2}+m_\rho^{\,2}}{2k_f^{\,2}}\,\ell n\,\frac{m_\rho^{\,2}+4k_f^{\,2}}{m_\rho^{\,2}} - 2\right\} \qquad (1\text{-}83)$$

where

$$\hat{f}_\rho^{\,2} = 0.4\ f_\rho^{\,2}, \qquad (1\text{-}84)$$

the 0.4 taking into account [17] in a crude way effects of short-range correlation, due mainly to ω-meson exchange. (This 0.4 is the ratio $A/4\pi$, where A will be given in (1-94) of the next section.) For $k_f=2m_\pi$

$$F_{1\rho} = -0.56\ m^*/m \tag{1-85}$$

and

$$F_{1\pi}+F_{1\rho} = -1.02\ m^*/m = -0.76, \tag{1-86}$$

where we have used (1-50), to compare with Sjoberg's -0.77 [29]. In fact, Sjöberg used the Reid potential, which has a smaller ρ-exchange component, but the second-order tensor interaction, which is then stronger, makes up for some of the difference. In any case, our model calculation is probably adequate for not only qualitative, but also semiquantitative considerations. Inclusion of ω-exchange should make our F_1 larger, because effects from a repulsive core grow with energy.

6.3 Symmetry Energy

It is well known that the neutron-proton interaction is substantially more attractive than the like particle one, resulting in a term which appears in the energy per particle

$$\delta E_\tau = \beta\left(\frac{N-Z}{A}\right)^2 \tag{1-87}$$

where β is given by

$$\beta = \frac{1}{3}\,\frac{k_f^{\,2}}{2m^*}\,(1+F_0{}') \tag{1-88}$$

Empirically [44], $\beta \cong 25$ MeV, which would give $F_0{}' \cong 0.4$, using $m^*=(3/4)m$.

If we go back to §2 to the nucleon-nucleon interaction, we see that second- and higher-order effects of the tensor force provide the main difference between neutron-proton and like-particle interactions.

From our model of the last section for F_1, we can relate $F_1{}'$ and F_1. Eq. (1-77) would give

$$F_1{}' \cong -\frac{1}{3}\,F_1. \tag{1-89}$$

This is only very roughly satisfied in Sjöberg's calculations [29,32], where $F_1{}'$ is more like $-\frac{2}{3}\,F_1$. Our argument should be sufficient to give the sign and general size of $F_1{}'$, however. Inclusion of the second-order tensor interaction leads to a ratio more like Sjöberg's.

6.4 The Spin-Dependent Interactions

We shall treat G and G$'$ together, since we believe them to arise mainly from a common agency, ρ-meson exchange. Furthermore, since the range of the ρ-exchange potential is short, we expect these to be well represented by a zero-range interaction, independent of nuclear density, as was used by MIGDAL [1].

Since $f(\underset{\sim}{k},\underset{\sim}{k}')$ in Landau theory represents direct and exchange terms in the interaction - to all orders, in principle-spin-independent interactions can contribute, e.g., through exchange terms, to the spin-dependent effective interaction. To make this clear, let us look back at the lowest-order particle-hole interaction shown in Fig. 24.

In order to pin down the spin-dependent interaction, we look at the spin waves; the wave functions of these can be written [6]

$$T = 0 : \frac{1}{2}[P\uparrow\overline{P\uparrow} + N\uparrow\overline{N\uparrow} - P\downarrow\overline{P\downarrow} - N\downarrow\overline{N\downarrow}]$$

$$T = 1 : \frac{1}{2}[P\uparrow\overline{P\uparrow} - N\uparrow\overline{N\uparrow} - P\downarrow\overline{P\downarrow} + N\downarrow\overline{N\downarrow}]$$

where $P\uparrow$ represents a particle describing a proton of spin up, $\overline{N\downarrow}$ a neutron hole with spin down.

Given a spin- and isospin-independent interaction, only the exchange term, Fig. 24b, can contribute, as is easily verified with the above wave functions. Thus, V_σ and V_ω, as we shall call the short-range repulsion, both contribute, the former given a repulsive contribution (so as to push the particle-hole state up in energy), the latter, an attractive one. On the average, V_σ will win out, as is known from the fact that nuclei are bound, so the net effect of V_σ and V_ω will be to push both $T=0$ and $T=1$ states up, an equal amount. Translated into <u>effective</u> interactions, this means an equal contribution to G and G'. A rough estimate [17] of the combined σ- and ω-contributions to G_0 and G_0' gives ~0.2, a small but nonnegligible part of these parameters. This fits in with general experience, that spin-dependent states are most easily moved up in energy by spin-dependent forces, and that spin-independent interactions have relatively small effects.

Exchange of π- and ρ-mesons give spin-dependent forces. In infinite nuclear matter, the tensor interaction in the OPEP brings down the $M_S = 0$, $S = 1$, $T = 1$ wave [45], making it into spin-isospin sound at densities somewhat higher than nuclear-matter density [46]. In finite nuclei, no indication of this tendency is found, probably because the effective momenta q involved in the state are rather small. In Pb^{208}, for example, the spin-spin and tensor interactions from the OPEP give contributions which nearly cancel each other for the low lying $S = 1$ states [17].

This leaves us with the ρ-exchange interaction <u>and</u> the spin-dependent term from the second-order pionic tensor force, (1-5). These contribute in the same way, the latter being somewhat longer range. This similarity is no coincidence; if one looks at the crossed channel in the iterated OPEP, then a $\vec{\tau}_1 \cdot \vec{\tau}_2$ interaction involving two pions can arise only if the pions are in a state of odd angular momentum; this follows from Bose statistics. The two pions will be mainly in a P-state, and thus look like a ρ-meson of predominantly low mass, ~2 - 4m_π. Thus, the term

$$\delta V_{eff} = \frac{4}{E} \underset{\sim}{\sigma}_1 \cdot \underset{\sim}{\sigma}_2 \; (\vec{\tau}_1 \cdot \vec{\tau}_2) \; V_{t\pi}^2(r) \tag{1-90}$$

has the same sign and characteristics as the $\sigma_1 \cdot \sigma_2 \; (\vec{\tau}_1 \cdot \vec{\tau}_2)$ term in $V_\rho(r)$, (1-28), but the V_{eff} has a somewhat longer range. The question of range is important, because both $V_\rho(r)$ and V_{eff} must be modified by short-range correlations in the nuclear wave functions induced by V_ω. In G-matrix calculations this is automatically included. For our present (semiquantitative) purposes it is sufficient to multiply the above potentials by g(r), where g(r) is the two-body correlation function that the short-range repulsion would introduce.

Thus, we end up with a spin-dependent effective potential

$$\hat{V}_{\sigma\tau}(r) = \underset{\sim}{\sigma}_1 \cdot \underset{\sim}{\sigma}_2 \; (\vec{\tau}_1 \cdot \vec{\tau}_1) \left\{ \frac{4}{E} V_{t\pi}^2(r) + \frac{2}{3} \frac{f_\rho^{\;2}}{4\pi} \frac{e^{-m_\rho r}}{r} - \frac{8\pi}{3} \frac{\delta(r)}{m_\rho} \right\} g(r) \tag{1-91}$$

In a reductio ad simplicitum absurdum, we can express this approximately as a zero-range potential, since the ranges involved in (1-91) are rather short compared with internucleon spacings. Thus

$$\hat{V}_{\sigma\tau}(r) \cong \left(\frac{C}{m_\rho^{\;3}}\right) m_\rho \; (\sigma_1 \cdot \sigma_2) \; \vec{\tau}_1 \cdot \vec{\tau}_2 \; \delta(\underset{\sim}{r}) \tag{1-92}$$

where

$$C = \frac{2}{3} \frac{f_\rho^{\;2}}{4\pi} A + \delta C \tag{1-93}$$

with

$$\frac{A}{m_\rho^{\;3}} = \int \left[\frac{e^{-m_\rho r}}{m_\rho r} - \frac{4\pi}{m_\rho^{\;3}} \delta(r) \right] g(r) \; 4\pi r^2 \; dr \tag{1-94}$$

and

$$\delta C = \frac{4}{m_\rho \bar{E}} m_\rho^{\;3} \int V_{t\pi}^{\;2}(r) \; 4\pi r^2 \; dr. \tag{1-95}$$

As noted earlier, $V_{t\pi}^{\;2}(r)$ is of range $\sim(2m_\pi)^{-1}$, and zero-range approximation may not be accurate for the term described by δC, but the largest contribution to C comes from $(2/3)(f_\rho^2/4\pi)A$. Evaluation of A using a g(r) from the Reid soft-core potential [17] gives $A = 5.1$.

The tensor-part of the ρ-exchange potential is assumed to be included as a short-range cutoff in the $V_{t\pi}(r)$ as discussed in §2. The ρ-exchange tensor interaction has very little effect in lowest order, because the tensor interaction has no diagonal term in S-states, mixing S- and D-states. The latter are small at short distances because of the centrifugal barrier.

We have now reduced the effective ρ-exchange force to a zero-range one, (1-92). This is an interaction in the usual shell-model sense; both direct and exchange terms should be calculated in the particle-hole interaction, Fig. 24. For the Landau f on the other hand, only the direct term should be used; it is an effective interaction which already includes effects of exchange terms, and higher order terms, by supposition.

Insofar as $\hat{V}_{\sigma\tau}$, (1-92), is a zero-range interaction, we can replace it by a spin-independent one

$$\hat{V}_{\sigma\tau}(r) = \left(\frac{C}{m_\rho^3}\right) m_\rho \, (\underset{\sim}{\sigma}_1 \cdot \underset{\sim}{\sigma}_2)(\vec{\tau}_1 \cdot \vec{\tau}_2) \, \delta(\underset{\sim}{r}) = -3\left(\frac{C}{m_\rho^3}\right) m_\rho \, \delta(\underset{\sim}{r}) . \qquad (1\text{-}96)$$

This is possible because if two particles are at the same place, they must, by Fermi statistics, be in an $S = 1$, $T = 0$ or $S = 0$, $T = 1$ state.

In the form (1-96), we see that $\hat{V}_{\sigma\tau}$ contributes only to the exchange term, the component $P\uparrow\overline{P\uparrow}$ of one particle-hole state connecting only with the $P\uparrow\overline{P\uparrow}$ part of the other one. Thus, use of (1-96) is equivalent to using

$$\frac{3}{4}\left(\frac{C}{m_\rho^3}\right) m_\rho \left[\underset{\sim}{\sigma}_1 \cdot \underset{\sim}{\sigma}_2 + \underset{\sim}{\sigma}_1 \cdot \underset{\sim}{\sigma}_2 (\vec{\tau}_1 \cdot \vec{\tau}_2)\right] \delta(\underset{\sim}{r}) \qquad (1\text{-}97)$$

as a Fermi-liquid interaction; i.e., only in the direct term, Fig. 24. (The change in sign between (1-96) and (1-97) comes from the change from exchange term in (1-96) to direct term in (1-97).) Consequently, we find

$$G_0 = G_0' = \frac{2k_f m^*}{\pi^2} \frac{3}{4} \frac{C}{m_\rho^2} . \qquad (1\text{-}98)$$

Calculations with the above $\hat{V}_{\sigma\tau}$ give [17]

$$G_0 = G_0' \cong 1.6, \qquad (1\text{-}99)$$

or, inclusive of the contribution from $V_\sigma + V_\omega$,

$$G_0 = G_0' = 1.8. \tag{1-100}$$

This may be somewhat larger than needed empirically for the Pb region [47]. The strength eq. (1-96) is, however, appropriate for a zero-range interaction used in a shell-model calculation [48]. It is seen in [48] that inclusion of the two-particle, two-hole states effectively renormalizes the interaction to be used in a purely one-particle, one-hole space. According to the estimate of [17] based on the calculation of [48], the interaction to be used in the one-particle, one-hole space would be obtained by multiplying the G_0 and G_0' of (1-100) by a factor of ~5/7. This is the effect of renormalization at the quasiparticle pole. We end up with $G_0 \simeq G_0' \simeq 1.3$. We should note that the tensor part of the one-pion interaction, in lowest order, splits $T=0$ and $T=1$ states, giving a net effect as if G_0' is larger than G_0, in Pb^{208} at least [17].

Let us ask now, in more detail, to what extent nature knows of our above considerations. Spectra and magnetic moments of nuclei around Pb^{208} should provide the best indications, but the experimental situation with respect to spectra is not at all clear. The components of the spin-flip state, $(h^{-1}_{11/2p}\ h_{9/2p})$ and $(i^{-1}_{13/2n}\ i_{11/2n})$ come at an unperturbed energy of ~5.5 MeV. No strength is seen in the 1+ spectrum up until 7 - 8 MeV, so the 1+ states made out of these components are pushed up at least 1½ - 2½ MeV; this displacement upwards can be accomplished by G_0 and $G_0' \sim 1.3$ [17]. From evaluation of configuration-mixing corrections (the so-called Arima-Horie effect) to magnetic moments, BAUER et al. [47] determined G_0 and G_0' to have the values

$$G_0 = 1.15,\ G_0' = 1.45.$$

However, in converting the dimensionless parameters they essentially used an effective mass of $m^* = 0.8$ m, and if we use $m^* = m$, as we have been doing, we would have

$$G_0 \approx 1.45, \quad G_0' \simeq 1.85,$$

which are in line with, but possibly somewhat larger than our above estimates.

Another line we have on the spin-dependent forces is from C^{12}, which may not be much of a nucleus, but at least the experimental situation is clear. The $T=1$, 1+ state comes at 15.1 MeV, the $T=0$ one at 12.7 MeV. Both are pushed up quite far from their unperturbed position, which is the spin-orbit splitting, ~6 MeV.

6.5 The Landau Sum Rule

From our discussion of the spin- and isospin-dependent parameters in §6.3 and §6.4, we see that

$$F_0' \simeq 0.4$$
$$F_1' > 0, \qquad (1\text{-}101)$$

where we assume the symmetry energy β to be ~25 MeV for nuclear matter, and take $m^* = 0.75\ m$ for the infinite system. Our values for G_0 and G_0' were in the region of 1.5, with G_0' probably somewhat larger than G_0. Our arguments that G and G' came from the short-range ρ-exchange potential made it plausible that G_ℓ and G_ℓ', $\ell > 0$, were negligible. Thus, the contribution of the spin- and isospin-dependent parameters to the Landau sum rule is approximately

$$\frac{F_0'}{1+F_0'} + \frac{F_1'}{1+F_1'/3} + \frac{G_0}{1+G_0} + \frac{G_0'}{1+G_0'} \gtrsim 1.5. \qquad (1\text{-}102)$$

In any case, all terms are positive. Thus, the sum rule must be satisfied by the F_ℓ being negative. Assuming m^* to be (3/4)m for nuclear matter, a value consistent with the calculations discussed in §6.2 and the connection with finite nuclei referred to in that section, F would equal -0.75 and

$$\frac{F_1}{1+F_1/3} = -1. \qquad (1\text{-}103)$$

With this value of m^* and an empirical compression modulus of ~220 MeV for nuclear matter, we would have $F_0 = -\ 1/3$, and

$$\frac{F}{1+F_0} = -0.5. \qquad (1\text{-}104)$$

Adding (1-102), (1-103) and (1-104), we arrive at a consistent picture. It is amusing and important that the spin- and isospin-dependent parameters put an important constraint, through the Landau sum rule, on the effective mass and compressibility.

In fact, just from the sum rule itself we can make the argument that m^*/m must decrease with increasing density (B. Friman, private communication), and this is very important in discussions of possible pion condensation [43]. The spin-dependent terms involving G_0 and G_0', which give the largest contributions to the sum rule (1-102), will be relatively independent of density, since they come from ρ-meson exchange, which gives a density-independent zero-range interaction within the approximations discussed in the previous section. The density of states factor $2k_f m^*/\pi^2$ entering into the Landau parameters would tend to grow with k_f, but m^* decreases, as we shall see below, so that initially between densities of nuclear matter density $\rho_0 \cong 0.16/\mathrm{fm}$ and $2\rho_0$, this factor is roughly constant.

The above argument implies that

$$\frac{F_0}{1+F_0} + \frac{F_1}{1+F_1/3} \simeq -1.5, \tag{1-105}$$

at least for the range of densities $\rho_0 \lesssim \rho \lesssim 2\rho_0$.

Now it is clear that the compression modulus K increases as nuclear matter is compressed and the nucleons experience each others repulsive cores more and more. The increase in K can be accomplished either by an increase in F_0, or a decrease in the (negative) F_1, or by both of these. If F_0 increases, then F_1 must decrease in order that (1-105) be satisfied. Thus, either way, F_1 decreases with increasing density, and m^*/m must go down.

Since m^* prefixes the density of states at the Fermi surface, a decrease in m^* will make phenomena like pion condensation more difficult. Similar arguments can be applied for dense neutron matter, where a decrease in m^* makes superfluidity more difficult.

References

1. A.B. Migdal, Theory of Finite Fermi Systems and Applications to Atomic Nuclei (Interscience, New York) 1967.
2. C. de Tar, Phys. Rev. D17 (1978) 323.
3. H. Feshbach, private communication.
4. G.E. Brown and A.D. Jackson, The Nucleon-Nucleon Interaction, North-Holland Publ. Co., 1976.
5. T.T.S. Kuo and G.E. Brown, Phys. Letters 18 (1966) 40. A factor of 2 error is corrected in ref. 6.
6. G.E. Brown, Unified Theory of Nuclear Models and Forces, North-Holland Publ. Co., 3rd Ed., 1971.
7. L. Rosenfeld, Nuclear Forces, North-Holland Publ. Co., (1948).
8. R. Vinh Mau, "Mesons in Nuclei", Vol. I, edited by M. Rho and D. Wilkinson, North-Holland Publ. Co. (1979).
9. J. Durso, M. Saarela, G.E. Brown and A.D. Jackson, Nucl. Phys. A278, (1977) 445.
10. M. Gell-Mann and M. Lévy, Nuovo Cimento 16 (1960) 705.
11. J. Durso, A.D. Jackson and B. Verwest, to be published.
12. G. Breit, Proc. Natl. Acad. Sci. USA 46 (1960) 746; Phys. Rev. 120 (1960) 287.
13. R.M. Woloshyn and A.D. Jackson, Nucl. Phys. B64 (1973) 269.
14. G.E. Brown and M. Rho, Phys. Letts. 82B (1979) 177.

15. M.M. Nagels, et al., Nucl. Phys. B109 (1976) 1.

16. G. Höhler and E. Pietarinen, Nucl. Phys. B95 (1975) 210.

17. M.R. Anastasio and G.E. Brown, Nucl. Phys. A285 (1977) 516.

18. G.E. Brown, S.O. Bäckman, E. Oset and W. Weise, Nucl. Phys. A286, (1977) 191.

19. F. Iachello, A.D. Jackson and A. Lande, Phys. Letts. 43B (1973) 191.

20. A.A. Abriksov and I.M. Khalatnikov, Reports on Progress in Physics 22 (1959) 329.

21. D. Pines and P. Noziéres, The Theory of quantum Liquids, Vol. I, W.A. Benjamin, N.Y. and Amsterdam, 1966.

22. L.D. Landau, Zh. Eksper Teor. Fiz. 30 (1956) 1058; ibid. 32 (1957) 59; ibid. 35 (1958) 97.

23. G.E. Brown, Rev. Mod. Phys. 43 (1971) 1.

24. S. Bäckman, O. Sjöberg and A.D. Jackson, Nucl. Phys. A321 (1979) 10.

25. S.O. Bäckman, Nucl. Phys. A120 (1968) 593.

26. N.F. Berk and J.R. Schrieffer, Phys. Rev. Lett. 17 (1966) 433.

27. S. Doniach and E. Engelsberg, Phys. Rev. Lett. 17 (1966) 750.

28. S. Babu and G.E. Brown, Ann. Phys. 78 (1973) 1.

29. O. Sjöberg, Ann. Phys. 78 (1973) 39.

30. S.O. Bäckman, Nucl. Phys. A130 (1969) 427.

31. J. Bardeen, G. Baym and D. Pines, Phys. Rev. 156 (1967) 207.

32. O. Sjöberg, Nucl. Phys. A209 (1973) 363.

33. V.R. Pandharipande and R.B. Wiringa, Nucl. Phys. A266 (1976) 269.

34. O. Sjöberg, private communication.

35. N. Marty, M. Morlet, A. Willis, V. Comparat, R. Frascaria and J. Källne, Nucl. Phys. A, to be published.

36. D.H. Youngblood, C.M. Rosza, J.M. Moss, D.R. Brown and J.D. Bronson, Phys. Rev. Letts. 39 (1977) 1188.

37. J.P. Blaizot, D. Gogny and B. Grammaticos, Nucl. Phys. A265 (1976) 315.

38. J.P. Blaizot and B. Grammaticos, Nucl. Phys. A, to be published.

39. G.E. Brown, J.H. Gunn and P. Gould, Nucl. Phys. 46 (1963) 598.

40. G.F. Bertsch and T.T.S. Kuo, Nucl. Phys. A112 (1968) 204.

41. J.P. Jeukenne, A. Lejeune and C. Mahaux, Phys. Reports 25C (1976) 83.

42. I. Hamamoto and P. Siemens, Nucl. Phys. A269 (1976) 119.

43. Y. Futami, H. Toki and W. Weise, Phys. Lett. 77B (1978) 37.

44. A. Bohr and B.R. Mottelson, Nuclear Structure, Vol. I, W.A. Benjamin, Inc., N.Y., Amsterdam (1969).

45. A.B. Migdal, Zh ETF 61 (1971) 2210; Jetp (Sov. Phys.) 34 (1972) 1184.

46. W. Weise and G.E. Brown, Phys. Letts 48B (1974) 297.

47. R. Bauer, J. Speth, V. Klemt, P. Ring, E. Werner and T. Yamazaki, Nucl. Phys. A209 (1973) 535.

48. J.S. Dehesa, S. Krewald, J. Speth and A. Faessler, Phys. Rev. C15 (1977) 1858.

Chapter II

THE NUCLEAR SHELL MODEL

Dieter Kurath
Argonne National Laboratory
Argonne, IL

1. Introduction

The nuclear shell model is an attempt to describe the many-body nuclear system in terms of a single-particle Hamiltonian representing the average effect of the strong nucleon-nucleon interactions on a given nucleon, plus residual interactions among a smaller number of valence nucleons near the Fermi surface.

$$H = H_o(i) + v_{RES}(i,j) \tag{2-1}$$

The model was presented in this form some thirty years ago by Mayer and Jensen as a means of interpreting the observation of shell-closures at nucleon numbers 20, 50, 82 and 126 as well as providing the observed angular momentum of the ground state for a large number of odd-A nuclei throughout the periodic table. The main ingredient was the inclusion of a strong spin-orbit coupling term in the single-particle Hamiltonian so that the energy of the $j=l+\frac{1}{2}$ orbital is lowered appreciably. In addition v_{RES} was represented by coupling rules based on an attractive short-range pairing interaction so that a j^n configuration would have a lowest state which had $J=0$ for even n and $J=j$ for odd n.

Since that time much has been done both experimentally and theoretically to establish the validity of the shell model. Elastic scattering of nucleons on spherical nuclei like Ca, Ni, Sn and Pb has been measured for a wide range of energies and analyzed in terms of an optical potential. The real part of this potential is well represented by a Woods-Saxon distribution consistent with the nuclear charge distribution measured by electron scattering. The polarization measured in nucleon-nucleus scattering can be described by including a spin-orbit potential having the sign proposed by the shell model and being proportional to the derivative of the nuclear distribution and hence peaked at the surface. Under the plausible assumption that upon extrapolating to zero incident energy the optical potential will approach the single-particle potential of the shell model, we obtain a form for the first term of (1) as a function of mass-number A. The single-particle level spectrum for neutrons in such a potential is given in Fig. 1, taken from BOHR and MOTTELSON [1].

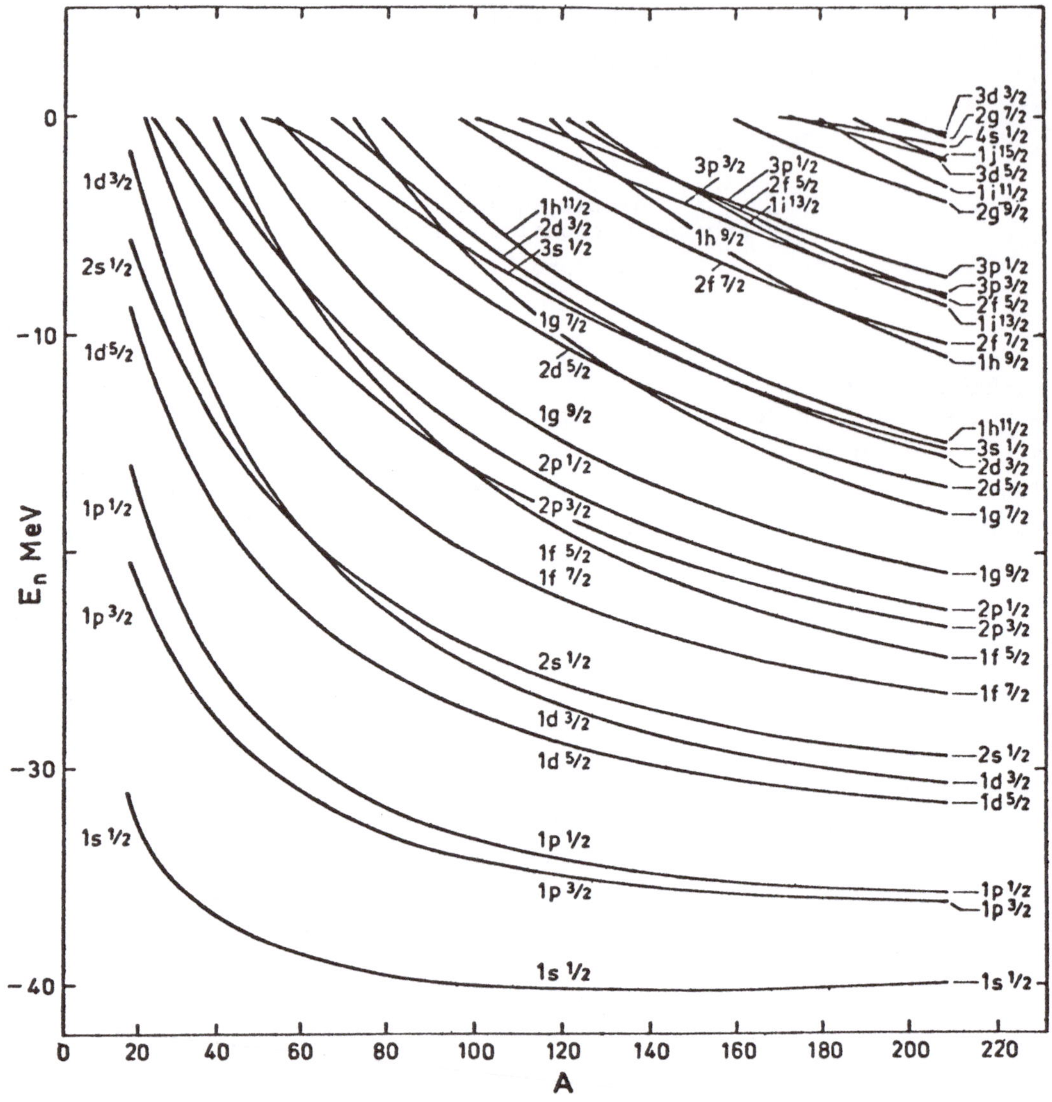

Fig. 1 Single particle levels in a potential well consistent with optical model analysis of neutron scattering [1].

The presence of magic numbers is apparent from the level spacing, so this result from elastic scattering and polarization measurements offers strong confirmation of the single-particle Hamiltonian.

A large theoretical effort has been devoted to showing why nuclei exhibit features of independent particle behavior despite the fact that in nucleon-nucleon scattering the interactions are seen to be very strong. One aspect is the treatment of infinite nuclear matter where translational invariance requires that the uncorrelated single-particle functions be plane waves. The objective is to see whether one can start with a two-body interaction determined by fitting nucleon-nucleon scattering up to an energy of 300 MeV, and obtain the binding energy per nucleon and density found in the interior of heavy nuclei. The basic ingredient which enables one to treat the strong repulsive core found at short distances in the interaction is the use of the Brueckner G-matrix. The G-matrix is an effective operator whose matrix elements between pairs of plane-wave states takes account of the high-energy two-particle excitations produced by the hard core. Up to the present, the results of such calculations are that if they produce saturation at the empirical density the binding energy per nucleon is 80% of that observed; if they agree with the observed binding the calculated density is too high. There is one important result from these calculations which gives basic support to the shell model. If one compares the correlated wavefunction which describes the relative motion of two nucleons to the uncorrelated one for the motion when no interaction is present, one finds that the major differences occur at short range, generally two fermis or less. Thus the correlation produced by the G-matrix affects the relative motion of two nucleons only when they are at distances small compared to nuclear dimensions and a large part of their motion is as though they were independent particles. This is mainly due to the Pauli Principle which inhibits scattering at low relative momenta in nuclear matter.

A related effort is the calculation of double-closed-shell nuclei starting with the G-matrix approach and a harmonic oscillator basis. This is a more difficult problem since the single-particle states must be determined as linear combinations of the oscillator states in a Hartree-Fock procedure while at the same time evaluating G-matrix elements with these states in a self-consistent fashion. Results obtained for ^{16}O, ^{40}Ca and ^{208}Pb exhibit features similar to what was found in the nuclear matter calculations. Relative to experiment one tends to have not enough binding energy and too great a density. Nevertheless, considering the nature of these hard-core potentials, the fact that one calculates nuclei of roughly the observed size and binding energy must be considered a qualitative success.

There remains the problem of determining the residual interaction v_{RES} of (1), which is to be used in calculations wherein the model space consists of a number of nucleons in a few nearly-degenerate single-particle levels near the Fermi surface. The pioneer calculation was done by Kuo and Brown who treated ^{18}O and ^{18}F assuming that v_{RES} should be the G-matrix modified by the fact that the ^{16}O core can be excited by various particle-hole vibrations. After calculating such an interaction they then calculated the energy spectra for the A=18 systems obtaining results reasonably close to observation. Attempts to obtain better agreement by including higher order terms have shown this problem to be much more difficult than was originally believed. The present procedure for determining v_{RES} in a given region of the periodic table is to start with G-matrix elements but make modifications determined by fitting the spectra of the simplest nuclei in the region so that v_{RES} is semi-phenomenological.

An alternative procedure which has been carried out for the 1p-shell and the (2s-1d) shell is to fit the spectra of a range of nuclei using the matrix elements of v_{RES} as parameters. The general result is that only a few linear combinations of matrix elements are well-determined, so that only certain aspects of v_{RES} are important in fitting the low-lying energy levels. Nevertheless the resulting wave functions have been tested by calculating many observable properties such as gamma-decays and nucleon transfer probabilities and are found to give generally good agreement with experiment.

For a thorough exposition of the techniques and problems involved in establishing these foundations of the shell model, the reader is referred to lectures by MACFARLANE [2] at the Scottish summer school in 1977. After this brief review these lectures will be concerned with the basic execution of shell model calculations and how one extracts information from them for comparison with experiment.

2. Basic Language of the Shell Model

2.1 Occupation Number Representation

Since one is dealing with systems of fermions, the many-nucleon wavefunctions must be antisymmetric under exchange of any two nucleons. For two nucleons the wavefunction is a Slater determinant

$$\psi_{\alpha\beta} = \frac{1}{\sqrt{2}} \begin{vmatrix} \phi_\alpha(1) & \phi_\beta(1) \\ \phi_\alpha(2) & \phi_\beta(2) \end{vmatrix} = \frac{1}{\sqrt{2}}[\phi_\alpha(1)\phi_\beta(2)-\phi_\alpha(2)\phi_\beta(1)]$$

where the labels α, β refer to the quantum numbers of the state. For instance in the spherical representation one has $\alpha = (nljmt_3)$ with t_3 as the neutron or proton label. Clearly the only relevant information is what states are occupied, so an equivalent form in terms of creation operators is

$$\psi_{\alpha\beta} = a_\alpha^\dagger a_\beta^\dagger|0\rangle \qquad \text{with} \qquad a_\alpha^\dagger a_\beta^\dagger = -a_\beta^\dagger a_\alpha^\dagger$$

where $|0\rangle$ is the vacuum for states of the type α, β. Clearly ψ is zero for $\alpha = \beta$ just as in the Slater determinant. We also introduce the operator a_α which destroys a particle in the state α,

$$a_\alpha a_\alpha^\dagger |0\rangle = |0\rangle \quad ; \quad a_\alpha |0\rangle = 0$$

together with the commutation relation

$$a_\alpha a_\beta^\dagger = \delta_{\alpha,\beta} - a_\beta^\dagger a_\alpha \quad \text{so that} \quad a_\alpha a_\beta^\dagger |0\rangle = \delta_{\alpha,\beta} |0\rangle .$$

The Hermitean adjoint of ψ is

$$\psi_{\alpha\beta}^\dagger = \langle 0| a_\beta a_\alpha \quad \text{and} \quad \psi_{\alpha\beta}^\dagger \psi_{\alpha\beta} = 1$$

since $\alpha \neq \beta$. The commutation relations can be summarized as

$$[a_\alpha^\dagger, a_\beta^\dagger]_+ = 0 = [a_\alpha, a_\beta]_+ \; ; \; [a_\alpha^\dagger, a_\beta]_+ = \delta_{\alpha,\beta} . \tag{2-2}$$

Since one writes the wavefunction for n nucleons in different states as the product of n creation operators applied to the vacuum, this is called the occupation number representation. It is widely used in shell model computer codes since it is well suited to the bit structure of computer words. One assigns an α value to each position in the word and then has a 1 or 0 at each position depending on whether or not the state is occupied. Bit manipulation is then easily encoded.

One-body operators are written in this representation as

$$Q = \sum_{\alpha,\beta} \langle \phi_\alpha | \, q \, | \phi_\beta \rangle a_\alpha^\dagger a_\beta \tag{2-3}$$

where the bracket is the single-particle matrix element of the operator Q. The simplest example is the number operator for which the single-particle operator is just $\delta_{\alpha,\beta}$. Two-body operators such as $v_{RES}(i,j)$ which is symmetric under exchange of i and j are written as

$$v = \sum_{\alpha<\beta,\gamma<\varepsilon} \langle [\phi_\alpha \phi_\beta] | v | [\phi_\gamma \phi_\varepsilon] \rangle a_\alpha^\dagger a_\beta^\dagger a_\varepsilon a_\gamma \tag{2-4}$$

where the wave functions inside the matrix element are antisymmetrized so that

$$\langle [\phi_\alpha \phi_\beta] | v | [\phi_\gamma \phi_\varepsilon] \rangle = \int \phi_\alpha^*(1) \phi_\beta^*(2) v(1,2) \{\phi_\gamma(1)\phi_\varepsilon(2) - \phi_\gamma(2)\phi_\varepsilon(1)\} d^3r_1 d^3r_2 \tag{2-5}$$

The matrix elements change sign under exchange of α with β or of γ with ε, and this property has been used to restrict the summations in (4) assuming some chosen order for the indices. If desired, one can write (4) as an unrestricted sum with a factor ¼ before the summation. With these definitions we can write the shell model Hamiltonian in the occupation number representation as

$$H=\sum_{\alpha,\beta}\langle\phi_\alpha|h_0|\phi_\beta\rangle a^\dagger_\alpha a_\beta+\sum_{\substack{\alpha<\beta\\ \gamma<\varepsilon}}\langle[\phi_\alpha\phi_\beta]|v|[\phi_\gamma\phi_\varepsilon]\rangle a^\dagger_\alpha a^\dagger_\beta a_\varepsilon a_\gamma\ . \tag{2-6}$$

2.2 Angular Momentum in the Spherical Basis

While the formalism of the previous section is applicable in general, for example in a deformed basis such as the Nilsson model, for nuclei near closed shells a spherical basis is used. In this case the α label refers to $(nljmt_3)$ of the orbital and a certain amount of angular momentum algebra is necessary. While such algebraic manipulations are internal to a shell model code, a certain minimal amount is necessary to understand what is going on. The basic requirement is that the operators transform under rotation like tensors of definite angular momentum j and z-component m. While this is true for the creation operator $a^{j\dagger}_m$, the destruction operator must be modified so we introduce the operator $\tilde{a}^j_m$ where

$$\tilde{a}^j_m = (-1)^{j+m}\, a^j_{-m} \tag{2-7}$$

If both neutrons and protons are filling the same levels it is convenient to use the isospin notation, where the nucleon has isospin ½ and component t_3 (+½ for neutrons) on the z-axis of isospin space. With an analogous requirement on behavior under rotation in isospin space the creation operator is written $a^j_{mt_3}$ and the hole operator is

$$\tilde{a}^j_{mt_3} = (-1)^{j+m+\frac{1}{2}+t_3}\, a^j_{-m-t_3} \tag{2-8}$$

Vector coupling of two such operators to a resultant J, component M, is achieved with Clebsch-Gordan coefficients (CGC). For $j_1\neq j_2$ we get

$$A^{J\dagger}_M(j_1t_{31}j_2t_{32})=\sum_{m_1m_2}(j_1j_2m_1m_2|JM)\,a^{j_1\dagger}_{m_1t_{31}}a^{j_2\dagger}_{m_2t_{32}}\equiv(a^{j_1\dagger}_{t_{31}}\times a^{j_2\dagger}_{t_{32}})^J_M\ . \tag{2-9}$$

The orthogonality properties of the CGC are

$$\sum_{m_1m_2}(j_1j_2m_1m_2|JM)(j_1j_2m_1m_2|J'M') = \delta_{J',J}\delta_{M',M}$$
$$\sum_{JM}(j_1j_2m_1m_2|JM)(j_1j_2m_1'm_2'|JM) = \delta_{m_1',m_1}\delta_{m_2',m_2}\ . \tag{2-10}$$

Applying (9) to the vacuum produces a normalized two-nucleon wave function. For $(nljt_3)_1=(nljt_3)_2$ the Pauli Principle, as embodied in the commutation relations (2), leads to the requirement that only even J is allowed. In this case the normalized two-nucleon state results from applying the operator

$$A_M^{J\dagger}\,(jt_3jt_3) = \sqrt{\tfrac{1}{2}}\;\delta_{J,\mathrm{even}}(a_{t_3}^{j\dagger}\times a_{t_3}^{j\dagger})_M^J. \tag{2-11}$$

In a similar way one obtains states with isospin $\Gamma=1$ or $\Gamma=0$ by vector coupling in isospin space. The general coupled operator then becomes

$$A_{M\Gamma_3}^{J\Gamma\dagger}\,(j_1j_2) = \gamma(j_1j_2J\Gamma)\;(a^{j_1\dagger}\times a^{j_2})_{M\Gamma_3}^{J\Gamma} \tag{2-12}$$

where $\gamma=1$ except in the case of identical orbitals where $\gamma=\sqrt{\tfrac{1}{2}}\,\delta_{J+\Gamma,\mathrm{odd}}$. Equation (11) is clearly included since for $t_{31}=t_{32}$ we have $\Gamma_3=\pm1$ so that Γ must equal 1, and the isospin CGC is unity. By inverting (12) with the help of (10) and using the fact that the nuclear residual interaction is generally a scalar in both ordinary space and isospin space, one can rewrite the two-body part of the Hamiltonian (6) in coupled form:

$$\sum_{J\Gamma}\;\sum_{\substack{j_1\leq j_2\\ j_3\leq j_4}} \langle\Phi^{J\Gamma}(j_1j_2)|v|\Phi^{J\Gamma}(j_3j_4)\rangle \sum_{M\Gamma_3} A_{M\Gamma_3}^{J\Gamma\dagger}(j_1j_2)A_{M\Gamma_3}^{J\Gamma}(j_3j_4). \tag{2-13}$$

This form has the advantage that there are many fewer coupled matrix elements of the interaction than in the uncoupled form. The nl label has been suppressed in (13) but it is implied by the j value.

A final necessary mathematical ingredient is the Wigner-Eckart theorem which states that the dependence on magnetic quantum numbers in a matrix element is contained in a CGC, namely

$$\langle\phi_M^I|Q_\mu^\lambda|\phi_{M_o}^{I_o}\rangle = (I_o\lambda M_o\mu|IM)\langle\phi^I||Q^\lambda||\phi^{I_o}\rangle. \tag{2-14}$$

The quantity on the right is called the reduced matrix element. One must take care since there are several definitions of RME in the literature. The quantity in (14) equals $(-1)^{2\lambda}$ times the RME of BRINK and SATCHLER [3] so it differs only for half-integral λ; it equals $\sqrt{2I+1}$ times the RME in [1] and [2]; it equals $(-1)^{2\lambda}\sqrt{2I+1}$ times the RME of Racah's original definition used in deSHALIT and TALMI [4]. The Wigner-Eckart theorem can be applied both in ordinary space and in isospin space, so we define a doubly-reduced matrix element by dividing the ordinary matrix element by a space CGC and an isospin CGC.

2.3 Spectroscopic Factors

One of the earliest means to identify which shell model orbitals are near the Fermi surface was the discovery that in single-nucleon transfer the angular distribution is determined by the ℓ value of the transferred nucleon. Later polarization measurements established the j value. Thus stripping reactions such as (d,p), (d,n) or (^{3}He,d) and pickup reactions (p,d), (d,t) or (d,^{3}He) are widely used in determining the degree of occupation of single-particle levels in a given target.

For a target having A nucleons which has quantum numbers $I_0T_0\alpha_0$ and generally $T_{03}=T_0$, the result of adding a nucleon can be expanded in states of the A+1 nucleon system as

$$a^{j\dagger}_{mt_3}\,\psi^{I_oT_o\alpha_o}_{M_oT_o}(A) = \sum_{IT\alpha,MT_3} \langle \psi^{IT\alpha}_{MT_3}|a^{j\dagger}_{mt_3}|\psi^{I_oT_o\alpha_o}_{M_oT_o}\rangle\, \psi^{IT\alpha}_{MT_3}(A+1) \qquad (2\text{-}15)$$

Note that T_3 must equal T_0+t_3 so that if t_3 is a neutron T can only equal $T_0+\frac{1}{2}$. We now apply (14) so the right hand-side of (15) becomes

$$\sum_{IT\alpha} \langle \psi^{IT\alpha}_{T_o+t_3} || a^{j\dagger}_{t_3} || \psi^{I_oT_o\alpha_o}_{T_o} \rangle \sum_{M} (I_ojM_om|IM)\psi^{IT\alpha}_{MT_o+t_3} .$$

The I_fM_f component can be projected out of both sides by multiplying with $(jI_0mM_0|I_fM_f)$ and summing over m and M_0. With the help of the orthogonality conditions (10) we have

$$(a^{j\dagger}_{t_3} \times \psi^{I_oT_o\alpha_o}_{T_o})^{I_f}_{M_f} = (-1)^{I_o+j-I_f} \sum_{T\alpha} \langle \psi^{I_fT\alpha}_{T_o+t_3} || a^{j\dagger}_{t_3} || \psi^{I_oT_o\alpha_o}_{T_o} \rangle\, \psi^{I_fT\alpha}_{M_fT_o+t_3}$$

(2-16)

where the phase factor comes from changing the order of I_0 and j in the CGC above. The reduced matrix element on the right indicates the degree to which the strength for transfer is spread over the final states, and the cross section for reaching a particular state by transfer of a j,t_3 nucleon is

$$\frac{d\sigma}{d\Omega}^{j}(I_oT_o\alpha_o\rightarrow I_fT_f\alpha_f) = \left(\frac{2I_f+1}{2I_o+1}\right) \langle \psi^{I_fT_f\alpha_f}_{T_o+t_3} || a^{j\dagger}_{t_3} || \psi^{I_oT_o\alpha_o}_{T_o} \rangle^2\, \sigma(n\ell j) \qquad (2\text{-}17)$$

Here $\sigma(n\ell j)$ is the DWBA cross section for adding such a nucleon to a core nucleus in which this orbital is empty. The definition of the spectroscopic factor S^j is

$$(T_o 1/2 T_o t_3 | T_f T_o + t_3)^2 \; S^j(I_oT_o\alpha_o, I_fT_f\alpha_f) = \langle \psi^{I_fT_f\alpha_f}_{T_o+t_3} || a^{j\dagger}_{t_3} || \psi^{I_oT_o\alpha_o}_{T_o} \rangle^2$$

(2-18)

that is, S^j is the square of the matrix element of a creation operator reduced in both ordinary space and isospin space. If it is possible to connect the states I_o and I_f by adding nucleons in more than one orbital, the total stripping cross section is obtained by summing (17) over such j values.

Sum-rules for stripping can be derived by taking the product of (15) with its adjoint as

$$<\psi^{I_oT_o\alpha_o}_{M_oT_o} | a^j_{mt_3} a^{j\dagger}_{mt_3} | \psi^{I_oT_o\alpha_o}_{M_oT_o}> = \sum_{IT\alpha} < \psi^{IT\alpha}_{T_o+t_3} || a^j_{t_3} || \psi^{I_oT_o\alpha_o}_{T_o}>^2 \sum_M (I_o j M_o m | IM)^2 . \quad (2\text{-}19)$$

If we sum over m on both sides the CGC sum becomes $(2I+1)/(2I_o+1)$ and on the left

$$\sum_m a^j_{mt_3} a^{j\dagger}_{mt_3} = \sum_m (1-a^{j\dagger}_{mt_3} a^j_{mt_3}) = (2j+1) - n^j_{t_3} \equiv h^j_{t_3}$$

which is the number of t_3 holes in the j orbital. With the help of (18) we obtain

$$<h^j_{t_3}>_{I_oT_o\alpha_o} = \sum_{IT\alpha} \left(\frac{2I+1}{2I_o+1}\right) (T_o 1/2 T_o t_3 | T T_o+t_3)^2 \; S^j(I_oT_o\alpha_o, IT\alpha) . \quad (2\text{-}20)$$

Further sum rules are obtained by summing over j or t_3 or both.

The nucleon removal problem can be treated in analogous fashion starting by applying the destruction operator to the target wavefunction and expanding in states of the A-1 nucleon system. The equation for the pickup cross section becomes

$$\frac{d\sigma}{d\Omega}(I_oT_o\alpha_o \to I_fT_f\alpha_f) = <\psi^{I_oT_o\alpha_o}_{T_o} || a^{j\dagger}_{t_3} || \psi^{I_fT_f\alpha_f}_{T_o-t_3}>^2 \sigma(n\ell j) \quad (2\text{-}21)$$

where here $\sigma(n\ell j)$ is the DWBA cross section for removing such a nucleon from a core nucleus where this orbital is full. The reduced matrix element in (21) is

$$<\text{RME}>^2 = (T_f\; 1/2\; T_o - t_3 t_3 | T_o T_o)^2 \; S^j(I_fT_f\alpha_f, I_oT_o\alpha_o) . \quad (2\text{-}22)$$

Note that the spectroscopic factor is always defined as the doubly-reduced matrix element of a creation operator, so that the same S is involved when one measures pickup from the A+1 ground to the A ground state as for the inverse stripping reaction. The pickup sum rule analogous to (20) gives the number of $n^j_{t_3}$ nucleons in the target.

$$<n^j_{t_3}>_{I_oT_o\alpha_o} = \sum_{IT\alpha} (T 1/2 T_o - t_3 t_3 | T_o T_o)^2 \; S^j(IT\alpha, I_oT_o\alpha_o) . \quad (2\text{-}23)$$

Experimentally it is found that the strength for transfer involving orbitals near the Fermi surface is generally confined to a few states quite localized in excitation energy. Such transfer measurements have provided a large amount of information on the degree of occupation of orbitals throughout the periodic table.

In the development of the shell model, coefficients of fractional parentage (CFP) were introduced in order to simplify the calculation of the effect of one-body operators in configurations of n active nucleons. This technique was adopted from Racah's treatment of atomic configurations and the nuclear formulation is presented in [4]. The CFP are defined by the expansion

$$\psi^{IT\alpha}(1\ldots n) = \sum_{I_oT_o\alpha_o, j} \langle IT\alpha\{|I_oT_o\alpha_o; j\rangle \left[\psi^{I_oT_o\alpha_o}(1\ldots n-1) \times \phi^j(n)\right]^{IT} \qquad (2\text{-}24)$$

wherein the wavefunctions for n and n-1 nucleons are antisymmetric. Then the elements for a one-body operator can be calculated for the nth nucleon and the final answer is just n times this result.

The CFP are simply related to the reduced matrix elements of creation operators in the occupation number representation. An equation similar to (24) is obtained by applying the number operator for the n nucleons in the active orbitals to ψ,

$$\sum_{jmt_3} a^{j\dagger}_{mt_3} a^j_{mt_3} \psi^{IT\alpha}(1\ldots n) = n\, \psi^{IT\alpha}(1\ldots n)\,.$$

By inserting a complete set of wavefunctions for n - 1 nucleons between $a^{j\dagger}$ and a^j one can write

$$n\,\psi^{IT\alpha} = \sum_{I_oT_o\alpha_o, j} \langle \psi^{IT\alpha}||a^{j\dagger}||\psi^{I_oT_o\alpha_o}\rangle (-1)^{n-1} \left[\psi^{I_oT_o\alpha_o} \times a^{j\dagger}\right]^{IT} \qquad (2\text{-}25)$$

where the doubly-reduced matrix element, called the spectroscopic amplitude, is the square root of the spectroscopic factor appearing in (23). The phase factor in (25) comes from permuting $a^{j\dagger}$ through $\psi^{I_oT_o\alpha_o}$ to get the same order as in (24).

One cannot yet relate the coefficients since the coupled wavefunction on the right of (24) is not antisymmetric to exchange of the nth nucleon with the others as is true in (25). This is remedied by applying the antisymmetrizer A

$$A = \frac{1}{\sqrt{n}}\left[1 - \sum_{k=1}^{n-1} P_{kn}\right]$$ (where P_{kn} exchanges nucleons k and n) to both

sides of (24). On the left we get $\sqrt{n}\,\psi^{I}$ and on the right, since ψ^{I_o} is some linear combination of Slater determinants of n-1 nucleons, the effect of A is to produce Slater determinants of n nucleons. These Slater determinants have a one to one correspondence with the terms in the coupled wavefunction on the right of (25). Hence

$$\sqrt{n}\,\psi^{IT\alpha}(1\ldots n) = \sum_{I_oT_o\alpha_o,j} \langle IT\alpha\{|I_oT_o\alpha_o;j\rangle\, A\,[\psi^{I_oT_o\alpha_o}(1\ldots n-1)\times\phi^{j}(n)]^{IT} \qquad (2\text{-}26)$$

and the spectroscopic amplitude is simply related to the one-nucleon CFP as

$$\langle\psi^{IT\alpha}\,||\,a^{j\dagger}\,||\,\psi^{I_oT_o\alpha_o}\rangle = (-1)^{n-1}\sqrt{n}\langle IT\alpha\{|I_oT_o\alpha_o;j\rangle. \qquad (2\text{-}27)$$

In similar fashion, doubly-reduced matrix elements of the coupled operator A (12) are related to two-nucleon CFP as

$$\langle\psi^{IT\alpha}\,||A^{J\Gamma\,\dagger}(j_1,j_2)||\,\psi^{I_oT_o\alpha_o}\rangle = \sqrt{\frac{n(n-1)}{2}}\,\langle IT\alpha\{|I_oT_o\alpha_o;J\Gamma(j_1j_2)\rangle. \qquad (2\text{-}28)$$

Thus these doubly-reduced matrix elements are simply generalizations of Racah's original CFP definitions.

3. Current Shell Model Programs

3.1 Standard Programs

The development of the shell model since the 1950's has coincided with the increasing power of digital computers. Despite the tremendous increase in speed and core size of computers, the problem of treating many nucleons in many unfilled orbitals expands in size so rapidly that only modest gains have been made in the complexity of problems that can be treated today. It is of course true that complete 1p-shell calculations which took many months in the 1950's can be done in minutes today. However it is only recently that the treatment of nuclei in the middle of the ($2s_{1/2}\ 1d_{3/2}\ 1d_{5/2}$) shell has been carried without any restriction on the space. The main things we hope to learn from such complete treatments is how to make physically sensible approximations to the few low-lying eigenstates for which we have experimental counterparts.

There are two systems of shell model programs which start with a model space ($j_1^{N_1}\, j_2^{N_2}\ldots j_k^{N_k}$), produce a set of basis states for N nucleons in this space, and then construct and diagonalize matrices consisting of the single-particle Hamiltonian $\sum_j \varepsilon_j n_j$ plus the two-body term (13). The N_k may take on all values consistent with a total N or be restricted by assumption or necessity. The Rochester-Oak Ridge system and the Argonne system proceed in different ways, but each calculates and stores CFP and uses the power of Racah algebra to handle the involved angular momentum coupling. In this sense they are standard programs, since they have adapted techniques originally developed for hand-computation to the digital computer.

The Rochester-Oak Ridge shell model system is compartmentalized by first constructing and storing IT eigenfunctions and CFP (reduced matrix elements for $a^\dagger, a^\dagger a^\dagger, a^\dagger\tilde{a}$, etc.) for each j^N configuration. Multi-configuration basis states and energy matrices are then constructed by coupling results of this factorized representation. A thorough discussion of this procedure is given in the review by FRENCH, HALBERT, McGRORY AND WONG [5]. The system is available at many computing centers in both IBM and CDC versions.

Since the Argonne shell model system is less familiar in general and more familiar to me, I use it as a step-by-step example of a standard shell model calculation. The system has evolved through four increasingly sophisticated computers and now operates on IBM 360 or 370 computers, in a version designed by GLOECKNER [6]. Neutrons and protons are treated separately until they are coupled in the final energy matrix, so the Coulomb interaction of protons can be easily included. Of course if the single-particle energies and residual interaction are the same for neutrons and protons, the wavefunctions will be isospin eigenfunctions although each matrix contains all T values. The individual programs are described below in normal sequence.

BASIS The input specifies orbital (nlj) values, number of neutrons, possible restrictions and desired resultant I and parity values. The N-neutron basis is constructed by filling the orbitals in order for $I_z=I$ and then solving the $I^+\psi = 0$ equations to obtain eigenfunctions of I. This is done for both the N and N-2 neutron systems.

BDFP Constructs reduced matrix elements of $A^{J1\dagger}(j_1j_2)$ between the basis states of N and N-2 neutrons in the form of (28).

TBME Produces the interaction matrix elements of (13) either by calculating them for central interactions or by having them read in numerically.

ENEMAT Constructs the energy matrices of (13) for the N-neutron system, each matrix element being of the form

$$\langle I\alpha|v|I\alpha'\rangle = \sum_{\substack{J,j_1\leq j_2\\ j_3\leq j_4}} \langle\phi^{J1}(j_1j_2)|v|\phi^{J1}(j_3j_4)\rangle \quad \mathrm{x}$$

$$\sum_{I_0\alpha_0} <\psi^{I\alpha}||A^{Jl^\dagger}(j_1j_2)||\psi^{I_0\alpha_0}><\psi^{I\alpha'}||A(j_3j_4)^{Jl\dagger}||\psi^{I_0\alpha_0}>$$

DIAG Adds the single-particle contributions to the matrix of V and diagonalizes the resultant matrix for each I of the N-neutron basis.

Thus far we have carried out an identical-nucleon calculation. The sequence is then repeated for protons, including the Coulomb interaction if desired.

The proton-neutron interaction is handled in differing fashion from that for identical nucleons. Starting from (6) one couples the neutron creation operator with the neutron destruction operator in the form of (7) and does the same for the proton operators. Then one calculates reduced matrix elements of these coupled particle-hole operators separately in the neutron and proton bases. This procedure also requires the V_{np} matrix elements to be in the particle-hole form which means linear combinations of the interaction matrix elements of (13) weighted by Racah coefficients. This is achieved by the following programs.

BUFP Constructs reduced matrix elements of the coupled particle-hole operator for states of the N-neutron system:

$$<I\alpha||(a_\nu^{j_1\dagger} \times \tilde{a}_\nu^{j_2})^J||I'\alpha'>$$

This is also done for the proton basis. These matrix elements can also be used later on to calculate transitions between diagonalized states.

PNME Produces the neutron-proton matrix elements V_{np} as in TBME.

INTEMB This program produces the final energy matrix for a designated angular momentum by combining the DIAG matrices for protons with those for neutrons and then calculating and adding the contributions of the neutron-proton interaction.

HMAT This is a diagonalization routine which produces eigenvalues and eigenvectors expressed in the neutron-proton basis.

SPECTRA Arranges eigenvalues in order listing the energy with respect to the ground state.

After this basic shell model calculation, additional programs are used to calculate quantities of interest such as one and two nucleon spectroscopic amplitudes, electromagnetic transition probabilities and beta decays. A typical calculation of the normal parity states of ^{12}C and ^{13}C for $(1s)^4(1p)^{A-4}$ configurations (wherein 9 matrices, the largest of dimension 17 are constructed and diagonalized) takes 15 seconds at a cost of 88¢ on the IBM 370/195. The calculation expands rapidly for the positive parity states of ^{13}C where the basic configurations in a $1\hbar\omega$

approximation are $(1s)^3(1p)^{10}$ and $(1s)^4(1p)^8(2s,1d)^1$. Here four of the matrices have dimension greater than 100, the largest being 198 for $I=3/2^+$, and the calculation takes 440 seconds at a cost of $31.

Despite the fact that these two standard shell model systems use different methods of constructing basis functions and energy matrices, there is not much difference between them in speed of execution or size of problem that can be handled. The maximum complexity without restricting level occupation is something like 6 nucleons in the (2s1/2 1d3/2 1d5/2) levels. Nevertheless much detailed information has been obtained from such calculations. They have provided microscopic descriptions of the nature of electromagnetic transitions between low-lying states. They have also suggested which features of the residual interaction are important for making approximations in nuclei where standard shell model calculations are not feasible.

3.2 The SU_3 Representation

The classification of the spatial wavefunctions of many-particle states in a deformed harmonic oscillator potential in terms of the group SU_3 was demonstrated by ELLIOTT [7]. He also showed the connection between this classification and the separation of states into rotational bands. For light nuclei where the effect of spin-orbit coupling is relatively less important than elsewhere, it was found that the low-lying states resulting from diagonalization of the Hamiltonian including v_{RES} had a large overlap with certain states of the SU_3 classification. This result suggests that a system of shell model programs in such a representation would be very useful, offering a good way to truncate the model space. The early review of the merits of the SU_3 scheme was presented by HARVEY [8]. Since then computer programs have been developed and applied to many problems. The new ingredient is inclusion of the algebra of the SU_3 group (see [9] for example) which is analogous and supplementary to the Racah algebra of the rotation group.

The unique feature of the SU_3 representation is that one can adequately remove the effect of the excitation of the center of mass. This problem arises because 3 of the 3A coordinates in a shell model description are concerned with the motion of the center of mass of the system. In a harmonic oscillator well, as long as one does not put nucleons into the next major oscillator shell before the lower shells are filled, all solutions have the center of mass in its ground state and there is no problem. If only a single nucleon is in violation of this condition, as in our previous example of the positive parity states of ^{13}C, the spurious effects of excitation of the motion of the center of mass can be removed. The complete space required to describe such $1\hbar\omega$ excitation must be included, and then the standard shell model codes include the Hamiltonian of the center of mass with a large coefficient so that the spurious states are raised to very high excitation energy. However the model space required to implement the technique quickly becomes too large if several nucleons are promoted between shells. The SU_3 representation offers an easy way to remove spurious states since the center of mass raising operator has a simple effect in this classification. One can thus properly treat configurations wherein several major oscillator shells are only partially occupied, and this treatment has been applied to nuclei above and below ^{16}O with up to 3 such oscillator shells, (1p), (2s,1d) and (2p,1f).

Additional advantages of the SU_3 representation are simplicity in treating multinucleon transfer such as clusters with the spatial symmetry of alphas or tritons, and flexibility in extending the basis for particular types of states without a rapid explosion of the size of the space. Limitations are the increased complexity in treating nuclei for which spin-orbit coupling has a strong effect, and lack of documentation for existing systems.

3.3 The Glasgow Code

The most recent set of shell model programs is the Glasgow code which is best adapted of all to exploiting the power of modern digital computers. This code uses no CFP, does no angular momentum coupling, and has as its main ingredient the use of the Lanczos method of matrix construction and diagonalization. In this method one starts with an arbitrary vector, v_1, in a K-dimensional space and operates on it with the Hamiltonian H to obtain a new vector v_2 as indicated below, and repeats this sequence on v_2. A series of orthonormal vectors is generated as

$$Hv_1=\gamma_1 v_1+\beta_1 v_2$$

$$Hv_2=\beta_1 v_1+\gamma_2 v_2+\beta_2 v_3$$

$$Hv_3=\beta_2 v_2+\gamma_3 v_3+\beta_3 v_4$$

$$Hv_4=\beta_3 v_3+\gamma_4 v_4+\beta_4 v_5 \qquad \text{et cetera.}$$

This procedure terminates with the vector v_K since the space is spanned, and diagonalization of the tri-diagonal matrix formed by the coefficients of the v_i gives the eigenvalues of the problem. However an extremely important feature is that as the number of vectors k increases, the low eigenvalues of the (k x k) partial matrix converge rapidly to the low eigenvalues of the full (K x K) matrix. This feature is vital for shell model calculations since it is just these low energy eigenvalues which are of physical interest. A brief outline of the procedure is given below. It is described in clear detail in the review article by WHITEHEAD, WATT, COLE and MORRISON [10].

The calculations use as basis a complete set of Slater determinants for the N active nucleons in the form

$$\phi = a^{\dagger}_{\alpha_1} a^{\dagger}_{\alpha_2} a^{\dagger}_{\alpha_3} \cdots a^{\dagger}_{\alpha_N} |0\rangle$$

where $\alpha = (nljmt_3)$. No vector coupling is done and the only requirements are that $\Sigma m_i = M$ and $\Sigma t_{3i} = T_3$, so the basis functions are not eigenfunctions of J or T. To carry out the Lanczos method where V is in the uncoupled form of (5), many matrix elements must be stored even though one uses the symmetries mentioned after (5). The vectors are linear combinations of the Slater determinants

$$v_k = \sum_i a_{ki} \phi_i$$

and convergence is faster if v_1 contains ϕ_i which are most likely to occur in low-lying states. Clearly one must have efficient means of storing and accessing the ϕ_i, matrix elements of H and the a_{ki}. It is also necessary to maintain stringent numerical orthonormality in the v_k.

The matrices are set up and diagonalized after v_k has been created in the kth iteration. Numerical eigenvalues are noted as a function of k, and the convergence criterion is that the eigenvalue remain unchanged between iterations to 8 or 9 significant figures. Experience in the (2s,1d) shell has shown that irrespective of K, the size of the space, there is convergence to the lowest n eigenvalues, where $1 \leq n \leq 10$, in less than 100 iterations. This is remarkably fast since near the middle of this shell the size of the space is often greater than 10^4, the maximum size being $K = 9.3 \times 10^4$ for $M = 0 = T_3$ in ^{28}Si. Finally one must operate with J^2 and T^2 on each eigenfunction in order to identify these quantum numbers.

Some symmetries can be used to reduce the size of the calculation, such as separating even J from odd J in even A nuclei with M=0 by noting behavior under rotation of 180° and using only linear combinations of Slater determinants with the proper behavior. An example of the magnitude of calculation given in [10] is the case of the odd-J,T=0 states of ^{26}Al where the basis size is 17,500. The average time per iteration is 0.38 minutes and convergence was achieved for the lowest 2 states in about 50 iterations, the next 3 by about 75 iterations, and the next 3 by about 95 iterations. The identification of J and T took 1.32 minutes per state. The Glasgow code is clearly the most powerful technique in existence for carrying out shell model calculations in a large space.

4. One Body Transition Density

4.1 General Formulation for One Body Operators

After one has carried out a shell model calculation by selecting a basis of active single-particle levels, choosing a residual interaction and comparing the resultant energy spectra with experiment, there are still many ways to test whether the model is giving a good description of the nuclear states in question. One way mentioned earlier is the comparison of calculated spectroscopic factors with single-nucleon transfer experiments. This tests the occupation probabilities of the active levels. Most other tests involve transitions between nuclear states which are mediated by one body operators. In this category are electromagnetic transitions and moments, beta decays, muon capture and inelastic scattering of electrons, pions and nucleons. In the case of nucleons, exchange can be an important lowest order process and this requires a two body operator.

Since the nuclear states are eigenstates of J^2 one uses one body operators of definite multipolarity, λ, and z-component μ. Starting from (3), the one body operator for transitions in a given nucleus can be written as

$$Q^{\lambda}_{\mu}=\sum_{j_1 j_2 t_3} \langle \phi^{j_2}_{t_3} \| q^{\lambda} \| \phi^{j_1}_{t_3} \rangle \sum_{m_1 m_2} (j_1 \lambda m_1 \mu | j_2 m_2)\, a^{j_2 \dagger}_{m_2 t_3} a^{j_1}_{m_1 t_3} \tag{2-29}$$

where we have used the Wigner-Eckart theorem on the single-particle matrix element. By reordering the CGC and using (7) this becomes

$$Q^{\lambda}_{\mu} = \sum_{j_1 j_2 t_3} \left(\frac{2j_2+1}{2\lambda+1}\right)^{\frac{1}{2}} \langle \phi^{j_2}_{t_3} \| q^{\lambda} \| \phi^{j_1}_{t_3} \rangle \left(a^{j_2 \dagger}_{t_3} \times \tilde{a}^{j_1}_{t_3}\right)^{\lambda}_{\mu} \tag{2-30}$$

Therefore the reduced matrix element between two nuclear states is

$$\langle \psi^{I_f}_{T_3} \| Q^{\lambda} \| \psi^{I_i}_{T_3} \rangle = \sum_{j_1 j_2 t_3} \left(\frac{2j_2+1}{2\lambda+1}\right)^{\frac{1}{2}} \langle \phi^{j_2}_{t_3} \| q^{\lambda} \| \phi^{j_1}_{t_3} \rangle \langle \psi^{I_f}_{T_3} \| \left(a^{j_2 \dagger}_{t_3} \times \tilde{a}^{j_1}_{t_3}\right)^{\lambda} \| \psi^{I_i}_{T_3} \rangle \tag{2-31}$$

The quantity on the far right of (31) is a matrix element of the one body transition density. These same matrix elements appear for any one body process of multipolarity λ, but they are weighted by different single-particle matrix elements in the sum depending on the nature of the process involved. For example, in gamma transitions the operators q^{λ} are derived in the long wavelength limit for the electromagnetic interaction. For inelastic electron scattering with appreciable momentum transfer one must include the dependence on momentum transfer in the interaction and the q^{λ} are more complicated operators. The transition density matrix elements contain the dependence on the nature of the nuclear states. One can also formulate (31) in terms of isospin-coupled transition density since via (8)

$$(a_{t_3}^{j_2\dagger} x\tilde{a}_{t_3}^{j_1})^{\lambda}_{\mu} = \sum_{\Gamma} (-1)^{\frac{1}{2}+t_3} (\tfrac{1}{2}\tfrac{1}{2}t_3 - t_3|\Gamma o)(a^{j_2\dagger} x\tilde{a}^{j_1})^{\lambda\Gamma}_{\mu o}. \qquad (2\text{-}32)$$

However it is often of interest to see the separate contributions of neutrons and protons. Consider two M1 transitions to the $I=\frac{1}{2}=T$ ground state of ^{13}C, one from the first excited state $I=3/2$ and the other from an excited state $I=\frac{1}{2}$*. In a $1p_{3/2}$, $1p_{1/2}$ basis the calculated numerical values for

$<\frac{1}{2}\frac{1}{2}||(a_{t_3}^{j_2\dagger} x\ \tilde{a}_{t_3}^{j_1})^1||I\frac{1}{2}>$ are

	$j_2j_1=$	1/2 1/2	1/2 3/2	3/2 1/2	3/2 3/2
I=3/2,	Neutron	-.001	-.186	.819	-.095
	Proton	.050	-.018	.008	.080
I=1/2*,	Neutron	.087	.022	-.002	.039
	Proton	-.010	.302	.773	-.064

The corresponding single-particle contributions needed to evaluate (31) are

Neutron	.441	-1.246	1.246	-1.394
Proton	-.183	1.494	-1.494	2.763 .

Clearly the major contributions in both cases come from changing $p_{1/2}$ nucleons to $p_{3/2}$ nucleons and vice versa, but the transition from I=3/2 is mainly a neutron transition while that from I=1/2* is mainly a proton transition. The reduced matrix element is obtained by multiplying corresponding terms and adding all contributions as in (31). The reduced transition probability B(M1) is given by the square of (31) divided by $I+\frac{1}{2}$, and equals 1.21 $(nm)^2$ for I=3/2 and 0.85 $(nm)^2$ for I=1/2*, where (nm) stands for nuclear magnetons. These are both strong M1 transitions and although the calculated neutron or proton natures of the transitions cannot be confirmed by gamma decay measurements, other experimental means such as inelastic pion scattering, to be discussed later, can throw light on this question.

4.2 Varying Representations to Further Understanding

One of the primary functions of the shell model is to offer explanations of why particular transitions are favored or have a special nature such as the above transitions in ^{13}C. This requires casting the wave functions in some representation which makes the observed features obvious. Often this can be done by means of the CFP expansions (24) or (25). Then if one or two amplitudes with common parentage are dominant in the expanded wavefunctions, the nature of a transition between the two states is clear. In our ^{13}C cases this procedure is not helpful since the states involved have appreciable components to many states of the A=12 system which have both isospins T=0 and 1. However a useful representation is obtained by expressing these states as neutron holes coupled to states of ^{14}C. Furthermore this is an orthonormal basis because the 1p neutron shell is full in ^{14}C. A good approximation to the ^{13}C states is obtained by keeping only the lowest I=0,1 and 2 states of ^{14}C,

$$\psi 1/2 = .931[\tilde{a}_\nu^{p1/2} x \Phi^0 (^{14}C)]^{1/2} + \ldots.$$

$$\psi^{3/2} = .713[\tilde{a}_\nu^{p3/2} x \Phi^0 (^{14}C)]^{3/2} + .590[\tilde{a}_\nu^{p1/2} x \phi^2 (^{14}C)]^{3/2} + \ldots$$

$$\psi^{1/2*} = .921[\tilde{a}_\nu^{p1/2} x \Phi^1 (^{14}C)]^{1/2} + \ldots.$$

Thus the transition from I=3/2 to the ground state clearly comes from filling the $p_{3/2}$ neutron hole and making a $p_{1/2}$ neutron hole since there is no contribution from M1 transitions between the ^{14}C states. On the other hand, the transition from I=1/2* to the ground state is a transition from the I=1 state to the I=0 state of ^{14}C with the neutron hole as spectator, and this is a proton transition since the neutron shell is full. There must also be a simple relationship between this transition and the I=1 to I=0 transition in ^{14}C, which is confirmed by experiment. Thus the main features seen in the transition density matrices are given by these approximate wave functions and the remaining features arise from the small components.

A second example concerns the understanding of the gamma transitions in the self-conjugate odd-Z nucleus ^{10}B. Many transitions between low-lying states have been measured, and the strong M1 and E2 transitions are indicated in Fig. 2. In such a $T_3 = 0$ nucleus M1 transitions with $\Delta T = 0$ are expected to be weak since they must be isoscalar wherein the neutron and proton spin contributions interfere destructively. Therefore the strong M1 transitions are all $\Delta T = 1$ wherein these spin contributions are constructive; nevertheless there is clearly added selectivity evident in Fig. 2. The $T = 1$ states are on the right in Fig. 2 and the $T = 0$ states have been separated into two groups which have strong E2 transitons within each group but not between groups. One added feature not included in the figure is a selectivity among isoscalar M1 transitions; for example the transitions from 20 to 10 and from 10* to 10 are two orders of magnitude weaker than the transition from 20 to 10*. One would like to have an explanation for these various forms of selectivity.

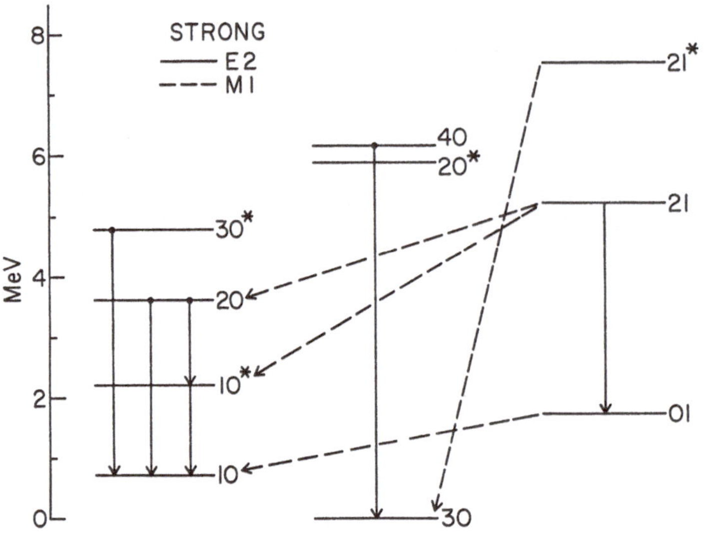

Fig. 2 Experimentally strong E2 and M1 transitions between ^{10}B states identified by IT.

Shell model calculations for ^{10}B have generally been carried out in the jj-representation of $1p_{3/2}$ and $1p_{1/2}$ orbitals. The early calculations assumed a purely central form for the residual two-body interaction while later calculations included two-body spin-orbit and tensor interactions as well. Both calculations fit the low-lying energy levels equally well, but the case with non-central terms gives better agreement with transition probabilities, mainly because the nature of the 10 and 10* states is exchanged relative to the purely central case. This is an example where one-body transitions can be useful in deciding between equally good energy level fits. The calculated transition probabilities in the better case show a reasonably good correlation with observed strong and weak transitions but no reason for selectivity is evident. Furthermore the calculated strong E2 transitions have values still about a factor of five weaker than measured values. While it is known that transitions within rotational bands in odd-A nuclei must be enhanced by effective charges, the band structure in this odd-odd nucleus is not evident in the jj-representation.

Clarification can be obtained by transforming the wavefunctions to the Wigner supermultiplet representation wherein the spatial wavefunction of the six 1p nucleons is classified according to its symmetry under permutation and then multiplied by a spin-isospin function of adjoint symmetry to obtain a totally antisymmetric wavefunction. The low-lying states of ^{10}B are found to belong predominantly to the most symmetric spatial classification [42] which contains states of orbital angular momentum L=0,2[2], 3 and 4. The adjoint spin-isospin function contains only (S=0,T=1) and (S=1,T=0). The [42] classification is equivalent to the SU_3 classification $(\lambda,\mu)=(2,2)$ and SU_3 has in addition the Elliott quantum number K_L, the projection of orbital angular momentum on the nuclear symmetry axis. This leads to a separation of L values according to $K_L=0$ with L=0 and 2; $K_L=2$ with L=2,3 and 4.

The selectivity of the transitions can be understood by keeping only the largest component in this classification for each state of Fig. 2. The T=0 states have S=1 and hence I can be L or L$\pm$1; the T=1 states have S=0 so that I=L. The corresponding dominant components are:

IT =	01	21	21*	30	20*	40	10	10*	20	30*
LK_L =	00	20	22	22	22	32	00	20	20	20.

Therefore the approximately valid quantum number K_L is what distinguishes the two groups of T=0 states, those on the left of Fig. 2 having $K_L=0$ while those in the center have $K_L=2$. One would then expect enhanced E2 transitions only within each K_L band.

Within the [42] classification the M1 transitions with $\Delta T=1$ must also have $\Delta S=1$, so the M1 operator is proportional to $(\mu_n-\mu_p)\sigma$ and hence has the selection rules $\Delta L=0$ and $\Delta K_L=0$. This explains the selectivity of the strong M1's in Fig. 2 since, aside from those indicated, other possible $\Delta T=1$ M1's are either forbidden by one of these selection rules or inhibited by having a small difference in energy. Finally the difference in strengths for isoscalar M1's is also explained since the operator is $(\mu_n+\mu_p-1/2)\sigma$ which has the same L and K_L selection rules. Hence the 20 to 10* transition is allowed while the 20 to 10 and 10* to 10 transitions have $\Delta L=2$ which accounts for their being weaker by two orders of magnitude.

Casting the wave functions into the SU_3 representation thus offers a simple qualitative explanation of the relative magnitudes of gamma transitions in ^{10}B. A more detailed description of this procedure is given in [11].

4.3 Stretched States in Inelastic Pion Scattering

The excitation of states of high angular momentum with change of parity by a stretched particle-hole operator was suggested by DONNELLY and WALKER [12] for (e,e') experiments at large momentum transfer. The term "stretched" means that for a given nucleus the hole is made in the valence orbital of largest j and the particle is excited to the state of largest j in the next oscillator shell with the particle and hole coupling to the maximum resultant, $J = j_2 + j_1$. For a 1p shell target this means $(a^{d_{5/2}\dagger} x \tilde{a}^{p_{3/2}})^{4-}$; for a (2s,1d) target it means $(a^{f_{7/2}\dagger} x \tilde{a}^{d_{5/2}})^{6-}$.

In the case of (e,e') these are magnetic excitations and they are strongest for isovector transitions since in this case the neutron and proton single-particle spin contributions add constructively. A recent comparison of results for (e,e') and (p,p') [13] sheds light on the nucleon-nucleon interaction needed to analyze (p,p') scattering near 135 MeV. There is much current experimental investigation of such states.

Recent results on (π,π') scattering also show excitation of stretched states, and one sees both T=1 and T=0 states in nuclei having T=0 ground states. For stretched states only a single term contributes to the $j_1 j_2$ sum in (30), leaving just the sum of neutron and proton contributions. For (π,π') scattering at energies near the J=3/2, T=3/2 resonance in the π-nucleon interaction, the relative contributions of neutrons and protons depend mainly on the charge of the pion through the isospin CGC coupling the pion and nucleon to a resultant T=3/2. The resultant expressions for $\pi^{\pm}$ cross sections depend on the transition density matrix elements approximately as

$$\begin{aligned} \sigma(\pi^-) &\doteq K \frac{2I_f+1}{2I_i+1} [3X_n + X_p]^2 \\ \sigma(\pi^+) &\doteq K \frac{2I_f+1}{2I_i+1} [X_n + 3X_p]^2 \end{aligned} \qquad (2\text{-}33)$$

where X_n and X_p are abbreviations for the matrix elements.

Inelastic scattering on ^{13}C has been measured recently for both π^+ and π^- at 162 MeV [14]. An interesting state at 9.5 MeV, now believed to be $I=9/2^+$, T=1/2, is found to be an order of magnitude weaker in π^+ scattering compared to π^- scattering. The transition density matrix element for this stretched state relative to the $I=\frac{1}{2}^-$, $T=\frac{1}{2}$ ground state is

$$X_{t_3} = \langle\psi^{9/2^+1/2}(^{13}C)||(a^{d_{5/2}\dagger}_{t_3} \times \tilde{a}^{p_{3/2}}_{t_3})^{4^-}||\psi^{1/2^-1/2}(^{13}C)\rangle . \qquad (2\text{-}34)$$

To evaluate this quantity we insert a complete set of states, $I_o T_o \alpha_o$, of the A=12 system between a^+ and $\tilde{a}$; because of the stretched nature, only $I_o=2^+$ can bridge the gap and we can apply the Wigner-Eckart theorem to obtain an expression in terms of spectroscopic amplitudes to states of A=12,

$$X_{t_3} = \langle\psi^{9/2\ 1/2}_{9/2\ 1/2}|a^{d_{5/2}\dagger}_{5/2\ t_3}\left(-a^{p_{3/2}}_{-3/2\ t_3}\right)|\psi^{1/2\ 1/2}_{1/2\ 1/2}\rangle$$

$$= \frac{-1}{\sqrt{5}} \sum_{T_o \alpha_o} \langle\psi^{9/2\ 1/2} || a^{d_{5/2}{}^{\dagger}} || \psi^{2T_o \alpha_o}\rangle \langle\psi^{1/2\ 1/2} || a^{p_{3/2}{}^{\dagger}} || \psi^{2T_o \alpha_o}\rangle$$

$$x(T_o, 1/2, 1/2-t_3 t_3 | 1/2\ 1/2)^2 \qquad (2\text{-}35)$$

The dependence on t_3 is contained in the squared isospin CGC. For T_o=1 this factor is 2/3 for protons and 1/3 for neutrons. For T_o=0 the factor vanishes for protons and is unity for neutrons. Thus if the common parentage of the 9/2 and 1/2 states of ^{13}C is mostly to T_o=0 states, the experimental observation is understandable since from (33) $(\sigma(\pi^-)/\sigma(\pi^+))=9$ for $X_p=0$. Shell model calculation shows that while the 1/2 state has strong parentage to both T_o=1 and T_o=0 states, the 9/2 state is very closely represented as being a $d_{5/2}$ neutron coupled to the lowest I_o=2,T_o=0 state of ^{12}C. Such "weak-coupling" nature is very commonly found for low-lying states of non-normal parity. A complete distorted-wave calculation [15] of these cross sections including the amplitudes obtained from shell model calculations gives good agreement with measured cross sections. There are states at higher excitation in the data where π^+ scattering is favored as one would get from (35) for T_o=1 parentage.

A similar situation is expected in ^{11}B where an $I=11/2^+$ state at 14 MeV can be reached by $\lambda=4^-$ transition from the $I=3/2^-$ ground state. The calculated common parentage is mainly to the I_o=3, T_o=0 ground state of ^{10}B, so again excitation by π^- scattering should be strongly favored. This sensitivity of pion scattering to isospin parentage is a good new test for nuclear structure.

References

1. A. Bohr and B.R. Mottelson, Nuclear Structure, Vol. I, (Benjamin, New York) 1969, pg. 239.
2. M.H. Macfarlane, Nuclear Structure Physics, (Ed. S.J. Hale and J.M. Irvine, Scottish Universities Summer School in Physics) 1978.
3. D.M. Brink and G.R. Satchler, Angular Momentum, (Clarendon Press, Oxford) 1962.
4. A. deShalit and I. Talmi, Nuclear Shell Theory, (Academic Press, New York) 1963.
5. J.B. French, E.C. Halbert, J.B. McGrory, and S.S.M. Wong, Advances in Nuclear Physics, Vol. 3, (Ed. M. Baranger and E. Vogt, Plenum Press, New York) 1969.
6. D.H. Gloeckner, Argonne Report ANL-8113, 1974 (Unpublished).
7. J.P. Elliott, Proc. Roy. Soc. A245, 128 and 562 (1958).

8. M. Harvey, Advances in Nuclear Physics, Vol. 1, (Ed. M. Baranger and E. Vogt, Plenum Press, New York) 1968.

9. D.J. Millener, J. Math. Phys. 19, 1513 (1978).

10. R.R. Whitehead, A. Watt, B.J. Cole and I. Morrison, Advances in Nuclear Physics, Vol. 9, (Ed. M. Baranger and E. Vogt, Plenum Press, New York) 1977.

11. D. Kurath, Nucl. Phys. A317, 175 (1979).

12. T.W. Donnelly and G.E. Walker, Ann. of Phys. 60,209(1970).

13. R.A. Lindgren, W.J. Gerace, A.D. Bacher, W.G. Love and F. Petrovich, Phys. Rev. Lett. 42, 1524 (1979).

14. D. Dehnhard, S.J. Tripp, M.A. Franey, G.S. Kyle, C.L. Morris, R.L. Boudrie, J. Piffaretti and H.A. Thiessen, Phys. Rev. Lett. 43, 1091 (1979).

15. T.S-H. Lee and D. Kurath (unpublished).

Chapter III

Nuclear Vibrations

G. Bertsch†
Michigan State University
East Lansing, MI 48824

1. Introduction

In these lectures I will present some of the theoretical tools for the description of nuclear transitions. I first discuss the variables and the important operators, and find their matrix elements in Sect. 2. In Sect. 3 I derive sum rules which must be satisfied in any reasonable theory, based on the continuity equation and translational invariance. The only theory which is both guaranteed to satisfy the sum rules and is presently computible, is the RPA. I derive RPA from the time-dependent Hartree-Fock equations in Sect. 4, and make the connection with the various formulations that are given to the RPA. Of the various formulations, the coordinate-space representation best exhibits the consistency with sum rules. This representation also facilitates derivation of classical formulas for the frequencies of giant vibrations. The formulation as Landau theory, which emphasizes the properties of the interaction at the Fermi surface, is suited for compressional vibrations in large systems. The Green's function formulation is most useful for simplified interactions. A remarkably successful approximation uses a separable interaction whose form is determined by translational consistency. In the last section I compare the results of RPA with experiment, and discuss some open questions.

2. Operators and Matrix Elements

The most important operators we deal with are the density and the current, and we shall now derive their matrix elements in the Fock space representation, which was discussed in Chapter II.

2.1 Density Operator

The density operator is

$$\hat{\rho}(\vec{r}) = a_r^\dagger a_r \tag{3-1}$$

$$= \sum_{\kappa\lambda} \phi_\kappa^*(r)\phi_\lambda(r)a_\kappa^\dagger a_\lambda \tag{3-2}$$

†Supported by the National Science Foundation under grant PHY-7620097.

where $a_r^\dagger$ creates a particle at position r. In the second representation, the ϕ_λ are a complete set of single-particle states. The complications of spin and angular momentum coupling are an inevitable part of useful algebraic formulas. We will include these by coupling orbital angular momentum ℓ, m' and spin $\frac{1}{2}, s_z$ to total angular momentum j,m. In the usual phase convention this is,

$$\phi_{jm}(\vec{r},s_z) = R_j(r) \sum_{m',s_z} Y^\ell_{m'}(\hat{r}) \left|\frac{1}{2},s_z\right><\ell\ m'\frac{1}{2}\ s_z\left|j\ m\right> \tag{3-3}$$

$$= R_j(r) \sum_{m',s_z} Y^\ell_{m'}(r) \left|\frac{1}{2},s_z\right> \sqrt{\frac{(\ell+(-)^{s_z+j-\ell}m'+\frac{1}{2})}{2\ell+1}} (x(-1) \text{ if } j=\ell-\frac{1}{2} \text{ and } s_z=\frac{1}{2}).$$

There is another representation of ϕ_j that is more convenient for evaluating matrix elements. This is the helicity representation [1], in which the axis of quantization for the spin is along the radius vector $\vec{r}$ rather than in the z-direction,

$$\hat{r}\cdot s|h> = h|h> \qquad h = \pm\frac{1}{2}$$

It is simple to show that a wavefunction of angular momentum (j,m) is proportional to a linear combination of the rotation matrices† $D^j_{mh}(\hat{r})$. If this wavefunction is rotated by R, the transformed wavefunction is

$$D^j_{mh}(R\hat{r})|h> = \sum_{m'} D^j_{mm'}(R) D^j_{m'h}(\hat{r})|h>$$

which is just how (j,m) must transform. To get the phases to agree with (3), and insure a definite parity, the precise definition is

$$\phi_{jm}(\vec{r},h) = \sqrt{\frac{2j+1}{8\pi}} \left(D^j_{m,-\frac{1}{2}}(\hat{r})\left|-\frac{1}{2}\right> + (-)^{j-\ell-\frac{1}{2}} D^j_{m,\frac{1}{2}}(\hat{r})\left|\frac{1}{2}\right>\right) \tag{3-4}$$

The normalization is obtained from the orthonormality conditions

$$\int D^{*j}_{mh}(\hat{r},0) D^j_{mh}(\hat{r},0) d\hat{r} = \frac{4\pi}{2j+1} \tag{3-5}$$

If we want to recover the old representation (3) it is only necessary to express $|h>$ in terms of $\left|\frac{1}{2},s_z\right>$,

$$|h>_{\hat{r}} = \sum_{s_z} D^{*\frac{1}{2}}_{s_z h}(\hat{r}) \left|\frac{1}{2},s_z\right>. \tag{3-6}$$

We use the creation operator $a^+_{j,m}$ to add a particle with wavefunction

† There are many conventions for the rotation matrices; we follow here Bohr and Mottelson.

(3) or (4) to the system. As discussed in Chapter II, the adjoint operators $a_{j,m}$ transform in the usual way under rotations when given a phase,

$$\tilde{a}_{j,m} = (-1)^{j+m} a_{j,-m} \tag{3-7}$$

It is easy to see the necessity of this m-dependent phase. The number operator is

$$n_j = \sum_m a^\dagger_{jm} a_{jm} \tag{3-8}$$

It is a scalar, and so must be expressible in the angular momentum algebra as

$$n_j \sim (a^\dagger_j \tilde{a}_j)^0 = \sum_m \langle jm\ j{-m}|00\rangle a^\dagger_{jm} \tilde{a}_{j,-m} \tag{3-9}$$

$$= \sum_m \frac{(-)^{j-m}}{\sqrt{2j+1}} a^\dagger_{jm} \tilde{a}_{j,-m}$$

The relation between a_{jm} and $\tilde{a}_{jm}$ follows immediately from (8) and (9).

We now derive the matrix elements of the density operator in j-j coupling. We first write the density operator in coordinate space as

$$\hat{\rho}(\vec{r}) = \sum_h a^\dagger_{\vec{r},h} a_{\vec{r},h} \tag{3-10}$$

From (4), the coordinate operators are related to the ℓjm coupled operators by

$$a^\dagger_{\vec{r},h} = \sum_{\ell jm} R^*_j(r) \left(\frac{2j+1}{8\pi}\right)^{\frac{1}{2}} D^{*j}_{mh}(\hat{r}) a^\dagger_{\ell jm} \begin{cases} 1, & h=-\frac{1}{2} \\ (-)^{j-\ell-\frac{1}{2}}, & h=+\frac{1}{2} \end{cases} \tag{3-11}$$

We substitute (11) in (10) and simplify. Let us first consider the $h = -\frac{1}{2}$ term,

$$a^\dagger_{\vec{r},-\frac{1}{2}} a_{\vec{r},-\frac{1}{2}} = \sum R^*_{j'}(r) R_j(r) \left(\frac{2j'+1}{8\pi}\right)^{\frac{1}{2}} \left(\frac{2j+1}{8\pi}\right)^{\frac{1}{2}} D^{*j'}_{m',-\frac{1}{2}} D^j_{m,-\frac{1}{2}} a^\dagger_{\ell' j' m'} a_{\ell jm} \tag{3-12}$$

The D-functions are combined using the relations

$$D^{*j}_{mh} = (-1)^{m-h} D^j_{-m,-h} \tag{3-13}$$

$$D^{j_1}_{m_1 h_1}(R) D^{j_2}_{m_2 h_2}(R) = \sum_L (j_1 m_1 j_2 m_2|LM)(j_1 h_1 j_2 h_2|LK) D^L_{MK}(R)$$

We also use the symmetry property of the vector coupling coefficient,

$$(j_1m_1\ j_2m_2|JM) = (-1)^{j_1+j_2-J}(j_1-m_1\ j_2-m_2|J-M) \qquad (3\text{-}14)$$

and the relation of the D to spherical harmonics: $Y^L_M(\hat{r})=(\frac{2L+1}{4\pi})^{\frac{1}{2}}D^L_{MO}(\hat{r})$.

Eq. (12) then becomes (3-15)

$$a^\dagger_{\vec{r},-\frac{1}{2}}\, a_{\vec{r},-\frac{1}{2}} = \sum R_j(r)R^*_{j'}(r)\sqrt{\frac{(2j'+1)(2j+1)}{(2L+1)16\pi}}\,(-)^{j'+j-L+m'+\frac{1}{2}}\ \mathrm{X}$$

$$\langle j'm'\ j\text{-}m|L\ m'\text{-}m\rangle\langle j'\tfrac{1}{2}\ j\text{-}\tfrac{1}{2}|LO\rangle\ Y^L_{m-m'}(\hat{r})\,a^\dagger_{j'm'}a_{jm}\ . \qquad (3\text{-}16)$$

Next we express the sum over m, m' as the J-coupled product of operators using (7),

$$a^\dagger_{\vec{r},-\frac{1}{2}}\, a_{\vec{r},-\frac{1}{2}} = \sum R_j(r)R^*_{j'}(r)\sqrt{\frac{(2j'+1)(2j+1)}{(2L+1)16\pi}}\,(-1)^{\frac{1}{2}+j'-L}(a^\dagger_{j'}\tilde{a}_j)^L_M\ Y^{*L}_M(\hat{r})\ \mathrm{X}$$

$$(j'\tfrac{1}{2}\ j\text{-}\tfrac{1}{2}|LO)\ . \qquad (3\text{-}17)$$

The other helicity in (10) gives an identical contribution except for sign, as may be seen from (11). Summing the two terms gives the final result,

$$\hat{\rho} = \sum_{j'j}\sum^{\text{nat}}_{L}\rho^L_{j'j}(r)\ Y^{*L}_M(\hat{r})\,(a^\dagger_{j'}\tilde{a}_j)^L_M \qquad (3\text{-}18)$$

where $\rho^L_{j'j}(r) = R_j(r)R^*_{j'}(r)\sqrt{\frac{(2j'+1)(2j+1)}{4\pi(2L+1)}}\,(-1)^{j'-L-\frac{1}{2}}\langle j'\tfrac{1}{2}\ j\text{-}\tfrac{1}{2}|LO\rangle$ (3-19)

and the sum over L is restricted by $(-1)^{L+\ell+\ell'} = 1$.

As an example of the use of this formula, the matrix element of $\hat{\rho}$ between a closed shell state $|0\rangle$ and a particle-hole state $|JM\rangle = (a^\dagger_{j'}a_j)^J_M|0\rangle$ is

$$\langle JM|\hat{\rho}|0\rangle = \begin{cases} 0 & \ell'+\ell-J\ \ \text{odd} \\ \rho^J_{j'j}(r)Y^{*J}_M(r) & \ell'+\ell-J\ \ \text{even}\ . \end{cases}$$

The nondiagonal matrix elements of $\hat{\rho}$ are called transition densities.

There are two simple models for the transition density that are useful for comparison purposes and unsophisticated calculations. The first is the deformed model of Bohr and Mottelson, in which it is imagined that the surface moves a slight amount without changing the intrinsic density. The transition density is then related to the derivative of the ground state density with some proportionality constant. The conventional definition, for a transition from a spherical ground state to an excited state of angular momentum L,M, is

$$\langle LM|\hat{\rho}(\vec{r})|0\rangle = \frac{d}{\sqrt{2L+1}}\frac{d\rho_0}{dr}Y_M^{*L}(\hat{r}) = \frac{\beta R}{\sqrt{2L+1}}\frac{d\rho_0}{dr}Y_M^{*L}(\hat{r}).$$

The parameter d is the deformation length, and is the product of the Bohr-Mottelson β and the nuclear radius R. Such properties as the electromagnetic transition strength can be related to the β-moment between the states as follows:

$$B(EL\downarrow) \equiv \left(\int r^L Y^L \langle L|\hat{\rho}|0\rangle d^3r\right)^2$$

$$\approx \frac{(\beta R)^2}{2L+1}\left(\int r^{L+2}\frac{d\rho_0}{dr}dr\right)^2 = \frac{(\beta R)^2}{2L+1}(L+2)^2\left(\int r^{L+1}\rho_0 dr\right)$$

$$\approx \frac{9(\beta R)^2}{2L+1}R^{2L-2}\frac{Z^2}{(4\pi)^2}$$

where in the last step a uniform charge density was assumed.

Another macroscopic model, proposed by TASSIE [2], has a superior functional form for the transition density. The model is

$$\langle L|\hat{\rho}(r)|0\rangle \sim Y^L r^{L-1}\frac{d\rho_0}{dr} \qquad L \neq 0$$

$$\sim 3\rho_0 + r\frac{d\rho_0}{dr} \qquad L = 0.$$

We will see in detail later how this model can be justified by sum rules and the assumption that the smoothest motions remain most coherent. It will also turn out that the Tassie model is remarkably accurate in describing the radial form of the transition densities associated with the strongest states.

2.2 Spin Density

The same technique as was used to derive (18), can be used to express the spin density operator in the shell representation. In the helicity representation, the spin density $\sigma(\hat{r})$ is

$$\hat{\sigma}_0(\vec{r}) = \sum_h (-)^h a^{\dagger}_{\vec{r},h} a_{\vec{r},h}$$

$$\hat{\sigma}_{\pm}(\vec{r}) = \sqrt{2}\, a^{\dagger}_{\vec{r},\pm\frac{1}{2}} a_{\vec{r},\mp\frac{1}{2}} \; . \tag{3-20}$$

The operator $\hat{\sigma}_0(\vec{r})$ is the same as $\hat{\rho}$ except for the change in sign of the two helicities. This eliminates the natural multipoles instead of the unnatural ones,

$$\hat{\sigma}_0(\vec{r}) = \sum_{jj'} \sum_{L}^{\text{unnat}} \rho^{L}_{jj'}(r) Y^{*L}_{M}(\hat{r}) (a^{\dagger}_{j'}, \tilde{a}_j)^{L}_{M} \; . \tag{3-21}$$

For the operators $\hat{\sigma}_{\pm}$, we can follow exactly the same technique as used in the derivation of (18) to obtain

$$\hat{\sigma}_{\pm}(\vec{r}) = \Sigma\, R_j(r) R^{*}_{j'}(r) \sqrt{\frac{(2j+1)(2j'+1)}{32\pi^2}}\, (-)^{j+j'-\ell'} (j'\tfrac{1}{2}\; j\tfrac{1}{2}|J1)\, D^{*J}_{M\pm 1}(\hat{r})\; X$$

$$(a^{\dagger}_{j'}, \tilde{a}_j)^{J}_{M} \begin{Bmatrix} 1 \\ (-)^{\ell+\ell'-J} \end{Bmatrix} \; . \tag{3-22}$$

This is further reduced using the following relation between Clebsch-Gordan coefficients [1],

$$(j'\tfrac{1}{2}\; j\tfrac{1}{2}|J1) = ((-)^{j+j'-J} e_{j'} + e_j) \frac{(j'\tfrac{1}{2}\; j-\tfrac{1}{2}|J0)}{\sqrt{J(J+1)}} \tag{3-23}$$

where $e_j = j + \frac{1}{2}$.

The final result is then

$$\hat{\sigma}_{\pm}(\vec{r}) = \sum^{\text{nat}} \rho^{J}_{j'j}(r) \sqrt{\frac{2J+1}{J(J+1)8\pi}}\, D^{*J}_{M\pm 1}(\hat{r}) (a^{\dagger}_{j'}, \tilde{a}_j)^{J}_{M} ((-)^{j+j'-J} e_{j'} + e_j) \tag{3-24}$$

$$X\; (-)^{j'-\ell'+J+\frac{1}{2}} \begin{Bmatrix} 1 \\ (-)^{-\ell+\ell'-J} \end{Bmatrix} \; .$$

We next derive the j-coupled formulas for the current operator, which is given in a single-particle space as

$$j = \frac{\vec{\nabla} - \overleftarrow{\nabla}}{2i} \; .$$

This can be expressed as the following limit with coordinate space Fock operators,

$$\hat{x}\cdot j(\vec{r}) = \lim_{x\to 0} \frac{a^{\dagger}_{r-\frac{x}{2}}\, a_{r+\frac{x}{2}} - a^{\dagger}_{r+\frac{x}{2}}\, a_{r-\frac{x}{2}}}{2i|x|} \;. \tag{3-25}$$

The helicity zero component is nothing more than the derivative in the radial direction, $\hat{r}\cdot\vec{\nabla} \equiv \nabla_0 = \partial/\partial r$. This only acts on the radial wave-functions, leaving the angular part the same as for the density operator,

$$j_0 = \sum_{j'j}\sum_{J}^{nat} \rho^J_{j'j}\left(\frac{dR_{j'}}{R_{j'}dr} - \frac{dR_j}{R_j dr}\right)(a^{\dagger}_{j'},\tilde{a}_j)^J_M\, Y^{*J}_M(\hat{r}) \;. \tag{3-26}$$

Calculation of the $\pm$ helicity components of the current has a subtlety if the helicity representation of the particles is used. When performing the derivative limit in (25), the spin must not be reoriented. To avoid this problem we calculate $\hat{j}_{\pm}$ by first noting that the coefficient of $(a^{\dagger}_j,\tilde{a}_j)^J_M$ in $j_+(\vec{r})$ must transform under rotations as $D^{+J}_{MK}(\hat{r})$. Then we need only determine its value on the z axis, and make use of the transformation property to find it elsewhere,

$$j_{\pm}(\vec{r}) = \sum (-)^{j-k}\langle\phi_{j'k'}|j_+(z)|\phi_{j,-k}\rangle\langle j'k'\; jk|J1\rangle D^{*J}_{M\pm 1}(\hat{r},0)\,(a^{\dagger}_j,\tilde{a}_j)^J_M \,. \tag{3-27}$$

The only matrix elements that are nonvanishing on the z-axis have $k,k' = \pm 1/2,\ \pm 3/2$. We evaluate these using the behavior of the spherical harmonics in the vicinity of the z-axis,

$$\langle Y^{\ell'}_1|j_+(z)|Y^{\ell}_0\rangle = \frac{-1}{2\pi i r}\sqrt{(2\ell+1)(2\ell'+1)\ell'(\ell'+1)/2} \tag{3-28}$$

together with the representation of j-coupled wavefunctions in terms of Y^{ℓ} functions, (2).

The result is simplified with the help of (23) and the further relation between Clebsch-Gordon coefficients,

$$(j'\tfrac{3}{2}j-\tfrac{1}{2}|J1) = \sqrt{\frac{J(J+1)}{(j'-\frac{1}{2})(j'+\frac{3}{2})}}\;(j'\tfrac{1}{2}j-\tfrac{1}{2}|J0)\left(1 - \frac{e_j((-)^{j'+j-J}e_{j'}+e_j)}{J(J+1)}\right). \tag{3-29}$$

The final result is

$$j_+(\vec{r}) = \sum \frac{(-)^{j+\frac{1}{2}}}{8\pi i r}\sqrt{\frac{e_j e_{j'}}{2J(J+1)}}\;(j'\tfrac{1}{2}j-\tfrac{1}{2}|J0)\,[\,((-)^{\alpha}+(-)^{\beta})J(J+1)$$

$$- 2((-)^{\beta}e_{j'}+e_j)(e_{j'}+(-)^{\alpha}e_j-(-)^{\delta})\,]\,D^{*J}_{M1}(\hat{r})\,(a^{\dagger}_j,\tilde{a}_j)^J_M \tag{3-30}$$

where $e_j = j+\frac{1}{2}$, $\alpha = j+j'-\ell-\ell'-1$, $\beta = j+j'-J$ and $\delta = j'-\ell'-\frac{1}{2}$.

3. Sum Rules

There is a relation between density and current which must be satisfied in any reasonable theory. This is the quantum version of the equation of continuity, which is derived by evaluating the commutator of the Hamiltonian H and the density $\hat{\rho}(r)$. If the interaction is a function of the local densities $\hat{\rho}(r)$, $\vec{\sigma}(r)$, etc., as is the case for interactions based on meson exchange, then the interaction in H commutes with $\hat{\rho}(r)$. We need only consider the kinetic energy in the Hamiltonian,

$$T = -\sum_{xyz} \lim_{x\to 0} \int \frac{(a^{\dagger}_{\vec{r}} a_{\vec{r}+x} - 2a^{\dagger}_{\vec{r}} a_{\vec{r}} + a^{\dagger}_{\vec{r}+x} a_{\vec{r}})}{2mx^2} d^3r . \tag{3-31}$$

Now using the coordinate Fock space representations of $\hat{\rho}(r)$ and $\vec{j}(r)$, (1) and (25), it is a simple exercise to show that

$$[H,\hat{\rho}(\vec{r})] = \frac{i\nabla\cdot j(\vec{r})}{m} . \tag{3-32}$$

Taking the expectation of this in a time-varying wavefunction, we have

$$\langle\psi|[H,\hat{\rho}(\vec{r})]|\psi\rangle = -i\frac{\partial}{\partial t}\langle\hat{\rho}(r)\rangle = i\frac{\vec{\nabla}\cdot\langle j(r)\rangle}{m} . \tag{3-33}$$

This is just the equation of continuity,

$$\dot{\rho} + \vec{\nabla}\cdot\vec{v}\rho = 0 . \tag{3-34}$$

With the operator relation (32) satisfied, all its matrix elements must obey the relation, and we can write down relations between transition densities and transition currents,

$$(E_f-E_i)\langle f|\hat{\rho}(r)|i\rangle = \frac{i\nabla\cdot\langle f|\vec{j}(r)|i\rangle}{m} . \tag{3-35}$$

The left-hand side is obviously closely related to experiment, but what about the current? We can get a further relation by taking the commutator of (32) with an arbitrary external potential field,

$$[V,[H,\hat{\rho}(r)]] = \frac{+i\nabla\cdot[V,\vec{j}(r)]}{m} \tag{3-36}$$

where

$$V = \int d^3r V(r)\hat{\rho}(r) .$$

The right-hand side is evaluated using the commutator relation,

$$[\vec{j}(\vec{r}),\hat{\rho}(r')] = \frac{\hat{\rho}(r)}{i}\vec{\nabla}_r\,\delta(r-r')\,. \tag{3-37}$$

We now take the expectation value of both sides of (36) in a state i, and write out explicitly the sum over intermediate states on the left-hand side.

$$\sum_f (E_f-E_i)\langle f|\hat{\rho}(r)|i\rangle\langle i|V|f\rangle = -\frac{1}{2m}\vec{\nabla}\cdot\langle i|\hat{\rho}(r)|i\rangle\vec{\nabla}V(r)\,. \tag{3-38}$$

This is the sum rule first utilized by FALLIEROS [3] and NOBLE [4]. If the potential field V should happen to connect only a single eigenstate, then there is only one term in the sum and it can be solved for $\langle f|\hat{\rho}(r)|i\rangle$,

$$\langle f|\hat{\rho}(r)|i\rangle = \frac{-\vec{\nabla}\cdot\rho_0\vec{\nabla}V}{2m(E_f-E_i)\langle i|V|f\rangle}\,. \tag{3-39}$$

We can now derive the Tassie model by demanding that (39) be satisfied for the smoothest potential fields. These fields are the harmonic polynomials for $L\neq 0$,

$$V(r) = r^L Y^L(\hat{r})\,. \tag{3-40}$$

For L=0, the simplest field, a constant, gives no transitions and we take the next smoothest,

$$V(r) = r^2 \quad \text{for} \quad L = 0\,. \tag{3-41}$$

It is easy to see in physical terms what the sum rule is telling us Let us imagine a projectile passing quickly by a nucleus, so that its potential field acts for only a very short time,

$$V(r,t) = V(r)\delta(t)\,.$$

A nucleon at position r will receive an impulse, changing its momentum by $\Delta p = \vec{\nabla}V$. The nucleon started with zero velocity expectation value so its final velocity is

$$\vec{v} = \frac{\vec{\nabla}V}{m}\,. \tag{3-43}$$

This is the velocity field created by the projectile. The equation of continuity, (34), relates this to the rate of change of density at $t = 0_+$. Now if the field V happens to excite only a single normal mode, the time varying density has to have the same r-dependence at all times. This gives the Tassie model. Of course, the assumption of coherence of a single normal mode is a very strong one, and it is not obvious that it should be a reasonable approximation in the many-body system.

The conventional energy-weighted sum rule is derived by multiplying (38) by V(r) and integrating over r. The result is

$$\sum_f (E_f - E_i) \langle i|V|f\rangle^2 = \int d^3r \rho_0 \frac{(\vec{\nabla}V)^2}{2m} . \tag{3-44}$$

This sum rule also has a nice physical interpretation. Again we consider a fast-moving projectile with its perturbative field $V(r)\delta(t)$. The energy transferred to a nucleon at some position r is $(\Delta p)^2/2m$, so the right hand side of (44) is the total energy transferred to the nucleus. When the quantum mechanics of a sudden perturbation is calculated, one finds that the probability of exciting a state f is given by $\langle i|V|f\rangle^2/\hbar^2$. Thus the left hand side also measures the average energy absorbed by the nucleus. This is independent of the dynamics because the energy is absorbed before the nucleus is much disturbed from equilibrium.

When the specific field $r^L Y^L$ are inserted in (44), the right side can be evaluated in terms of the expectation value of a power of r. This is LANE's sum rule [5],

$$\sum_f (E_f - E_i) \langle i|r^L Y^L|f\rangle^2 = \frac{AL(2L+1)\langle r^{2L-2}\rangle}{8\pi m} \tag{3-45}$$

This sum rule is a valuable tool for both experimenters and theoreticians. When results for transition strengths are quoted as a fraction of the sum rule, there is no ambiguity as to the definition of the quoted matrix elements.

The final sum rule I wish to discuss was applied extensively by SATCHLER [6] and is based on a mixed use of the macroscopic model and Lane's sum rule. We earlier evaluated the matrix element of $r^L Y^L$ in the macroscopic model, assuming a uniform matter distribution,

$$\langle i|r^L Y^L|f\rangle = \frac{\beta_{fi} R}{\sqrt{2L+1}} \frac{3}{4\pi} AR^{L-1} .$$

We now use the uniform model to evaluate $\langle r^{2L-2}\rangle$ on the right-hand side of (45),

$$\langle r^{2L-2}\rangle_{\text{uniform}} = \frac{3}{2L+1} R^{2L-2}. \tag{3-46}$$

Inserting the above into (45), we obtain Satchler's sum rule,

$$\sum_f (E_f-E_i)(\beta_{fi}R)^2 = \frac{2\pi L(2L+1)}{3Am} . \tag{3-47}$$

This sum rule is not exact, but it has proved very convenient for discussing transition strengths associated with nuclear projectiles. This is because it is very easy to extract from experimental data a value for the deformation length associated with the strong inelastic transitions.

There is one more important sum rule in the theory of isoscalar vibrations, involving the gradient operator. The gradient commutes with a translationally invariant Hamiltonian, so the right-hand side is zero. The only way this condition can be met on the left-hand side is for all of the strength of the operator to be concentrated at zero energy. This is of course the spurious state. We shall see that it is helpful in dealing with low lying collective states to have this translation invariance built into the theory.

Sum rules for spin excitations and isovector excitations are necessarily more complicated, because the interaction does not commute with the corresponding densities. Stated differently, the mesons carry spin and isospin, and must therefore be considered explicitly when discussing spin- and isospin-densities. They do not carry baryon number, and so their effect can be ignored on the isoscalar densities.

4. RPA

There are many ways to formulate RPA and to derive the equations of motion. The most convenient version depends on the particular application. I will start from the time-dependent Hartree-Fock approximation, to derive the RPA equations in a coordinate space representation. This formulation is particularly useful for deriving simple formulas for the giant vibrational frequencies. I will also make the equivalence of this formulation with the particle-hole representation in configuration space, and with the Landau kinetic equation for an infinite medium. Finally, I will derive the response function for RPA. This last formulation is the most efficient for treating simplified interactions.

4.1 RPA in Coordinate Space

We begin with the time-dependent Hartree-Fock equations,

$$i \frac{\partial}{\partial t} \phi_i(t) = [- \frac{\nabla^2}{2m} + V[\rho]]\phi_i(t). \tag{3-48}$$

Here the $\phi_i(t)$ are single-particle wavefunctions which depend on $\vec{r}$ and t. I write the potential as a functional of ρ to suggest that V can be nonlocal, depending on the full density matrix. We start from an equilibrium solution $\phi_i^0(t)$ having time dependence

$$i \frac{\partial}{\partial t} \phi_i^0(t) = \varepsilon_i \phi_i^0(t). \tag{3-49}$$

The RPA is the small amplitude limit of vibrations about this equilibrium. Let us write the perturbed wavefunction as

$$\phi_i(t) = e^{-i\varepsilon_i t}(\phi_i^0 + \lambda\phi_i'(t)) \tag{3-50}$$

where λ is a small parameter. Choosing ϕ_i^0 to be real, we find for the density and potential,

$$\rho = \rho_0 + \lambda\delta\rho + O(\lambda^2) \tag{3-51}$$

where $\delta\rho = 2\sum_i^A \phi_i^0 \mathrm{Re}\phi_i'(t)$. The time-varying potential is

$$V[\rho] = V(\rho_0) + \lambda \frac{\delta V}{\delta\rho} \delta\rho + O(\lambda^2) . \tag{3-52}$$

We now insert this in (48) and extract the coefficient of λ,

$$i \frac{\partial}{\partial t} \phi_i'(t) + \varepsilon_i\phi_i'(t) = [-\frac{\nabla^2}{2m} + V(\rho_0)]\phi_i'(t) + \frac{\delta V}{\delta\rho} \delta\rho(t)\phi_i^0 . \tag{3-53}$$

Next we separate ϕ_i' into its real and imaginary parts. Eq. (53) then gives two equations,

$$- \frac{\partial}{\partial t} \mathrm{Im}\phi_i'(t) = [H_0 - \varepsilon_i]\mathrm{Re}\phi_i'(t) + \frac{\delta V}{\delta\rho} \delta\rho(t)\phi_i^0 \tag{3-54}$$

$$\frac{\partial}{\partial t} \mathrm{Re}\phi_i'(t) = [H_0 - \varepsilon_i]\mathrm{Im}\phi_i'(t) \tag{3-55}$$

where $H_0 = -\frac{\nabla^2}{2m} + V(\rho_0)$.

If (55) is multiplied by ϕ_i^0, we immediately recognize the left-hand side as the time derivative of the density of particle i, except for a factor 2λ. The right-hand side can be rewritten

$$\phi_i^0(H_0 - \varepsilon_i)\mathrm{Im}\phi_i'(t) = (H_0 \mathrm{Im}\phi_i'(t))\phi_i^0 - (H_0\phi_i^0)\mathrm{Im}\phi_i'(t) \tag{3-56}$$

$$= (\frac{\nabla^2}{2m}\phi_i^0)\,\mathrm{Im}\phi_i'(t) - (\frac{\nabla^2}{2m}\,\mathrm{Im}\phi_i'(t))\,\phi_i^0 \tag{3-56}$$

$$= + \frac{\vec{\nabla}}{2m}\cdot((\vec{\nabla}\phi_i^0)\,\mathrm{Im}\phi_i' - (\vec{\nabla}\mathrm{Im}\phi_i')\,\phi_i^0)\,.$$

This is just the divergence of the current associated with particle i, up to a factor (-2λ). Thus the equation of continuity follows directly from (55). The RPA conserves current and will obey the sum rules. This key fact makes RPA the most useful theory for vibrations.

As discussed in Chapter I, Hartree-Fock theory is not adequate for describing single particle excitations, because of the correlations. This leaves a challenge to the theorist to construct a theory of vibrations which incorporates moderate frequency correlations and still satisfies the sum rules.

I now want to make the physics of RPA more concrete and describe the situation where the vibration has been excited by a potential field $V(r)\delta(t)$. Then at $t = 0_+$, there is a velocity field but no change in density,

$$\mathrm{Re}\phi_i'(0_+) = 0, \quad \mathrm{Im}\phi_i'(0_+) = V(r)\phi_i^0. \tag{3-57}$$

If the vibration has a frequency ω, these functions will vary in time as

$$\mathrm{Im}\phi_i'(t) = \cos\omega t\,\mathrm{Im}\phi_i' \qquad \mathrm{Re}\phi_i'(t) = \sin\omega t\,\mathrm{Re}\phi_i' \tag{3-58}$$

where the symbol ϕ_i' without a time label denotes a time-independent function of position. With the substitution (3-58) the equations of motion (54) and (55) become

$$\omega\mathrm{Im}\phi_i' = [H_0 - \varepsilon_i]\mathrm{Re}\phi_i' + \frac{\delta V}{\delta\rho}\,\delta\rho\phi_i^0 \tag{3-59}$$

$$\omega\mathrm{Re}\phi_i' = [H_0 - \varepsilon_i]\mathrm{Im}\phi_i'\;. \tag{3-60}$$

The fact that there must be a spurious state at zero frequency is seen in this representation by considering the excitation to be a static translation:

$$\mathrm{Im}\phi_i' = 0 \qquad \mathrm{Re}\phi_i' = a\cdot\nabla\phi_i^0\;. \tag{3-61}$$

Eq. (60) is automatically satisfied for $\omega = 0$, while (59) is satisfied if

$$[H_0 - \varepsilon_i]a\cdot\nabla\phi_i^0 + \frac{\delta V}{\delta\rho}\,\delta\rho\phi_i^0 = 0\;. \tag{3-62}$$

The first term is manipulated to obtain

$$[H_0-\varepsilon_i]a\cdot\nabla\phi_i^0 = -a\cdot\nabla V(\rho_0)\phi_i^0 \tag{3-63}$$

$$= -\frac{\delta V}{\delta\rho}(a\cdot\nabla\rho_0)\phi_i^0$$

which cancels the second term in (62) because

$$\delta\rho = 2\sum_i \phi_i^0 \mathrm{Re}\phi_i^! = 2\sum_i \phi_i^0 a\cdot\nabla\phi_i^0 = a\cdot\nabla\rho_0 \; .$$

4.1.1 Reduction of RPA to Quadrature

The RPA can be reduced to simple formulas if we assume that the motion initiated by a potential field, (57), is an eigenstate with frequency ω. Then from (60) we may solve for $\mathrm{Re}\phi_i^{'}$,

$$\mathrm{Re}\phi_i^! = (u\cdot\nabla + \tfrac{1}{2}(\nabla\cdot u))\phi_i^0 \tag{3-64}$$

$$\text{where } \vec{u} = \frac{\vec{\nabla}V}{m\omega} \; . \tag{3-65}$$

In fact we could allow more general motions [7], and start with an arbitrary field $\vec{u}$. If we go back to the representation with a time varying wavefunction $\phi_i(r,t)$, we find that (64) represents a displacement of the wavefunction by a distance $\vec{u}$. The $(\nabla\cdot u)$ term is the change in amplitude due to the stretching or squeezing of the wavefunction under a (nonuniform) displacement. Substituting (64) into (59), we find an equation to be satisfied by the field V,

$$\omega V\phi_i^0 = [H_0, u\cdot\nabla + \tfrac{1}{2}(\nabla\cdot u)]\phi_i^0 + \frac{\delta V}{\delta\rho}\,\delta\rho\;\phi_i^0. \tag{3-66}$$

If we multiply this by $\mathrm{Re}\phi_i$, sum over i and integrate over r, the equation assumes a very nice form [8],

$$\omega^2 = \frac{I[\vec{u},\vec{\nabla},\rho]}{\frac{m}{2}\int\vec{u}\cdot\vec{u}\rho_0 d^3r} \tag{3-67}$$

where the numerator is an integral over the ground state depending on $\vec{u}$. This has the form of Rayleigh's variational principle, and does in fact give an upper bound on the frequency of the lowest excitation. A differential equation for $\vec{u}$ can be derived in the usual way from such a variational principle. We will not discuss this any further except to mention that the differential equation is that of vibrations in an elastic medium rather than in a fluid. In the case that V does not excite a single eigenstate, the meaning of (67) is the average frequency associated with V, obtained by dividing the ω^3-weighted strength by the ω-weighted strength [9,10].

The giant quadrupole vibration is described by a field $\vec{u}=\vec{\nabla}r^2Y^2$. Applying (64) with such a field, we find that a short-range potential in Hamiltonian gives no contribution to I, for the same reason that (62) is satisfied. Thus the restoring force in the giant quadrupole is the single particle kinetic energy. The actual formula for the quadrupole frequency is found to be

$$\omega_Q^2 = \frac{2\langle T\rangle}{m\langle r^2\rangle} \tag{3-68}$$

where $\langle T\rangle$ is the average single-particle kinetic energy. In the harmonic oscillator model, this gives the famous result [11]

$$\omega_Q^2 = 2\omega_{osc}^2 . \tag{3-69}$$

The expected coefficient of 4, for noninteracting particles in an oscillator well, is cut in half by throwing away the potential energy associated with the quadrupole distortion. The formula (68) is however in no way based on the oscillator model; substitution of Fermi gas parameters in (67) gives an equally good account of the empirical giant quadrupole state.

If the potential is velocity-dependent, its effects do not cancel out completely. In the effective mass approximation, (67) remains true with $\langle T\rangle = 3/5\ k_F^2/2m^*$. To get agreement of the theoretical quadrupole frequency with experiment, it is necessary to have $m^* \approx 1$. Since Brueckner theory requires $m^* \approx 0.7$, there is a contradiction here. A recently proposed resolution [12] involves considering the velocity dependence of the interaction going beyond the effective mass approximation.

4.2 Configuration Space Representations

Starting from (59) and (60), it is a simple matter to write down the RPA equations in the familiar matrix representation. We add and subtract the two equations, and integrate the resulting equations with an arbitrary Hartree-Fock wavefunction ϕ_i^0 . The amplitudes are expressed as

$$\begin{aligned} \frac{\lambda}{2} \int \phi_j^0(\mathrm{Re}\phi_i' + \mathrm{Im}\phi_i')d^3r &= X_{ij} \\ \frac{\lambda}{2} \int \phi_j^0(\mathrm{Re}\phi_i' - \mathrm{Im}\phi_i')d^3r &= Y_{ij} . \end{aligned} \tag{3-70}$$

In terms of these amplitudes the equations (59)(60) become

$$\omega X_{ij} = (\varepsilon_j-\varepsilon_i)X_{ij} + \sum_{k\ell}(X_{k\ell} + Y_{k\ell})V_{ik,j\ell}$$

$$-\omega Y_{ij} = (\varepsilon_j-\varepsilon_i)Y_{ij} + \sum_{k\ell}(X_{k\ell} + Y_{k\ell})V_{ik,j\ell} \tag{3-71}$$

where $V_{ij,i'j'}$ is the matrix element of $\frac{\delta V}{\delta \rho}$. For nonlocal potentials this is given by

$$V_{ik,j\ell} = \int d^3r_1 d^3r_2 d^3r_3 d^3r_4 \phi_i^0(r_1)\phi_k^0(r_2)\phi_j^0(r_3)\phi_\ell^0(r_4) \frac{\delta V(r_1,r_3)}{\delta\rho(r_2,r_4)} . \tag{3-72}$$

The normalization of the X and Y amplitudes can be determined by requiring that the energy associated with the vibration equals $\hbar\omega$. This gives the usual condition

$$\sum_{ij}(X_{ij}^2 - Y_{ij}^2) = 1. \tag{3-73}$$

As expressed in (71), the RPA is solved by diagonalizing a matrix. Matrix diagonalization is the only practical method of solution if the full nonlocality of the potential, as expressed in (72), is important for the physics. The continuum must be treated by a discrete set of states, and then the number of configurations required to give an adequate approximation to the true solution is of the order of 20 for excitations in ^{16}O to 300 for excitations in ^{208}Pb.

4.3 Landau Theory

The RPA assumes a simple form in infinite systems, when only long wavelength excitations are considered. Then the orbits are labelled by momentum $\vec{k}$, and the excitation can be characterized by a momentum $\vec{q}$. We now go back to (53), and instead of considering real and imaginary parts of $\phi_i'(t)$, we divide it into positive and negative frequency components,

$$\phi_k' = (x_k e^{iq\cdot r-i\omega t}+y_k^* e^{-iq\cdot r+i\omega t})\phi_k^0 . \tag{3-74}$$

The equation of motion for the coefficients x, y, are

$$\omega x_k = (\frac{k\cdot q}{m} + \frac{q^2}{2m})x_k + \sum_{k'}(x_{k'}+y_{k'})n_k^0 \tag{3-75}$$

$$-\omega y_k = (-\frac{k\cdot q}{m} + \frac{q^2}{2m})y_k + \sum_{k'}\frac{\delta V_k}{\delta\rho_{k'}}(x_{k'}+y_{k'})n_k^0 \tag{3-76}$$

where $n_k^0 = \theta(k_F - k)$. We now use a trick to write the equations in a form that only involves the amplitudes in the vicinity of the Fermi surface. In (76), change the variable from k to k+q, and subtract the equation from (75),

$$\omega(x_k + y_{k+q}) = \frac{k\cdot q}{2m}(x_k+y_{k+q}) + \sum_{k'}(x_{k'}+y_{k'})\frac{\delta V}{\delta\rho}(n_k^0-n_{k+q}^0). \quad (3\text{-}77)$$

Thus $(x_k + y_{k+q})$ is only nonvanishing in the region about the Fermi surface where $n_k^0 - n_{k+q}^0 \neq 0$. A differential equation equivalent to (77) can be made by the replacements

$$(x_k+y_{k+q}) \to \delta n \qquad \omega \to \frac{\partial}{i\partial t} \qquad \frac{k\cdot q}{m} \to \frac{v\cdot\nabla}{i}$$

$$(n_k^0-n_{k+q}^0) \to q\cdot\nabla_k n^0 \quad (3\text{-}78)$$

$$\sum_{k'}\frac{\delta V}{\delta\rho_{k'}}(x_{k'}+y_{k'}) = \delta V \to \frac{\vec{\nabla}U}{iq}.$$

Then (77) becomes the linearized Vlasov equation

$$\frac{\partial}{\partial t}\delta n + v\cdot\nabla\delta n - \nabla U\cdot\nabla_k n^0 = 0. \quad (3\text{-}79)$$

For the Landau Fermi liquid equation, we write in (77)

$$n_k^0 - n_{k+q}^0 \approx q\cos\theta_{kq}\delta(k-k_F), \qquad (x_k+y_{k+q}) \approx \delta n_k(n_k^0-n_{k+q}^0). \quad (3\text{-}80)$$

Then the sharply peaked function $(n_k^0 - n_{k+q}^0)$ factors out of (77) and we are left with

$$\omega\delta n_k - v_F q\cos\theta\delta n_k = \int\frac{d^3k'}{(2\pi)^3}\frac{\delta V_k}{\delta\rho_{k'}}\delta n_{k'}q\cos\theta_{kq}\delta(k-k_F) \quad (3\text{-}81)$$

$$= \frac{k_F^2 q}{(2\pi)^3}\int d\Omega\,\frac{\delta V_k}{\delta\rho_{k'}}\cos\theta_{kq}\delta n_{k'}.$$

We next make a multipole expansion of the interaction

$$\frac{\delta V_k}{\delta\rho_{k'}} = \frac{(2\pi)^3 v_F}{4\pi k_F^2}\sum_L F_L P_L(\cos\theta_k)\cdot P_L(\cos\theta_{k'})$$

to get the final equation for δn_k

$$\delta n_k = \frac{1}{\omega/v_F q - \cos\theta}\sum_L F_L P_L(\cos\theta)\int d\cos\theta'\cos\theta' P_L(\cos\theta')\delta n_{k'}. \quad (3\text{-}82)$$

4.4 Green's Functions and the Density Response

To find the vibration induced by an external field in RPA, we must solve coupled inhomogeneous differential equations based on the homogeneous equations (59), (60). The standard technique for dealing with such situations utilizes Green's functions. To solve the one-particle problem

$$(H_0 - \varepsilon_i - \omega)\phi_i' = V_{ext}\phi_i^0 \tag{3-83}$$

we define the inverse operator

$$g(\varepsilon_i + \omega) = (H_0 - \varepsilon_i - \omega)^{-1}. \tag{3-84}$$

The brute force way to construct this operator is to find the eigenstates ϕ_j^0 and eigenvalues ε_j of H_0, and then the Green's function is

$$g(r,r',\varepsilon_i+\omega) = \sum_j \frac{\phi_j^0(r)\phi_j^0(r')}{\varepsilon_j-\varepsilon_i-\omega} . \tag{3-85}$$

The more clever way is to use the solutions of $(H_0 - \varepsilon_i - \omega)\phi = 0$ that satisfy each boundary condition,

$$g(r,r',\varepsilon_i+\omega) = \frac{\phi^-(r_<)\phi^+(r_>)}{(\phi^+ \frac{d}{2mdr}\phi^- - \phi^- \frac{d}{2mdr}\phi^+)_R} . \tag{3-86}$$

To solve RPA with the Green's function, we substitute (59) into (60) and solve for $\mathrm{Re}\phi_i'$,

$$\mathrm{Re}\phi_i' = [\omega^2 - (H_0 - \varepsilon_i)^2]^{-1}[H_0 - \varepsilon_i](\frac{\delta V}{\delta\rho}\delta\rho\phi_i^0 + V_{ext}\phi_i^0). \tag{3-87}$$

The operator is simplified using

$$(A^2-B^2)^{-1}B = \frac{1}{2}[(A-B)^{-1} - (A+B)^{-1}]. \tag{3-88}$$

The equation for $\mathrm{Re}\phi_i'$ is then

$$\mathrm{Re}\phi_i' = -\frac{1}{2}(g(\varepsilon_i - \omega) + g(\varepsilon_i + \omega))(\frac{\delta V}{\delta\rho}\delta\rho + V_{ext})\phi_i^0. \tag{3-89}$$

Since the equation involves $\delta\rho$ on the right-hand side, let us make an equation with $\delta\rho$ on the left by multiplying (89) by ϕ_i^0 and summing over i. The result can be expressed compactly in terms of the free density response function,

$$G^0(r,r',\omega) = \sum_i^A \phi_i^0(r)\,(g(r,r',\varepsilon_i-\omega)+g(r,r',\varepsilon_i+\omega))\,\phi_i^0(r'). \qquad (3\text{-}90)$$

Then (89) becomes

$$\delta\rho = -G^0\left(\frac{\delta V}{\delta\rho}\,\delta\rho + V_{ext}\right). \qquad (3\text{-}91)$$

With another operator inversion we finally arrive at

$$\delta\rho = \left[1+G^0\,\frac{\delta V}{\delta\rho}\right]^{-1} G^0 V_{ext} \equiv G^{RPA} V_{ext}\;. \qquad (3\text{-}92)$$

We only need the free density response G^0 in (92) because I assumed that $\delta V/\delta\rho$ depends only on the local density, $\hat{\rho}(r)$. The theory can be generalized to dependence on $\vec{j}(r)$, $\vec{\sigma}(r)$, etc. which requires then additional Green's functions. It does not seem feasible to generalize to an arbitray dependence on the full density matrix $\rho(r,r')$, for then G^0 would be a function of four coordinate variables.

There are several ways G^{RPA} can be used. Its poles identify the eigenmodes, according to the representation

$$G^{RPA}(r,r',\omega) = \sum_f \frac{\langle i|\hat{\rho}(r)|f\rangle\langle f|\hat{\rho}(r')|i\rangle}{\omega-(\omega_f-\omega_i)}\;. \qquad (3\text{-}93)$$

The probability of exciting the system by an external field is given by

$$W = \frac{1}{\pi}\int d^3r\,d^3r'\; V_{ext}(r)\,\mathrm{Im}G(r,r')\,V_{ext}(r). \qquad (3\text{-}94)$$

Here G is given an imaginary part either by adding a small $i\eta$ to the energy, or, in the case of continuum states, using outgoing wave boundary conditions in the definition of ϕ^+ in (86).

To actually solve (92), we first make a spherical harmonic decomposition of the dependence on $\hat{r}$. Using the notation of (19), G^0 is expanded

$$G^0(\vec{r},\vec{r}',\omega) = \sum_L G_L^0(r,r',\omega)\,Y^L(\hat{r})\cdot Y^L(\hat{r}')$$

$$\text{where } G_L^0(r,r',\omega) = \sum_j \sum_i^A \frac{\varepsilon_j-\varepsilon_i}{\omega^2-(\varepsilon_j-\varepsilon_i)^2}\,\rho_{ji}^L(r)\,\rho_{ji}^L(r'). \qquad (3\text{-}95)$$

Expanding $\delta V(r)/\delta\rho(r')$ in the same way, the angle-decomposed Green's function is

$$G_L^{RPA}(r,r') = [1+G_L^0 v_L]^{-1} G_L^0\;. \qquad (3\text{-}96)$$

This still looks like a formal equation, but it can be easily implemented on the computer, representing the operators on a mesh in coordinate space. The unit operator is then the unit matrix, and $G^0_L(r,r')$ is computed at each pair of mesh points.

For large enough systems, (96) is bound to be superior to the configuration representation, because the dimensionality is directly proportional to the size of the system, rather than to the number of configurations, which grows at a much faster rate. As the practitioners of time-dependent Hartree-Fock theory have discovered, it is possible to get good accuracy with a very coarse mesh, up to 1 fm spacing.

4.5 Separable Interactions and Self-Consistency

If the interaction $\delta V/\delta\rho$ is separable,

$$v_L(r,r') = \kappa f(r) f(r') \tag{3-97}$$

the Green's function inversion becomes trivial. We make use of the following formula for the inverse of a dyadic operator,

$$[1 - |\eta\rangle\langle\xi|]^{-1} = 1 + |\eta\rangle \frac{1}{1-\langle\eta|\xi\rangle} \langle\xi| . \tag{3-98}$$

This may be verified by a power series expansion. Substituting (97) in (96) we find

$$\delta\rho_L(r) = G^0_L(r,r')V_{ext}(r')dr' - \frac{\kappa \int G^0_L(r,r')f(r')dr' \int f(r'')G^0_L(r'',r')V_{ext}(r')dr'dr''}{1+\kappa\int dr'dr''f(r')G^0_L(r',r'')f(r'')} . \tag{3-99}$$

It is then only necessary to solve the algebraic equation

$$1+\kappa\int fG^0_L f = 0 \tag{3-100}$$

to find the poles of G^{RPA}. Bohr and Mottelson's treatment of vibrations can be viewed as RPA with separable interactions that are constrained by consistency. If the motion had a $\delta\rho$ of the Tassie or macroscopic form,

$$\delta\rho = d\,\frac{d\rho_0}{dr} \tag{3-101}$$

then consistency demands that δV be related to the static potential in the same way,

$$\delta V = d\,\frac{dV_0}{dr} . \tag{3-102}$$

This can be achieved with the separable interaction (97) if we choose

$f = dV_0/dr$ and κ so that

$$v(r,r') = \frac{\left.\frac{dV_0}{dr}\right|_r \left.\frac{dV_0}{dr}\right|_{r'}}{r^2 dr \, \frac{dV_0}{dr} \, \frac{d\rho_0}{dr}} \, . \tag{3-103}$$

The theory with the interaction (103) used in the (100) is mathematically equivalent to the collective coordinate theory of BOHR and MOTTELSON [13].

5. Applications

I first wish to show that RPA is remarkably successful in reproducing the properties of low-lying vibrations. I then want to discuss the excitation of the vibrations by nuclear scattering, and their decay properties.

5.1 RPA Frequencies and Transition Densities

Fully self-consistent RPA requires the prior construction of a Hartree-Fock theory. This can be accomplished easily with zero-range interactions, such as given by the Skyrme parameterization. The nuclear size and binding energy constrains the interactions, but even so there is some freedom in the parameters. The saturation property of the interaction can come from a momentum-dependent term in the interaction or from a density-dependent term, or some combination. In Table 3-1 we show the results of RPA for two extremes, Skyrme I which has mainly density-dependent saturation and Skyrme II which has mainly momentum-dependent saturation. The nucleus is ^{208}Pb, and we give the comparison of theoretical and experimental energies and transition strengths from

Table 3-1 Low-lying vibrations in ^{208}Pb

J^π	E (MeV)	B(EL) (Weisskopf Units)	Theory
2^+	4.07	8	Empirical
	5.4	7	Skyrme I
	5.2	7	Skyrme II
	4.5	12	Harmonic oscillator self-consistency
3^-	2.61	40	Empirical
	2.7	29	Skyrme I
	3.2	50	Skyrme II

[14]. For the 2^+, the transition strength is very close but the frequency is somewhat high. We also quote the result of a much simpler calculation [15] using a self-consistent harmonic oscillator interaction, according to (103). This model lowers the frequency, but at the cost of increasing the transition strength. For the 3^- vibration, the frequency strength and transition strength for the Skyrme I interaction are quite close to experiment. For this transition, the self-consistent harmonic oscillator interaction has to be renormalized by ~10% to get agreement [16].

We next examine some transition charge densities. In Fig. 1 the

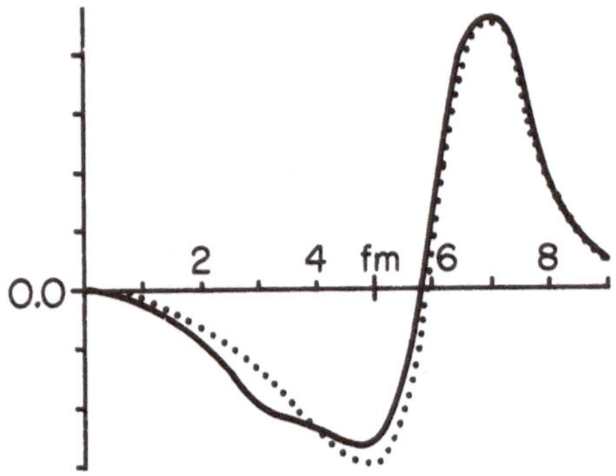

Fig. 1 Transition density, $r^2\langle f|\hat{\rho}|i\rangle$, for ^{208}Pb giant monopole state. The solid line is the RPA calculation with Skyrme I, and the dotted curve is the Tassie model.

transition density for the monopole giant vibration is compared with the Tassie model. The model is certainly a good approximation to the peak and width of the transition density. Finer details in the central region, associated with shell structure, are missed.

In Figs. 2 and 3 we compare RPA transition charge densities with densities measured by inelastic electron scattering. For the 3^- vibration in ^{208}Pb, shown in Fig. 2, all the features of the experiment are

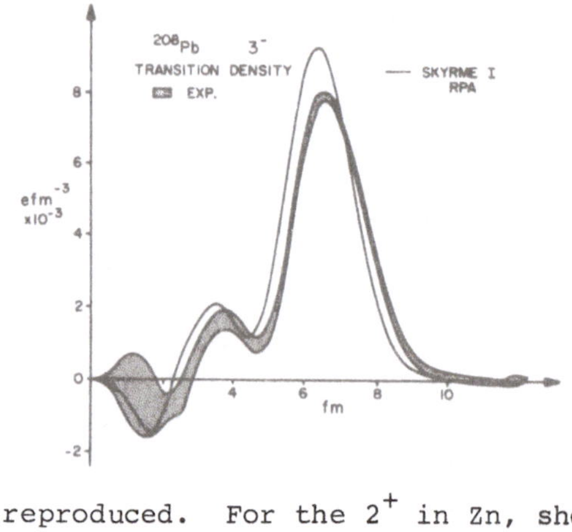

Fig. 2 Transition density for low octupole vibration in ^{208}Pb. Solid line is the RPA theory with the Skyrme I interaction, and the stippled curve is experiment [27].

reproduced. For the 2^+ in Zn, shown in Fig. 3, the surface peak is reproduced but there is predicted another lobe in the interior which is evidently not present.

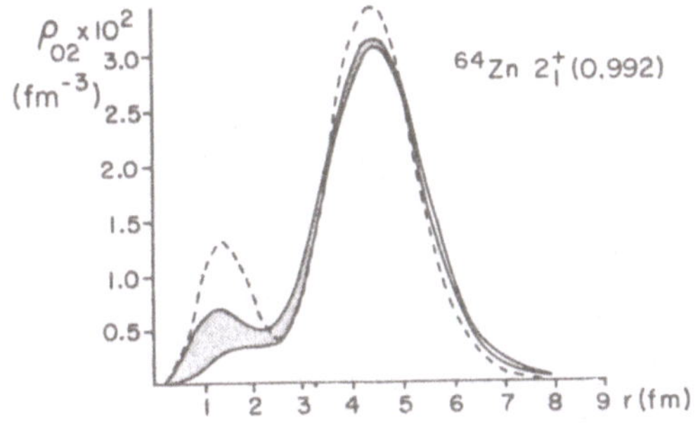

Fig. 3 Transition density for low quadrupole vibration in ^{64}Zn. Dashed line is RPA [28], and stippled curve is experiment.

5.2 Inelastic Scattering of Nucleons and Composite Projectiles

Inelastic scattering is conventionally described with the distorted-wave Born approximation using some optical potential that reproduces elastic scattering. It is then found that the cross sections to the strong vibrational states are very well described in the macroscopic model, in which the transition potential has the form (102). Turning this around, deformation lengths or β parameters can be extracted from the inelastic scattering, and the success of the description is that these parameters agree with those deduced by other means, for example B(EL).

In view of this success it is interesting to ask to what extent the total cross section can be accounted for by the excitation of vibrations. In this question we are not interested in individual states and the Green's function method is well suited. We shall now make an approximation and calculate only the β^2 strength associated with the response function, according to

$$\frac{d(\beta R)_L^2}{dE} = \frac{(2L+1)}{\pi} \, \frac{\mathrm{Im}\, \frac{d\rho_0}{dr} G_L^{RPA} \frac{d\rho_0}{dr}}{\left(\int \left(\frac{d\rho_0}{dr}\right)^2 r^2 dr\right)^2} \; .$$

The prediction of the theory [17] is compared with an experiment in Fig. 4. We see that the excitation of surface vibrations accounts for only ≈25% of the cross section, once we are above the lowest collective states and into the continuum region. There are two possible explanations of the discrepancy.

One is that knockout reactions may not be well described by the β parameterization of the response. Since the final nucleon is in the continuum, the transition density should peak farther out from the surface. This possibility was examined and rejected [17], but in the light of experimental data to be discussed in the next section, the knockout contribution should be reexamined.

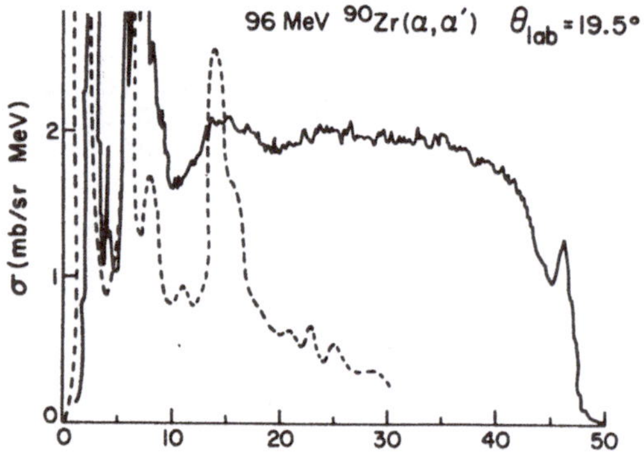

Fig. 4 Inclusive inelastic alpha scattering, in ^{90}Zr. The solid line is experiment [29], and the dashed line is the results of RPA [17]. The horizontal scale shows excitation energy of the residual nucleus in MeV.

The other possibility is that multistep reactions, not describable in DWBA, dominate the cross section in the continuum. Semiclassical calculations by BROGLIA, et al. [18] verify this hypothesis; for inelastic α scattering on ^{208}Pb, roughly half the cross section in the giant vibration region is due to multiple excitation of surface vibrations.

In either case, the study of the giant vibrations by inelastic scattering would benefit from a projectile that interacts less strongly with the target. Obviously multistep reactions would be reduced. Also, the projectile would probe more deeply into the target and thus be less sensitive to the knockout in the extreme surface. Possible projectiles that have been advocated are K^+ mesons, nucleons at 200 MeV bombarding energy, and pions of less than 50 MeV energy.

5.3 The Decay of the Giant Vibrations

The quadrupole giant vibrations have by now been thoroughly studied by inelastic scattering [19]. The quadrupole bump that appears on the continuum background has a width that varies from ~8 MeV in light nuclei to ~3 MeV in heavy nuclei. There are two sources to the width: decay by particle emission, and damping by mixing with the more complicated states of the nucleus.

From the theoretical side, only the decay by nucleon emission can be calculated in a straightforward way. The RPA includes the continuum damping automatically with the Green's function (86). It is found from the RPA calculations that nucleon emission accounts for only a fraction of the total width, of the order of 1 MeV in ^{16}O to ≈300 keV in ^{208}Pb [20].

The theory for the α-decay width is much less certain. It is possible to calculate the spectroscopic factor for an alpha particle in the giant vibration [21]. This gives the overlap of the state with the daughter state and an alpha particle in an oscillator wavefunction. To calculate the width of the state requires that the oscillator wavefunction be coupled to the continuum. A well-defined procedure for accomplishing this is the Feshbach projection operator technique [22]. Unfortunately, the results are quite sensitive to the optical potential that is assumed for the alpha particle. However, some idea of the width can be deduced from examining a simpler system, such as Be. The Be nucleus has a virtually 100% parentage for decomposing into two alpha particles. If the experimental L = 2 phase shift is fit to a resonance shape, the width is 1.5 MeV. We may regard this as an upper bound for the width in the larger nuclei. The total particle emission width is thus considerably smaller than the actual width. This leaves configuration mixing as the main damping.

The damping width is computed from the matrix elements of a residual interaction between the vibration, expressed in a particle-hole representation, and two-particle two-hole states. The damping that emerges is much larger than the decay width. The calculation is rather complicated, and it is instructive to compare the situation with the damping of single-particle motion by the mixing of one-particle states with 2p-1h states. The empirical damping is known from the optical model

$$\Gamma_i \approx 2\langle W\rangle_i$$

where W is the imaginary part of the optical potential. With standard optical potentials, the particle width is ~5 MeV. In the vibration there is additionally a hole, which also has a width. The widths do not add incoherently for a collective vibration. The interference is destructive, so the overall damping is reduced some from the single-particle value.

Detailed calculations in ^{208}Pb verify this relation between the single-particle damping and vibrational damping [23]. They also show that only certain final states, those containing a low-frequency collective vibration, are important in the damping. In effect, the particles damp by bouncing inelastically from the nuclear surface.

Experimentally, there are contradictions and paradoxes in the decay of the quadrupole vibrations. A particularly illuminating experimental study was the measurement of the decay of ^{16}O, excited by inelastic alpha scattering [29]. If the damping dominates the particle decay, one should expect decay patterns as in compound nucleus theory: final states populated according to penetrabilities, and symmetry in the decay angular distribution under a rotation by 180°. The experimental finding is that the α channels are preferred over proton channels, with the 2^+ final state particularly strong. This agrees with statistical interpretation, which favors final states that can be reached by several L values. It also agrees with the direct decay interpretation, with the spectroscopic factor of the 2^+ larger than the 0^+. This may be understood as a geometric effect: to have a spherical ^{16}O as a product of an alpha particle and ^{12}C, the ^{12}C must be oriented with the alpha along the symmetry axis. The oriented nucleus contains a superposition of L values, with weights going as (2L + 1) for the lower L-values.

Fig. 5 shows the angular distribution of decay α's to the ground state of ^{12}C, which should have a unique L. We see an asymmetry

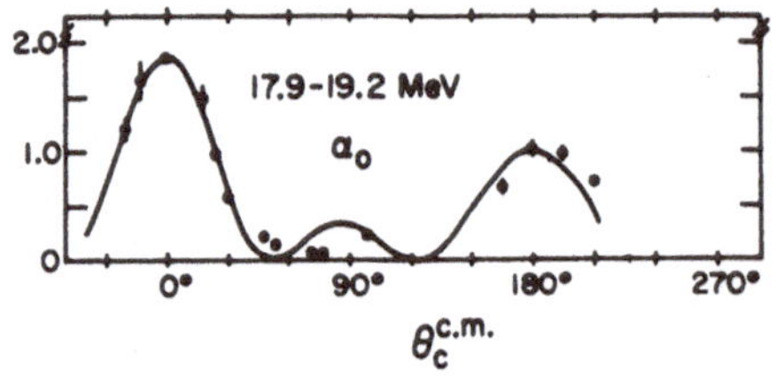

Fig. 5 Decay angular distribution α particles from giant quadrupole region of ^{16}O [24]. The solid line is a fit to a coherent mixture of L = 2 and L = 3.

about the recoil axis, which requires that different L mix coherently. The phase is just that required for a knockout reaction. In making a fit to the distribution, the authors of [24] find that only a small amount of L = 3 is necessary to make the lopsided distribution. Obviously, there is a challenge here for the theorist to understand these decay distributions and the coherence between different L-values.

While the lighter nuclei decay mostly by alpha particle emission, heavier nuclei decay preferentially by nucleon emission, with the crossover occurring near ^{40}Ca. This empirical finding is consistent with the statistical model, which disfavors alpha emission through larger Coulomb barriers. In the recent literature, there has been confusing and contradictory experimental evidence on this basic point. The inelastic alpha scattering experiments on ^{58}Ni show a decay branch in the quadrupole region similar to that outside, with neutrons favored and α-particles suppressed [25]. This was called into question by an electroexcitation experiment which indicated a greatly enhanced α-branch [26]; however recent work has revised the electroexcitation result downward [30]. The comparison of the decay branches of the vibration with the background has also been attempted with fissile nuclei. In the case of ^{238}U, the fission probability at moderately high excitation energies is ~20%. Unfortunately, the branching to fission channels has much structure, due to threshold effects, just in the region of the giant quadrupole. The best analysis compares the observed structure in the reaction (α,α'), which excites the quadrupole strongly, to the structure in the (t,p) reaction, which does not emphasize the quadrupole

[31]. No substantial difference was found.

As we have seen, theory predicts that the vibration should mainly damp by mixing with more complicated states. The more complicated states are indistinguishable from the background states, and decay in the same way as the background. Thus the experiments on decay branches support, or at least do not firmly contradict, this simple theoretical conclusion. The major problem now, which I leave you with, is to reconcile the obvious coherence in the decay exhibited by the angular distributions, with the apparent loss of memory of the system, seen by its choice of decay channels.

References

1. J. Raynal, Nucl. Phys. A97, 572 (1967).
2. L.J. Tassie, Aust. J. Phys. 9, 407 (1956).
3. T. Deal and S. Fallieros, Phys. Rev. C7, 1709 (1973).
4. J.V. Noble, Ann. Phys. (N.Y.) 67, 98 (1971).
5. A.M. Lane, Nuclear Theory (Benjamin, 1964), p. 80.
6. G.R. Satchler, Nucl. Phys. A195, 1 (1972).
7. G. Holzwàrth and G. Eckart, Z. Phys. A284, 291 (1978).
8. G. Bertsch, Nucl. Phys. A249, 253 (1975).
9. E. Lipparini, G. Orlandini, R. Leonardi, Phys. Rev. Letters 36, 660 (1976).
10. O. Bohigas, A.M. Lane, J. Martorell, Phys. Reports 51C, 267 (1979).
11. T. Suzuki, Nucl. Phys. A217, 182 (1973).
12. M. Kohno and K. Ando, Prog. Theoretical Phys. 61, 1065 (1979).
13. A. Bohr and B. Mottelson, Nuclear Structure Vol. II (Benjamin, 1975) p. 356.
14. G. Bertsch and S.F. Tsai, Physics Reports 18C, 127 (1975).
15. D.R. Bes, et al., Physics Reports 16C, 1 (1975).
16. I. Hamamoto, Physics Reports 10C, 64 (1974).
17. S.F. Tsai and G. Bertsch, Phys. Rev. C11, 1634 (1975).
18. R.A. Broglia, et al., Physics Letters 85B (1979).
19. F.E. Bertrand, Ann. Rev. Nucl. Sci., 26, 457 (1976).
20. S. Shlomo and G. Bertsch, Nucl. Phys. A243, 507 (1975).
21. K.T. Hecht, D. Braunschweig, Nucl. Phys. A295, 34 (1978).
22. N. Auerbach, et al., Rev. Mod. Phys. 44 (1972), eq. Al.51.
23. G. Bertsch, et al., Phys. Letters 80B, 161 (1979).
24. K.T. Knöpfle, et al., Phys. Letters 74B, 191 (1978).
25. M.T. Collins, et al., Phys. Rev. Letters 42, 1440 (1979).
26. E. Wolynec, et al., Phys. Rev. Letters 42, 27 (1979).
27. H. Rothhaas, et al., Phys. Letters 51B, 23 (1974).
28. S.F. Tsai and G. Bertsch, Phys. Rev. Letters 59B, 425 (1975).
29. J.M. Moss, et al., Phys. Letters 53B, 51 (1974).

30. E. Hayward, NBS preprint, 1979.

31. A. Schotter, et al., Phys. Rev. Letters 43, 569 (1979).

Chapter IV

COLLEÇTIVE DESCRIPTION OF DEFORMED AND TRANSITIONAL NUCLEI

Amand Faessler
Institut für theoretische Physik der Universität Tubingen
D-7400 Tübingen, West Germany

1. Introduction

In these lectures we discuss several topics of current interest in the theory of deformed nuclei. The anomaly of the moment of inertia called backbending (bb) is described in Section 2. We provide a theoretical framework which can encompass the various hypotheses put forth to explain the anomaly. It is found that the major cause of bb is the alignment of two $i_{13/2}$ neutrons, together with a general reduction of the pairing correlations. The next topic we take up, in Section 3, is the gamma decay of the systems resulting from heavy ion fusion reactions. There is a statistical decay toward the ground state band (gsb), the details of which are quite sensitive to the competition between collective and single-particle transitions. It is essential to include the strong collective transitions parallel to the gsb to describe the observed feeding of the gsb. This alternate cascade can even produce a second maximum in the gsb feeding intensities, as seems to occur in the reaction $^{26}Mg(^{136}Xe,4n)^{158}Dy$. In the last section, we discuss the spectrum of two particles coupled to a rotor. There are two situations to consider. In the peaceful case, the coupling for both particles is either weak or strong. In the conflicting case, one particle is strongly coupled to the core and the other particle is weakly coupled. Although there are many levels, we are able to find simple rules for how the spectra are built and how they vary with proton or neutron number. This description provides an explanation of the anomalies in the yrast spectrum of the even Hg and Pt isotopes.

2. Backbending in Deformed Nuclei

If one wants to study backbending which is an anomaly of the moment of inertia at high spin states, one has first to bring into the nucleus the high angular momentum. This is still best done by a heavy ion fusion reaction as indicated in Fig. 1.

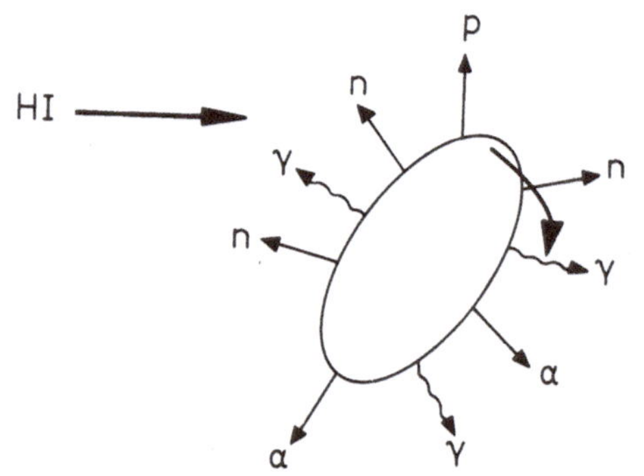

Fig. 1 A heavy ion fusion reaction is schematically indicated. The fused system is rotating fast and it is highly excited. The internal excitation is reduced by evaporating mainly neutrons but also protons, deuterons, α-particles ... If there is not enough excitation energy left to evaporate particles, the remaining internal energy of the residual nucleus is removed by γ-ray emission

If one bombards a nucleus with a heavy ion, there is a finite probability that the two nuclei fuse. The kinetic energy of the incoming projectile is transformed partially in rotational energy and partially in internal excitation of the compound system. Fig. 1 indicates that this internal excitation energy is removed by evaporating a few particles and, if not enough energy for particle emission is available, by the decay in γ-rays. If one bombards for example rare earth nuclei by 100 MeV α-particles, one evaporates in average 8 neutrons. But since the compound nucleus evaporates not only 8 neutrons but also 6, 7, 9, or even 10 neutrons, one measures with a γ-ray counter γ-rays in different residual nuclei. An important trick is therefore to measure the γ-rays in coincidence with a known γ-ray in the residual nucleus. Let us assume the γ-ray transition between the angular momentum 4^+ and 2^+ state has a transition energy of 159.37 keV. Then we are sure, if we use two GeLi's, that all γ-rays we measure in coincidence with 159.37 keV are transitions in this definite residual nucleus. Such coincidence spectra are shown in Fig. 2 [1]. If the rotational band follows a pure I(I+1) law, then the transition energies

$$\Delta E_{I,I-2} = (\hbar^2/\Theta)(2I-1)$$

are proportional to the angular momentum I. The reaction in the upper part of Fig. 2, ^{176}Yb(α,6n)^{174}Hf, shows therefore almost a normal behavior: the transition energy increases with increasing angular momentum. The big surprise found in 1971-72 first in Stockholm [2] and confirmed in Jülich [3], Brookhaven [4] and in many other laboratories are reactions like ^{162}Dy(α,8n)^{158}Er shown in the second part of Fig. 2. In ^{158}Er one finds that the transition energies are not increasing with angular momentum but that they are decreasing if one goes from the

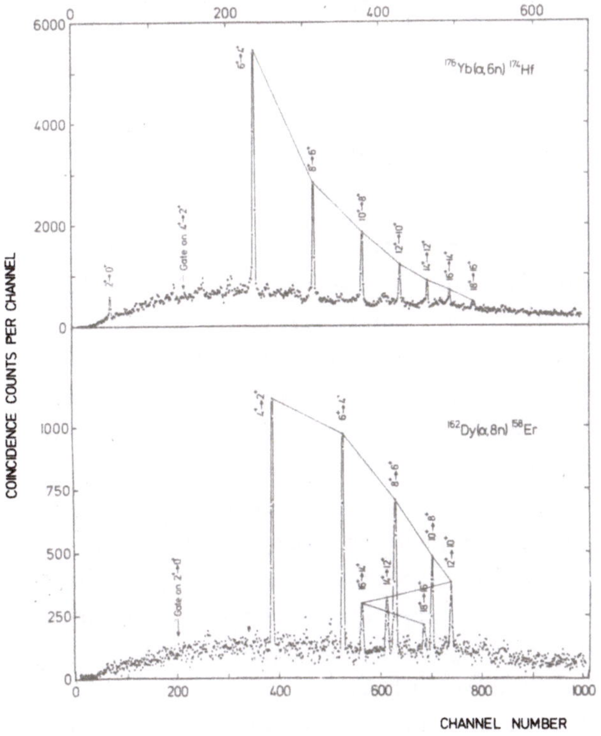

Fig. 2 γ-γ coincidence spectra for the reactions ^{176}Yb(α,6n)^{174}Hf and ^{162}Dy(α,8n)^{158}Er gated on the 4^+ to 2^+ transition in ^{174}Hf and gated on the 2^+ to 0^+ transiton in ^{158}Er

$12^+ \rightarrow 10^+$ to the $14^+ \rightarrow 12^+$ and $16^+ \rightarrow 14^+$ transitions. This anomaly, which can be directly seen in the data (lower part of Fig. 2), is called backbending. One normally prefers to present the data in different ways. The two most popular presentations are shown [1] in Fig. 3. The regular nucleus ^{174}Hf shows practically a straight line in the plot

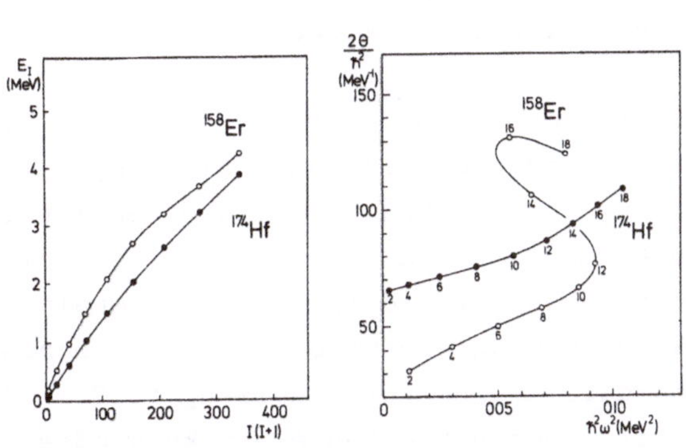

Fig. 3 On the left-hand side plot of excitation energy vs. I(I+1) for the gsb in ^{158}Er and ^{174}Hf, and on the right-hand side the plot of twice the moment of inertia against the square of the rotational frequency of the nucleus

of the energy vs. I(I+1). The anomaly found for ^{158}Er appears as a small kink. Therefore one prefers the so-called backbending plot where twice the moment of inertia is plotted against the square of the rotational frequency of the nucleus. The moment of inertia is defined by the excitation energy differences

$$(2\Theta/\hbar^2) = (4I - 2)/(E_I - E_{I-2}).$$

The rotational frequency is given by half the transition energy,

$$\hbar\omega = (E_I - E_{I-2})/2.$$

In this diagram a good rotor would be represented by a horizontal straight line indicating a constant moment of inertia. ^{174}Hf yields almost a straight line increasing with the square of the rotational frequency. This result is not surprising since we know it already qualitatively from molecular physics. If one has a hydrogen molecule consisting of two protons and two electrons, the moment of inertia is increasing with increasing angular momentum due to the centrifugal force. Due to time reversal symmetry, the increase of the moment of inertia with the rotational frequency can only depend on even powers of ω. A parametrization of the dependence of the moment of inertia up to ω^2 is given by the variable moment of inertia model [5]. The diagram in which the moment of inertia is plotted against the square of the rotational frequency is called the backbending plot.

The surprising result in 1971-72 was data as shown for ^{158}Er in Fig. 3. Up to angular momentum 10, the moment of inertia is increasing linearly with the square of the rotational frequency. But then the moment of inertia is increasing so fast that the rotational frequency is decreasing although the angular momentum is still increasing. With angular momentum 16 the moment of inertia reaches its highest value and decreases again if one goes to higher angular momenta. The question which we want to discuss here first is what happens inside the nucleus to increase the moment of inertia by a factor 2 between angular momentum 10 and 16.

Such a tremendous increase of the moment of inertia can be explained by the intersection of two rotational bands as discussed first by FAESSLER, GREINER and SHELINE already in 1965 [6]. The intersection of two bands and the resulting backbending plot are shown in Fig. 4. After the detection of backbending several groups tried to fit the anomaly of the moment of inertia by the intersection of two or even more bands. In this way one is able to fit the data if one introduces up to 7 intersecting bands. But this type of fits does not enhance our knowledge about the nucleus. What we want to know is the nature of the upper band. Up to 1976 several different explanations for the nature of the upper band have been discussed. I would like to mention here only the three which have been discussed very extensively at a large number of

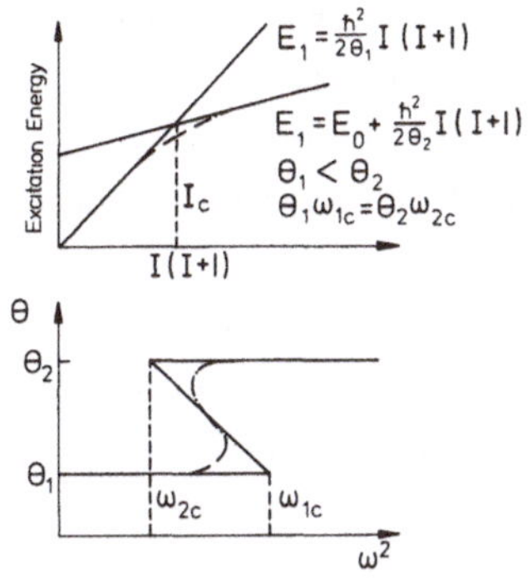

Fig. 4 The upper part shows the intersection of two bands in the diagram excitation energy vs. the square of the angular momentum I(I+1). The dashed-dotted line indicates the yrast energy if one assumes an interaction between the two crossing bands. The lower part gives the backbending plot for two intersecting bands without any interaction (solid line) and with interaction (dashed-dotted line)

international conferences.

(i) In the deformation jump (DEJ) one assumes that the deformation energy surface has a second excited minimum at a larger deformation as shown in Fig. 5. A rotational band is built on both minima. The excited rotational band has a larger moment of inertia and therefore intersects the ground state rotational band at around angular momentum 14 as indicated in Fig. 5.

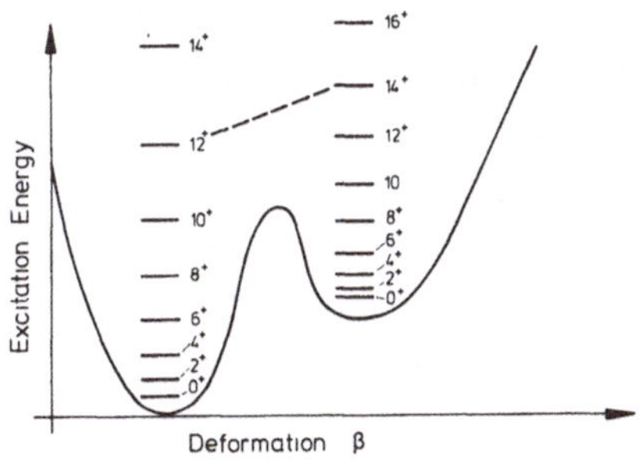

Fig. 5 Excitation energy against the deformation of the nucleus β. In the deformation jump (DEJ) one assumes that the deformation energy surface, indicated here by a solid line, has a secondary excited minimum at a larger deformation. On both minima a rotational band is built. The excited band has a larger moment of inertia and therefore forms the yrast line (lowest energy for a given angular momentum) starting with angular momentum 14^+

(ii) The Coriolis antipairing effect (CAP) has been first discussed by MOTTELSON and VALATIN [7]. The name CAP-effect has been coined by FAESSLER, GREINER and SHELINE [6,8]. Here one assumes that the lower band with the moment of inertia Θ_1 shown in Fig. 4 is the paired band, while the upper band is the unpaired band. The energy difference between the two 0^+ states is therefore the pairing correlation energy. If one goes up the yrast line (lowest energy for a given angular momentum) as a function of the increasing angular momentum, one has in this interpretation a phase transition from the paired into the unpaired state of the nucleus at the intersection of the two bands. The larger moment of inertia of the unpaired nucleus is explained by the fact that in this situation one allows for no correlation between the nucleons. The nucleons are therefore all coupled to the deformed potential. They have therefore to rotate all with the potential and one therefore gets the rigid body moment of inertia. If one has pairing correlations as in the gsb, the nucleons have additional degrees of freedom. Each pair can scatter from level to level coupled to angular momentum 0. In this way not all nucleons have to follow the rotation of the potential and therefore the moment of inertia is smaller by a factor 2 - 3 than the rigid body one. In the CAP-effect we have the following situation: At total angular momentum 0 all nucleon pairs are coupled to angular momentum 0. If the angular momentum of the nucleus is increasing, the Coriolis force starts to act on the pairs,

$$V_{Coriolis}(i) = -\frac{h^2}{\Theta} \vec{I}\cdot\vec{j}(i).$$

If no other force would act on the nucleons apart of the Coriolis force, all nucleons would be aligned with their single-particle angular momentum j along the total angular momentum I. Due to the minus sign in the Coriolis force, this would yield the minimum energy. But apart from the Coriolis force, other forces also act on the nucleons; for example, pairs of nucleons are bound to angular momentum 0 by the pairing correlations. One needs roughly 0.6 MeV to break a pair at the Fermi surface. At small angular momenta the Coriolis force is not able to do that. But if one comes to the critical angular momentum I_C, the pairs are broken roughly all near the same angular momentum, one has a "phase transition" (if one wants to use this expression of statistical mechanics for a system of only a few nucleons).

One has a close analogy to superconductivity in solid state physics with the Meissner-Ochsenfeld effect. In solid state physics one increases the magnetic field and here we increase the Coriolis field until superconductivity disappears in the whole system at once (superconductor of I-type).

(iii) STEPHENS and SIMON [9] looked closer at the Coriolis force and realized that it acts more strongly on single-particle states with a larger angular momentum j. They therefore requested that first the pair with the largest single-particle angular momentum near the Fermi surface is broken. Thus, in this rotational alignment (RAL) effect the upper band is a two quasi-particle band of $i_{13/2}$ neutrons aligned along the total angular momentum. In contrast to the Coriolis anti-pairing effect, where we have an analogy to a superconductor of I-type, we have here an analogy to a superconductor of II-type where the magnetic field breaks first only a few pairs.

One naturally would like to know what is the true nature of the upper band: Is it a rotational band built on a secondary minimum with a larger deformation as requested by the deformation jump (DEJ), or is it the unpaired band with the rigid body moment of inertia as indicated by the Coriolis antipairing effect (CAP), or is it a two quasi-particle $i_{13/2}$ neutron band with the single-particle angular momenta j aligned along the total angular momentum? These questions can be answered in the framework of the projected Hartree-Fock-Bogoliubov theory [10,11].

What do we have to request from the theory? (i) To describe the DEJ-effect one has to allow that the shape parameters of the nucleus β (here and in the rest of the paper the shape parameter β stands for the shape parameters β, γ and β_4 to simplify the notation) has to be determined for each angular momentum I separately.

(ii) To include the CAP-effect in our description, we have to allow the pairing gap parameters, Δ_P, Δ_N, to depend on the total angular momentum I.

(iii) To include the rotational alignment effect, the theory should be able to allow for the alignment of single-particle angular momenta j along the total angular momenta I. Therefore the description should contain the possibility that the expectation value $\langle J_x \rangle$ can be different from 0. This seems to be a trivial requirement. But, it is the most restrictive of all three. J_x is odd under time reversal symmetry. If the wavefunction has a good time reversal symmetry, then the expectation value of an operator which changes sign upon the action of the time reversal symmetry operator has to be zero. This is the same situation as with the electric dipole operator which is odd under the parity operation. Each state which has good parity yields an electric dipole moment zero. Only if one has parity mixing, the electric dipole moment may be different from zero. We have therefore to construct a wavefunction which has no good time reversal symmetry in the intrinsic system. This does not mean that we violate a natural conservation law. In the intrinsic system a wavefunction of the nucleus has also no good angular momentum. Only if one rotates into the laboratory system, one has to project either exactly or approximately on good angular momentum and good time reversal symmetry.

In Section 2.1 we shall formulate a theory which fulfills the above three requirements and is therefore able to describe on a fair equal footing all three possibilities for the character of the excited band which we have discussed above.

2.1 Formulation of the Theory

We shall formulate here a theory [10] which includes for the nature of the upper band the possibility that it is built on an excited secondary minimum with a larger deformation (deformation jump), that it is the unpaired band with a moment of inertia close to the rigid body one (Coriolis antipairing effect), and that it is a two quasi-particle band of aligned $i_{13/2}$ neutrons. We proceed in two steps, A and B. In step A we produce a class of wavefunctions which still depend on a few variational parameters. In step B we search from these wavefunctions the best ones by minimizing in a Ritz variational principle the expectation value of the many-body Hamiltonian. This two-step procedure allows restoration of some of the symmetries, after constructing the trial wavefunctions in step A, by projecting on good quantum numbers and then doing the final variation after projection. We especially will do a particle number projection on a definite proton and neutron number before the final variation. This turns out to be very essential since, as we shall see below, the backbending properties vary rapidly with the particle number.

To construct a class of trial wavefunctions, we start with a trial Hamiltonian

$$H_{trial} = \sum_{i=1}^{A} \{h_{Nil}(\beta,i) - \omega j_x(i)\} + H_{pair}(\Delta) \qquad (4\text{-}1)$$

The trial Hamiltonian consists for each nucleon i of the Nilsson Hamiltonian h_{Nil} which depdns on the deformation parameters β, γ and β_4 (represented in the formula only by β) of the shape of the single-particle potential and on a cranking term $-\omega j_x$ which violates time reversal symmetry and therefore allows for an expectation value $\langle J_x \rangle$ different from zero. In addition, we have a pairing term depending on the pairing gaps for the protons Δ_P and for the neutrons Δ_N, (characterized in the formula by Δ only). To construct a class of trial wavefunctions, we solved the Hamiltonian (1) with the Hartree-Fock-Bogoliubov (HFB) approach. The HFB method combines the advantages of the Hartree-Fock (HF) and the treatment of the pairing correlations by Bardeen, Cooper and Schrieffer (BCS). Fig. 6 indicates schematically the HFB method. The Hilbert-space consists of the combined HF and BCS degrees of freedom. Lines of equal expectation value $\langle H \rangle$ of the Hamiltonian are indicated. A Hartree-Fock calculation minimizes this expectation value along the abscissa and a BCS calculation along the ordinate. The HFB approach searches the minimum in the combined Hilbert-space and is therefore able to find an improved solution.

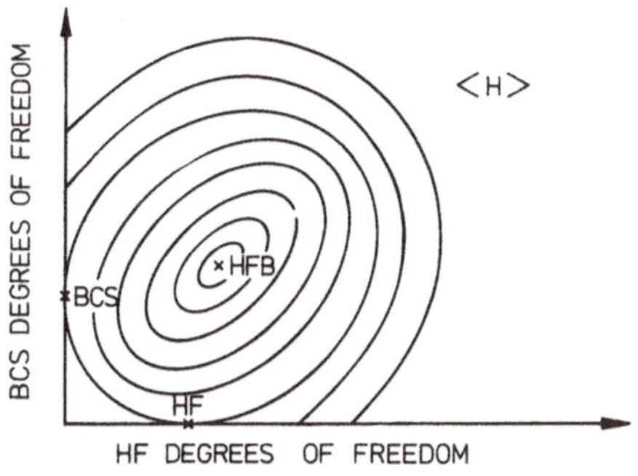

Fig. 6 Schematic representation of the Hartree-Fock-Bogoliubov (HFB) approach. The Hilbert-space in which the expectation value of the many-body Hamiltonian $\langle H \rangle$ is minimized, is the combined space of the Hartree-Fock (HF) and the pairing (BCS) degrees of freedom. The lines indicate equal values of the total energy $\langle H \rangle$. The HF approach minimizes along the abscissa. A pairing calculation minimizes $\langle H \rangle$ along the ordinate. The HFB approach finds the deepest minimum in this Hilbert-space

In the HF approach one constructs the wavefunction as the Slater determinant out of the best independent single-particle wavefunctions.

$$a_i^+ = \sum_a c_a^+ A_{ai} \qquad (4\text{-}2)$$

The operators c_a^+ create a complete set of independent single-particle wavefunctions (for example oscillator wavefunctions). In the BCS approach one assumes that the single-particle states are already good enough, but that one has to take into account that the pairing correlations smear out the Fermi surface. Therefore one introduces quasi-particle states b_a^+ which are linear combinations of a particle and a hole state. (The bar at the hole state indicates the time reversed state.)

$$b_a^+ = u_a c_a^+ + v_a c_{\bar{a}} \qquad (4\text{-}3)$$

In the HFB method we combine the degrees of freedom of (2) and (3) to define independent quasi-particle states which are linear combinations of particle and hole basis states.

$$b_i^+ = \sum_a \{c_a^+ A_{ai} + c_a B_{ai}\} \qquad (4\text{-}4)$$

The HFB wavefunction then is the Slater determinant built of all quasi-particle annihilation operators.

$$|\rangle \equiv |HFB, \beta, \Delta, \langle I \rangle\rangle \equiv \prod_{\text{all } i} b_i(\beta, \Delta, \langle I \rangle)|0\rangle \qquad (4\text{-}5)$$

The quasi-particle annihilation operators depend still on the shape parameters β and the pairing parameters Δ. They further depend on the cranking frequency ω which can be replaced by the average angular momentum due to the relation:

$$\langle \hat{J}_x \hat{P}_{P/N} \rangle = \langle I \rangle (\langle I \rangle + 1) \qquad (4\text{-}6)$$

The operator $\hat{P}_{P/N}$ projects on good proton and good neutron number [10].

In step B we take a more microscopic many-body Hamiltonian, one of KUMAR and BARANGER [12] which contains for the particle-particle interaction the monopole pairing force with the strength G_τ $(\tau = P,N)$ and for the particle-hole interaction the quadrupole-quadrupole force with the strength χ.

$$H = \sum_i \varepsilon a_i^+ a_i + V$$

$$V = -\frac{1}{4}\sum_\tau G_\tau \hat{P}^+_{00}(\tau)\hat{P}_{00}(\tau) - \frac{1}{2}\chi \sum_{\tau,\tau',m} \hat{Q}^+_{2m}(\tau)\hat{Q}_{2m}(\tau') \tag{4-7}$$

with $\hat{P}^+_{00}(\tau) = \sum_i a_i^+ a_{\bar{i}}^+ \delta_{\tau_i,\tau}$

$$\hat{Q}_{2m}(\tau) = \sum_{ij} \langle i|r^2 Y_{2m}|j\rangle a_i^+ a_j \delta_{\tau_i,\tau}\delta_{\tau_j,\tau}$$

The normalized and particle-number projected expectation value of this many-body Hamiltonian is minimized for each average angular momentum $\langle I\rangle$ as a function of the three shape parameters β and the two pairing parameters Δ,

$$E_I(\beta,\gamma,\beta_4;\Delta_P,\Delta_N) = \langle \hat{H}\, \hat{P}_{P/N}\rangle .$$

In practice, we take an inert core of ^{110}Zr and include for the protons the $N = 4,5$ and for the neutrons the $N = 5,6$ oscillator shells. The single-particle energies are taken from KUMAR and BARANGER [11]and the pairing and quadrupole force constants are identical to the ones in [10] and [13].

$$G_P = 23/A \text{ MeV};\; G_N = 18/A \text{ MeV}$$

$$\chi = 73 \cdot A^{-1/4} \text{ MeV};\; \hbar\omega_o = 41.2\, A^{-1/3} \text{ MeV}. \tag{4-8}$$

The quantity $\hbar\omega_o$ is the oscillator energy for the spherical single-particle basis.

2.2 First Backbending in Even Mass Nuclei

Figure 7 shows the backbending plot for ^{162}Er. The circles indicate the experimental data, while the triangles and squares give the theoretical results. The squares include also the variation of the deformation parameter γ. (One should keep in mind that even for $\gamma = 0$ the projection of the single-particle angular momentum on the symmetry axis Ω is not a good quantum number due to the cranking term.) Theory and experiment show a similar behavior, but theory backbends earlier than the experimental data. I shall come back to this problem later on, but for the

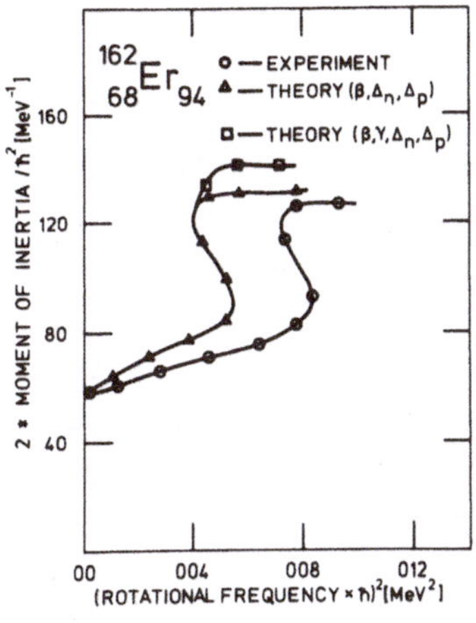

Fig. 7 Twice the moment of inertia of ^{162}Er as a function of the square of the rotational frequency. The circles indicate the experimental values from angular momentum 2^+ to 20^+. The triangles and squares give the theoretical values up to 18^+. For the triangles the asymmetric deformation γ was fixed to zero

moment we want to keep in mind that the theory presented in Fig. 7 contains the DEJ, CAP and RAL effects. We can now ask which effect is really producing backbending: The deformation β stays roughly constant up to angular momentum 20. Thus, bb cannot be produced by a DEJ. The deformation γ is zero up to around angular momentum 12 and increases slightly then to about 10° at angular momentum 20. The proton gap parameter is only slightly decreased by 10 - 20% up to angular momentum 20. The neutron gap parameter shows a more drastic decrease and one could speculate that bb is due to the CAP of the neutrons. A more detailed inspection shows that the general decrease of neutron pairing is only responsible for the linear increase of the moment of inertia in the bb plot. The important push for backbending comes from the alignment of an $i_{13/2}$ neutron pair along the x-axis which is the axis of the total angular momentum (due to the choice of the cranking term). This is demonstrated by the expectation value of j_x for the $i_{13/2}$, $\Omega = 5/2$ level, plotted as a function of <J> in Fig. 8.

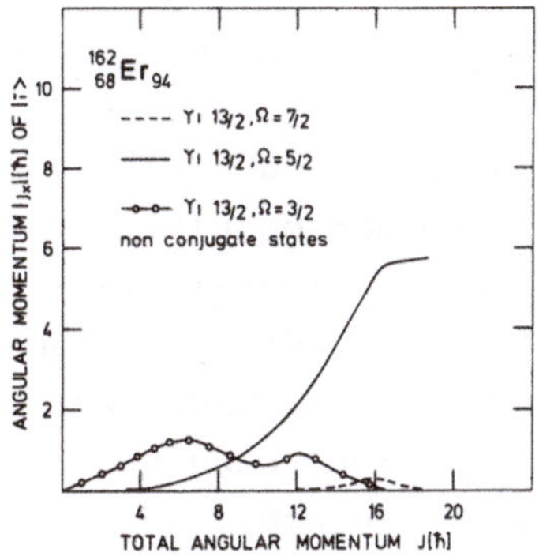

Fig. 8 Angular momentum projection along the rotational axis for different quasi-particle states in the canonical representation of the neutron $i_{13/2}$ levels as a function of the total angular momentum. The angular momentum projection is calculated with particle-number projection. The level which contains mainly an $i_{13/2}$ $\Omega = 5/2$ admixture for low total angular momenta shows alignment above the backbending region (RAL)

A corresponding diagram can also be drawn for the conjugate states. (Please keep in mind that the angular momentum projection Ω to the symmetry axis is only a good quantum number for very small total angular momenta.) At total angular momentum $I = 16$ the alignment of the conjugate pair of $i_{13/2}$ neutrons yields already 10.5 angular momentum units. We could also ask whether one still gets backbending after switching off the CAP effect (by keeping the pairing gap parameters constant). In Fig. 7 this yields a constant moment of inertia below backbending. The backbending behavior is also changed. We get no backbending but only a slight upbending. Backbending is therefore an interplay between the CAP effect and the RAL effect, where the important push for bb comes from the alignment of two $i_{13/2}$ neutrons along the total angular momentum (RAL).

In Fig. 7 one sees that the theory is backbending too early. This might be due to the fact that the many-body Hamiltonian of Kumar and Baranger contains only monopole pairing. That means that a pair of nucleons is only kept together when it couples to angular momentum zero. But realistic nucleon-nucleon forces show also a particle-particle attraction for higher multipolarities. Therefore they would also attract each other when they couple to angular momentum 2 although this force is definitely smaller than the monopole pairing force. If one neglects the higher multipole pairing forces it is obvious that one breaks the pairs at an already too low angular momentum, since if a pair is not any more coupled to angular momentum zero, there is nothing to prevent full alignment. WAKAI and FAESSLER [14] used the above description for bb but included also quadrupole pairing,

$$V^{(2)} = -\frac{1}{4} \sum_{\tau,m} 4\pi G_{\tau}^{(2)} \hat{P}_{2m}^{+}(\tau)\, \hat{P}_{2m}(\tau) \tag{4-9}$$

with $\hat{P}_{2m}^{+}(\tau) = \sum_{ij} \langle i|Y_{2m}|\bar{j}\rangle a_i^{+} a_{\bar{j}}^{+} \delta_{\tau_i,\tau}\, \delta_{\tau_{\bar{j}},\tau}$

This Hamiltonian has in addition to the parameters given in (8) the quadrupole pairing force constant $G_{\tau}^{(2)}$. Since we do not want to use a new parameter, we assume that the monopole pairing and the quadrupole pairing force are derived from the surface delta interaction [15]. This yields $G^{(0)} = G^{(2)}$. Since we use the same parameters as given in (8), we no longer have an optimal fit to the ground state properties in the rare earth nuclei as performed by KUMAR and BARANGER [12]. Thus, we do not expect to get such a good agreement as without including the quadrupole pairing for the low spin states. We want therefore to see only the qualitative tendency what happens to backbending if one includes quadrupole pairing.

The result of a calculation including also quadrupole pairing is shown in Fig. 9. One sees that the inclusion of quadrupole pairing with the projection to the symmetry axis $m=0$ and $m=1$ yields an improvement compared to only including monopole pairing. The backbending point is shifted to higher angular momenta and the slope of the moment of inertia as a function of the square of the rotational frequency is improved at lower angular momenta. MIGDAL [16] and BELIAEV [17] who discussed first quadrupole pairing argued that one needs only to include the $m=1$ part to calculate the moment of inertia. Indeed, we find that the inclusion of the $m=1$ part of quadrupole pairing increases the moment of inertia at angular momentum 2 by 20%, but it does not change the backbending plot starting from angular momentum 8 and higher. The $m=0$ quadrupole pairing part is reducing again the moment of inertia at angular momentum 2 by 20% (it therefore just cancels the increase by the $m=1$ part), but it affects the results especially near the backbending region.

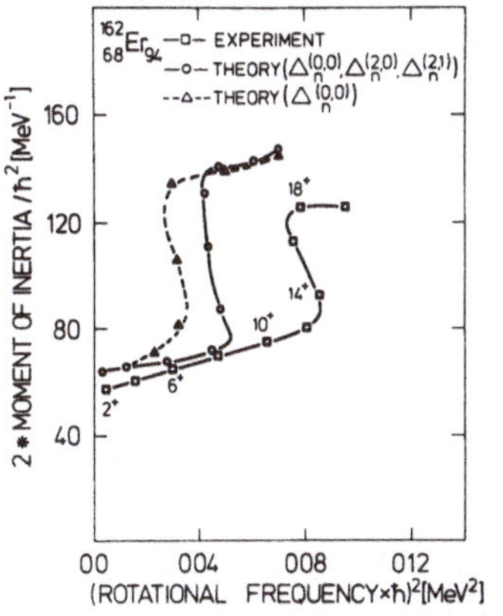

Fig. 9 Twice the moment of inertia of ^{162}Er as a function of the square of rotational frequency for spins from 2^+ to 20^+. The squares indicate the experimental values. The open circles give the theoretical values calculated with the Y_{20} and $Y_{2\pm1}$ pairing forces. The triangle gives the values calculated without any quadrupole pairing forces. The parameters employed in both calculations are $G_N^{(0)}=G_N^{(2)}=18/A$ MeV. It is seen that the effects of the Y_{20} and $Y_{2\pm1}$ pairing forces cancel each other for very low spins (2^+ and 4^+).

Quadrupole pairing improves therefore the agreement between theory and experiment for bb considerably. But the details of the theoretical results depend very strongly on the choice of the single-particle energies. Minor modifications of the $i_{13/2}$ Nilsson levels relative to the other single-particle states would affect strongly the quantitative description of backbending. Thus, a semi-quantitative description of bb is already quite satisfactory.

2.3 The Second Backbending

In the preceding section we have shown that the anomaly of the moment of inertia around angular momentum 12, which one normally calls backbending (bb), is well understood as the alignment of an $i_{13/2}$ neutron pair by the Coriolis force. A second anomaly around angular momentum 28 has been measured in ^{158}Er by LEE, et al. [18]. This anomaly has been explained by FAESSLER and PLOSZAJCZAK [19] as the alignment of an $h_{11/2}$ proton pair. Recently, the isotones $^{156}_{66}Dy_{90}$ and $^{160}_{70}Yb_{90}$ have been measured up to the angular momentum region of the second anomaly [20,21] in the sequence of the isotones ^{156}Dy, ^{158}Er and ^{160}Yb (see Fig. 10). One finds around angular momentum 28 no anomaly of the moment of inertia, upbending and bb, respectively. Here we want to

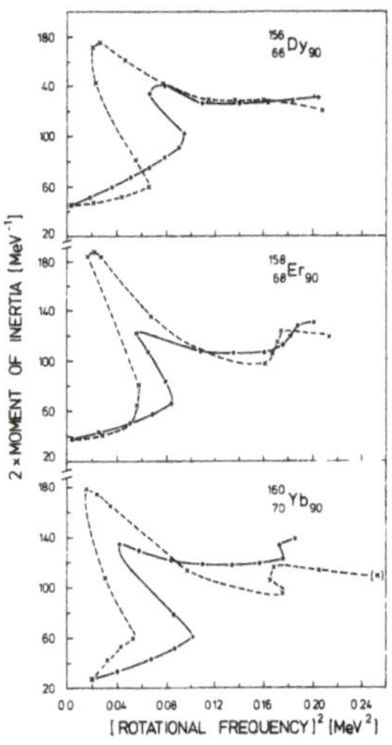

Fig. 10 Backbending plots for the N = 90 isotones $^{156}_{66}Dy_{90}$, $^{158}_{68}Er_{90}$ and $^{160}_{70}Yb_{90}$. The solid line gives the experiment [1,3,4] and the dashed line the theory. The theoretical results are calculated with the HFB approach [7] and the Hamiltonian of KUMAR and BARANGER [9] ($A \cdot G_P = 25$ MeV; $A \cdot G_N = 20$ MeV; $\chi = 72 \cdot A^{-1.4}$ MeV). The proton gap parameter has been varied to minimize the total energy for each angular momentum. The other shape and pairing parameters are kept fixed ($\beta = 0.26$; $\gamma = 0$; $\beta_4 = 0.08$; $\Delta_N = 0.9$ MeV)

discuss the explanation of this strange behavior. We use again the Hartree-Fock-Bogoliubov theory developed in Section 2.2. The parameters of the Hamiltonian are slightly modified and are given in the caption of Fig. 10. The expectation value of the total many-body Hamiltonian is in principle minimized as a function of the shape parameters β, γ and β_4 and the pairing gaps Δ_P and Δ_N. The calculation in which all these parameters are varied to find the minimum of the total energy is numerically not feasible. But extensive numerical studies [10,14,19] have shown that it is allowed to keep the shape parameters fixed with increasing angular momentum and to choose a constant value of the proton gap Δ_P if one studies the first bb and a constant value of the neutron gap Δ_N if one is interested in the second anomaly. The parameters chosen independently of the total angular momentum in the wavefunction have been chosen to minimize the energy in the intrinsic system [19]: $\beta = 0.2$, $\gamma = 0$, $\Delta_N = 0.9$ MeV. The choice of the hexadecapole deformation $\beta_4 = 0.08$ will be discussed below. The proton gap parameter Δ_P is varied to yield the minimum of the total energy for each average angular momentum <I>.

Fig. 10 shows the results for the three isotones in the bb plot. The variation of the second anomaly of the moment of inertia is nicely reproduced. This is essentially due to the choice of the β_4 deformation ($\beta_4 = 0.08$).

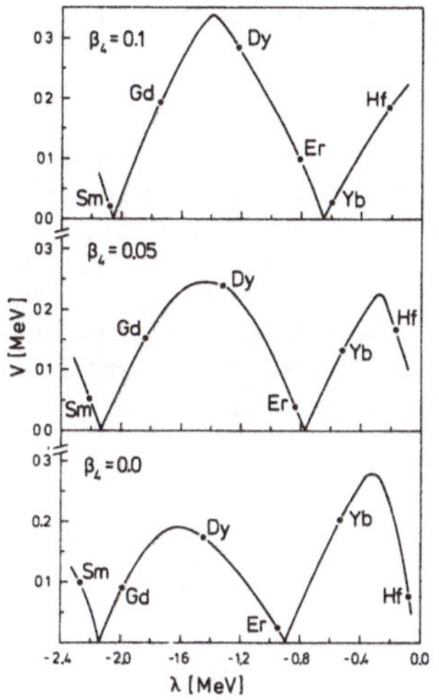

Fig. 11 Interaction V at the crossing with the proton 2qp $h_{11/2}$ band as a function of the proton Fermi surface λ for different β_4 values. The position of the Fermi surface for different N = 90 isotones is indicated. A small |V| yields a large second anomaly of the moment of inertia

Fig. 11 displays the interaction [22,23] V in MeV between the aligned two quasi-particle neutrons $i_{13/2}$ band and the band which in addition has aligned two $h_{11/2}$ protons. The interaction is defined as half the energy distance at the "crossing point" as a function of the cranking frequency ω. The interaction with the aligned $h_{11/2}$ proton pair shows the oscillations first discussed by BENGTSSON, HAMAMOTO and MOTTELSON [22] for the $i_{13/2}$ shell. For $\beta_4 = 0$ one obtains a strong second bb for ^{158}Er but none for ^{156}Dy and ^{160}Yb. The β_4 has to be chosen to be larger or equal than $\beta_4 = 0.08$ to find for a spherical single-particle energy of KUMAR and BARANGER [12] a strong second bb in ^{160}Yb, upbending in ^{158}Er and no anomaly in ^{156}Dy. The analysis of alpha-scattering data by HENDRIE, et al. [24] gives around Z = 62, N = 90 the value $\beta_4 = 0.05 \pm 0.01$. The theoretical calculations tend to give larger values. NILSSON and coworkers [25] obtain with the Strutinsky method for the N = 90 isotones $\beta_4 = 0.075$ (Sm), 0.07 (Gd), 0.075 (Dy), 0.08 (Er). (These values are partially extrapolated.) The value $\beta_4 = 0.08$ needed to obtain agreement with the data seems therefore slightly high but is still within the range of the theoretical results. One could obtain also the correct qualitative behavior of the second bb for $\beta_4 = 0.05$ if one increases the

$h_{11/2}$ proton single-particle energy. Such an increase has also recently been suggested by CHASMAN [26].

Fig. 11 shows that one expects for a hexadecapole deformation larger or equal than $\beta_4 = 0.08$ a very small interaction between the intersecting bands at the second anomaly for Yb, a larger interaction for Er, and a very large interaction for Dy. According to this we find in these three isotones strong backbending, upbending and no anomaly at all.

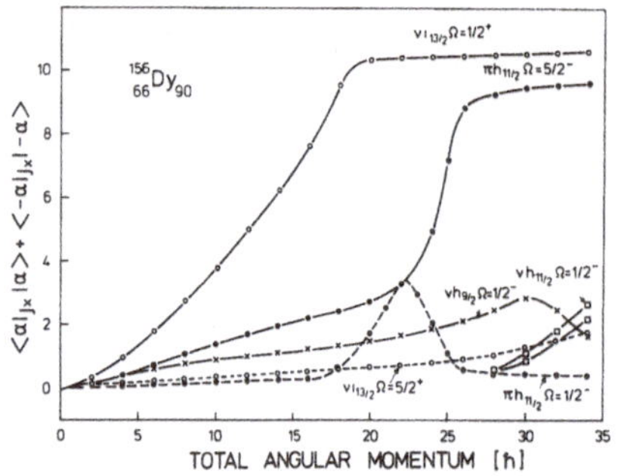

Fig. 12 Alignment of the single-particle angular momentum along the rotational-(x)-axis for conjugate single-particle states $|\alpha\rangle$ and $|-\alpha\rangle$ as a function of the total angular momentum I for ^{156}Dy. The quantum numbers assigned are only good for small I. At high I the main amplitude of $|\alpha\rangle$ may be characterized by a different Ω

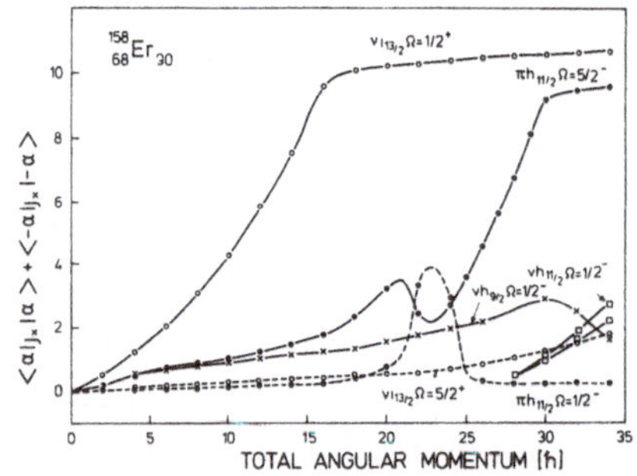

Fig. 13 Alignment of the single-particle angular momenta of a conjugate pair $|\alpha\rangle$ and $|-\alpha\rangle$ of nucleons along the rotational axis as a function of the total angular momentum for ^{158}Er

Figs. 12 and 13 show the alignment plots of different single-particle angular momenta for ^{156}Dy and ^{158}Er. It shows that the first anomaly is due to the alignment of an $i_{13/2}$ neutron pair, while the second anomaly is due to the alignment of an $h_{11/2}$ proton pair.

We are therefore able to reproduce the variation of the second anomaly with the proton number. This supports strongly the interpretation of the second anomaly as the alignment of two $h_{11/2}$ protons [19].

3. What is the Nature of the Yrast Traps?

3.1 Introduction

Before I come to the direct topic of this chapter, "What is the Nature of the Yrast Traps?", I will first study how to form a nucleus with large angular momentum and how such a nucleus decays. The most successful method of forming high angular momentum states is the fusion reaction, especially between two heavy ions. The kinetic energy of the incoming heavy ion is partially transformed in such a reaction into rotational energy and into internal excitations. The internal energy is mainly given away by evaporating a few particles. The large angular momentum is preferentially carried away by collective E2 γ-rays. We study here in the region where only γ-rays can be emitted, the competition between statistical γ-rays of E1, M1, E2 nature and collective E2 transitions. We assume that each level is member of a rotational band and can decay by a stretched E2 or by a statistical transition. Including this competition we will try to understand the side feeding pattern of the ground state rotational band and the measured γ-multiplicities especially for the reactions $^{150}Nd(^{16}O,4n)^{162}Er$ and $^{148}Nd(^{18}O,4n)^{162}Er$. The work reported here is by WAKAI and FAESSLER [27]. Similar conclusions, but based on less quantitative considerations, are obtained by LIOTTA and SORENSEN [28] and by NEWTON, et al. [29].

In Section 3.3 we then ask "What is the Nature of the Yrast Traps?" Do rare earth nuclei at higher angular momenta really rotate around an oblate symmetry axis as predicted by the liquid drop model [30,31]?

3.2 Excitation and Deexcitation of High Spins in Heavy Ion Fusion Reactions

The most successful method in exciting high spin states is the heavy ion fusion reaction. The kinetic energy of the heavy ion is partially transferred in angular momentum of the compound system and partially into internal excitation. The internal excitation is evaporated by emitting mainly neutrons, protons and alpha particles. After the energy is too low to emit particles, the rest of the excitation energy is carried away by γ-rays. The main purpose of this lecture is to study side feeding patterns and γ-multiplicities in the region where the energy is carried away by γ-rays only. Before we do this we have first to see how the compound nucleus is formed. Assuming spin zero for projectile and target, the compound nucleus formation cross section in a fusion reaction is proportional to:

$$\sigma_I \propto (2I+1)T_I \qquad (4\text{-}10)$$

Here σ_I is the compound nucleus cross section. T_I is the transmission coefficient shown in Fig. 14.

If one takes the fusion cross section as calculated from the transmission coefficients of the optical model, one finds a too large value. This is due to the fact that the transmission coefficients consider also

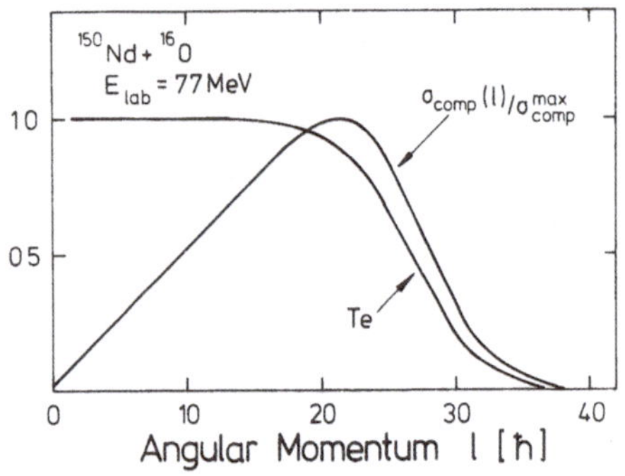

Fig. 14 Transmission coefficients T_ℓ and fusion cross section $\sigma_{comp}(\ell)$ for different incoming orbital angular momenta ℓ of the fusion reaction $^{150}Nd + ^{16}O \rightarrow ^{166}Er^*$. The transmission coefficients T_ℓ are calculated with the optical model potential, but the half value angular momentum ℓ_0 is adjusted to the total experimental fusion cross section (see discussion near (4-11))

other reactions than the fusion. We parametrized therefore the transmission coefficients T_I calculated from the optical model by a Fermi function,

$$T_I = [1 + \exp\{-(I - I_o)/d_o\}]^{-1}. \tag{4-11}$$

The diffuseness d_o is taken from the optical model calculations, but the half value angular momentum I_o is adjusted to reproduce the experimental fusion cross section. In this way we can describe the probability of forming compound nuclei for the different angular momenta I. The deexcitation of these compound nuclei into the ground states of different final nuclei is described by the Monte-Carlo technique. The computer code has the name "ICARUS" since the nucleus finally falls down into the ground state as Icarus fell into the sea. For the Monte-Carlo description of that part of the cascade where particles are emitted, we [27] follow closely the work of SARANTITES, et al. [32]: A random number weighted by the compound nucleus formation cross section of the fusion reaction determines the starting angular momentum. In a second step the partial widths for neutron, proton, alpha particle and γ-ray emission have to be calculated. The partial width for the particle emission is determined by the inverse fusion reaction using the corresponding optical model. With the help of detailed balance the final and the initial states are interchanged and the cross section is then summed up over all permitted energies and angular momenta of the new final state. Correctly normalized, this yields the partial width for the particle emission. The partial width for the γ-decay can be calculated from the transition probability to an energy interval dE_f at the final state

$$\tilde{R}_\lambda(E_i, I_i^{\pi_i}; E_f I_f^{\pi_f})dE_f = \tilde{C}(E_f - E_i)^{2\lambda+1}\Omega(E_f, I_f^{\pi_f})dE_f \tag{4-12}$$

Here $\Omega(E_f, I_f^{\pi_f})$ is the level density for which we use a modified Fermi gas level density in which the additional freedom of the rotational motion of nuclei is considered. Eq. (4-12) is modified if one takes into account that the electromagnetic transition operator is only a one-body operator and that a given final state can only be reached by initial states which distinguish themselves from the final state only by a one-particle-one-hole excitation. To include this effect, expression (4-12) has to be multiplied by the ratio of the initial states formed by one-particle-one-hole excitations from the final state over all the initial states [28]

$$R_\lambda(E_i, I_i^{\pi_i}; E_f I_f^{\pi_f})dE_f = C_\lambda(E_f - E_i)^{2\lambda+2}\Omega(E_f, I_f^{\pi_f})\Omega^{-1}(E_i, I_i^{\pi_i})dE_f \quad (4\text{-}13)$$

If all partial widths are calculated, a random number from the computer weighted by these partial widths selects if in this step of the cascade one emits a neutron, a proton, an alpha-particle or a γ-ray. In a third step a random number weighted by the corresponding transition rates selects the final energy, angular momentum and parity. If the energy is large enough to emit further particles, we return to step two. If the energy is too low to emit particles, we treat the γ-ray emission more carefully.

In region II, where only γ-rays are emitted, we include for the statistical γ-rays E1, M1 and E2 multipolarities. In addition we assume that each level is the member of a rotational band and can decay by a stretched E2 transiton along the band. These collective E2 transitions are independent of the energy above the yrast line. On the other hand, the rate of the E1 transition increases roughly with the fourth power of the energy above the yrast line. If one comes closer to the yrast line, the collective transitions are favored over the statistical ones. Formerly one assumed that the nucleus decays with a few statistical transitions near the yrast line and then it decays in a few parallel channels mainly with collective E2 transitions along the yrast region. The results of this work seem to indicate that one can only understand the side feeding pattern if one assumes that one has an irregular mixture already a few MeV above the yrast line of collective and statistical transitions. To understand the experimental data (see below), we have to assume that the collective E2 transitions are of about 150 single-particle units while the statistical E1 transitions are hindered by a factor 10^{-3}. The statistical M1 and E2 transitions do not seem to play a central role.

Fig. 15 shows the entry distribution of the nuclei as a function of the angular momentum for the reaction $^{150}Nd(^{16}O,4n)^{162}Er$ with a bombarding energy of 77 MeV in the lab system. The experimental values [33] are extracted from measurements of the γ-ray multiplicities by assuming that each γ-ray takes away a definite average angular momentum. The dashed line is the theoretical result calculated by ICARUS.

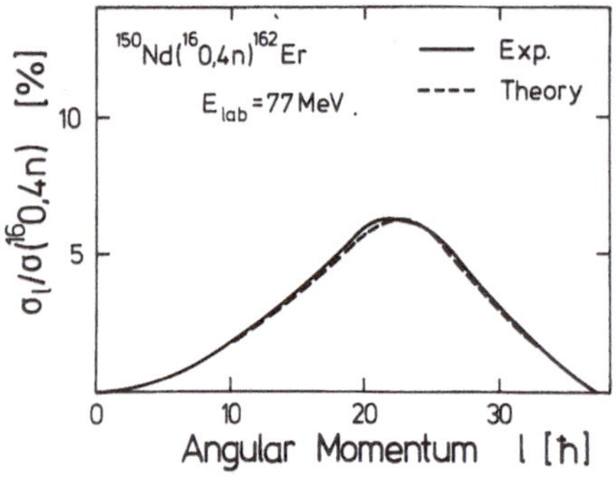

Fig. 15 Experimental and theoretical angular momentum distribution of the entry state in the reaction $^{150}Nd(^{16}O,4n)^{162}Er$ with $E_{lab} = 77$ MeV. The experimental values are taken from ref. 33

Fig. 16 shows the entry distribution for the same reaction as a function of the excitation energy and the angular momentum of the entry state. Such a distribution has not yet been measured. In principle, this distribution can be measured if one uses a subdivided 4π-counter to determine the total energy release and the multiplicity of one and the same γ-cascade.

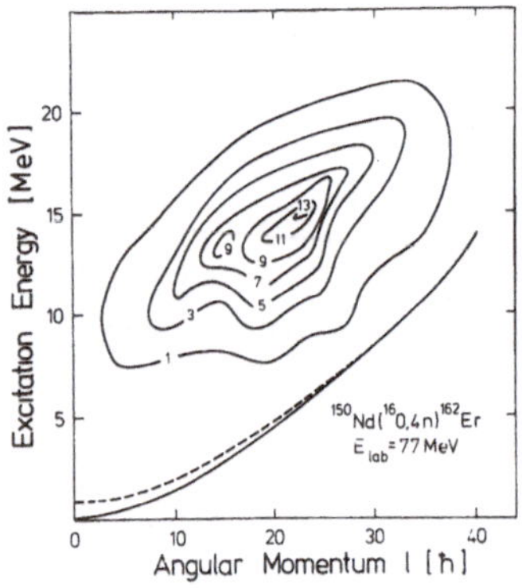

Fig. 16 Entry distribution of the reaction $^{150}Ne(^{16}O,4n)^{162}Er$ with $E_{lab} = 77$ MeV in the excitation energy and angular momentum plane. The solid line indicates the yrast energy. The dashed line shows the position of the yrast states with odd spins and with even spins and negative parity

Figure 17 shows the side feeding pattern for the same reaction with different bombarding energies. The percentage of the cascades which hit the ground state band at a definite angular momentum is plotted as a function of these angular momenta. One sees that the maximum percentage is shifted to larger angular momenta if the bombarding energy is increased. But one sees also that the very good agreement at lower bombarding energies gets worse if one increases the bombarding energy. This may be connected with a wrong extrapolation of the yrast energy or with a deteriorating quality of the level density formulas for higher angular momenta.

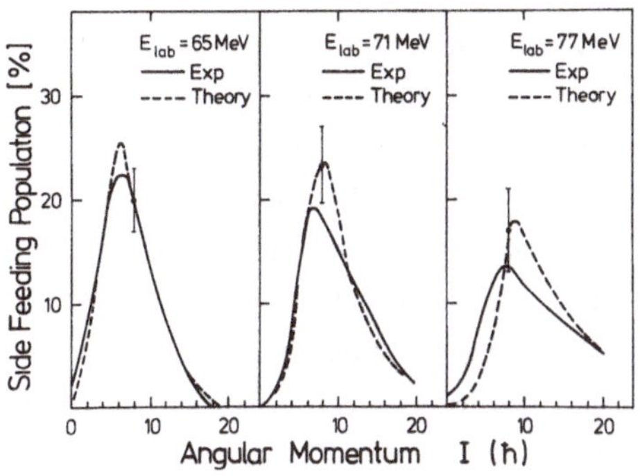

Fig. 17 Side feeding population of the ground state band in % as a function of the angular momentum of the ground state band into which γ-rays feed for different bombarding energies of the reaction $^{150}Nd(^{16}O,4n)^{162}Er$.

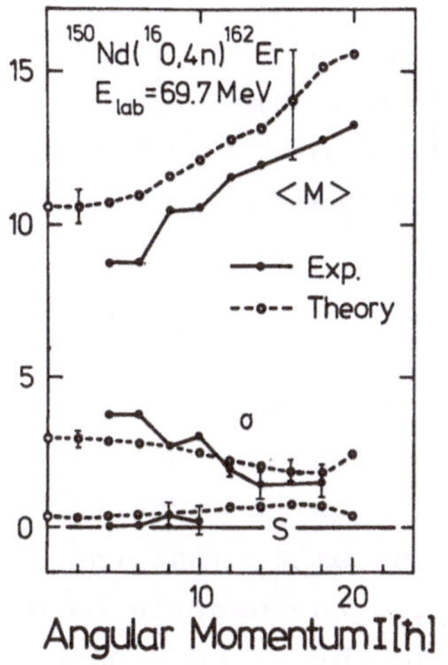

Fig. 18 γ-ray multiplicities $\langle M_{\gamma I} \rangle$, standard deviations σ_I and skewness s_T as defined in [27] and [33]. The experimental data are from the reaction $^{150}Nd(^{16}O,4n)^{162}Er$ with $E_{lab} = 69.7$ MeV. The theoretical results are calculated using (4-12) for the γ-ray transiton probabilities. Improved expression (4-13) modifies only the γ-multiplicities slightly by reducing the value by 1 to 1½ units. It therefore improves the agreement.

Figure 18 shows the γ-multiplicity $\langle M_\gamma \rangle$ as a function of the angular momentum in the ground state rotational band I through which the cascade has to go. The dashed theoretical curve shows also the statistical errors. It is calculated using (4-12) and therefore does not include the one-body nature of the electromagnetic transition operator. If this effect is included according to (4-13), the γ-multiplicities are reduced by 1 to 1½ units since the higher power on the transition energy prefers higher energy transitions. Thus the number of γ-rays in one cascade is slightly reduced. This reduction brings theory and experiment in close agreement. The figure shows in addition the standard deviation σ_I and the skewness s_I. Both agree with the experimental data [33]. The skewness is experimentally and theoretically zero within the statistical errors. The entry distribution as a function of angular momentum seems therefore to be symmetric.

Recently, EMLING, et al. [43] measured the reaction $^{26}Mg(^{136}Xe,4n)$ ^{158}Dy with a bombarding energy of 4.1 MeV per nucleon in the lab system. The surprising result as seen in Fig. 19 is the two maxima in the side feeding intensity into the yrast band. Even more surprising is that our pure statistical description yields two maxima in the side feeding intensities [44]. To test where these two maxima come from, we put the collective E2 transitions in the bands parallel to the yrast line equal to zero. In this case the lower maximum disappeared and the higher maximum increased in intensity. If we enlarged the E2 transitions in the collective band by a factor 10, we found a decrease of the upper maximum and an increase of the lower maximum and a shift of the lower maximum to smaller angular momenta. Thus, the explanation of the two maxima in our calculations is the following: The upper maximum is due to mainly pure statistical transitions which come from the entry states in a few steps down to the yrast line. The second maximum is due to γ-ray transitions which are captured in the collective bands parallel to the yrast line and slowly come down and hit the yrast line at a lower angular momentum.

In addition to this statistical explanation one could also think of a more structural explanation: backbending is due to the intersection of two rotational bands where the upper band has a larger moment of inertia. Since we know of two anomalies of the moment of inertia, one around angular momentum 12 and the other around angular momentum 28, we have three bands with increasing angular momentum forming the yrast line, the ground state band, the aligned $i_{13/2}$ two neutron band and the aligned $h_{11/2}$ two proton band. All of these three bands have a system of roughly parallel collective rotational bands higher up in energy. Therefore one expects that a large part of the γ-ray intensity hits the yrast line near the two intersection points. If the structural explanation is the correct one, one expects that the two maxima are not shifted if the bombarding energy is increased. But if the statistical explanation is the real reason for the two maxima, one would expect that with increasing bombarding energy and also increasing angular momentum for the entry states the two maxima shift to higher angular momenta. With this method one could distinguish experimentally between the two hypotheses for the side feeding maxima obtained in the $^{26}Mg(^{136}Xe,4n)$ ^{158}Dy reaction.

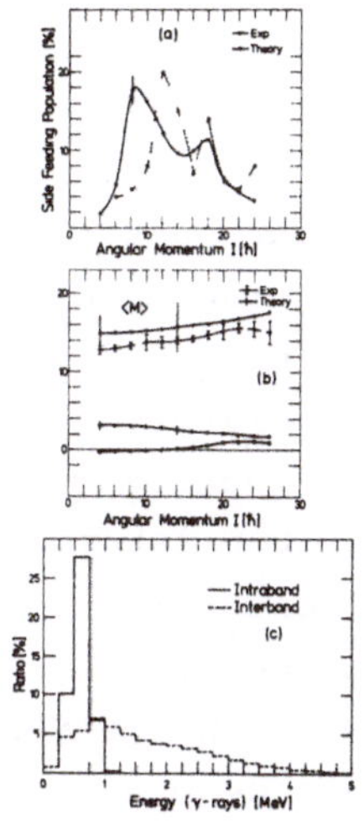

Fig. 19 Side feeding population of the ground state band in % as a function of the angular momentum of the ground state band into which γ-rays feed for the reaction $^{26}Mg(^{136}Xe,4n)^{158}Dy$ with a bombarding energy of 4.1 MeV per nucleon in the laboratory system. The two theoretical curves are calculated for slightly different level densities near the yrast line. The data are from [43]. The middle part shows the γ-multiplicities $\langle M_{\gamma I}\rangle$, the width σ_I and the skewness s_I. The lower part gives the theoretical γ-ray spectrum for the intra- and interband transitions.

3.3 Yrast Traps

For the behavior of nuclei at high and very high angular momenta COHEN, PLASIL and SWIATECKI [30] used the liquid drop model. For rare earth nuclei the liquid drop model predicts for angular momenta below $80\hbar$ a rotation around an oblate symmetry axis (at angular momentum 80 the nucleus is twice as wide than high). Such a collective rotation around a symmetry axis is quantum mechanically not possible. But even if such a rotation would be possible, such a state could not decay by an E2 transition since the electric field of the nucleus rotating around the symmetry axis does not show in lowest order a modification if the rotational frequency slows down. Since a collective rotation around the symmetry axis is not possible, the high angular momentum is built up by independent single-particle motion. We therefore have a similar situation like near doubly-closed shell nuclei. From there we know that we have a high probability of finding isomeric states. BOHR and MOTTELSON [31] predicted therefore that the rare earth nuclei should show yrast isomers, the yrast traps, at angular momenta between 30 and 80.

The liquid drop model can naturally only be a first guide. But already for the ground state of the rare earth nuclei the liquid drop model gives the wrong prediction. Due to the shell corrections the rare earth nuclei are deformed and not spherical. Along the yrast line we expect similar strong shell corrections as in the ground state. We are therefore forced to calculate the deformation energy surface including the shell corrections to see if the rare earth nuclei prefer at high angular momenta to rotate around a symmetry axis of the nucleus with a strongly oblate deformed shape.

To calculate the deformation energy surface two methods have been developed:

(i) The microscopic method developed in Jülich [34] is in principle based on cranked Hartree-Fock single-particle wavefunctions constrained to a definite shape of the nucleus. In practice, we use cranked Nilsson single-particle wavefunctions for different shape parameters β, γ and β_4. With these single-particle wavefunctions we build a Slater determinant for a definite shape of the nucleus with an average angular momentum determined by the cranking frequency ω. With this Slater determinant we calculate the expectation value of a many-body Hamiltonian (usually the pairing plus quadrupole force and sometimes also hexadecapole force Hamiltonian). The quality of this method depends on the many-body Hamiltonian and the single-particle wavefunctions. Since we use only one-center Nilsson wavefunctions and a limited single-particle basis in our applications, our results are restricted to deformation less than $\beta = 0.5$ and angular momenta below $I = 60$.

(ii) To overcome deficiencies for larger deformations in the deformation energy surface, the Strutinsky method has been invented. This method has been extended to higher angular momenta by the Lund-Warsaw group [35]. Similar calculations have been performed at Dubna using also Nilsson single-particle wavefunctions. These calculations have serious difficulties with the asymptotic moment of inertia which does not agree with the rigid body value. We therefore used in Jülich [36,37] the deformed Saxon-Woods potential for the Strutinsky method and we could overcome this difficulty. It is essentially connected with the ℓ^2-term of the Nilsson potential. Using the Saxon-Woods single-particle energies from Jülich, Strutinsky calculations with the Saxon-Woods potential have also been performed in Copenhagen [38].

All these calculations, microscopic and Strutinsky with Nilsson and Saxon-Woods potential, agree in the fact that typical rare earth nuclei do normally not prefer to rotate around an oblate symmetry axis. But one finds in the nuclei before the rare earth region that they prefer to rotate around a symmetry axis of the nucleus slightly oblate deformed. On the other hand, nuclei after the rare earth region rotate around a symmetry axis of nuclei which are slightly (typical $\beta = 0.05$ to 0.1) prolate deformed. In both regions one finds isomeric states [37].

This behavior, which seems to be strange on the first view, can be explained by the MONA effect (MONA = Maximization of the Overlap of Nucleonic wavefunctions by Alignment of single-particle angular momenta) [39]. We have to distinguish three situations:

(1) If we have a few particles outside a doubly closed shell nucleus

($^{146}_{64}Gd_{82}$) we need all the nucleons to create a large angular momentum. Therefore all these nucleons have to rotate in the same direction around the equator of the nucleus. This forms at the beginning of the shell a slightly oblate deformed nucleus "rotating" around the oblate symmetry axis. For the beginning of the shell we have therefore a strong correlation between the high angular momentum and the shape of the nucleus. The nuclei prefer to rotate around a slightly deformed oblate symmetry axis. Thus one finds yrast traps in these nuclei.

(2) In the middle of the shell one has enough nucleons to produce the deformation of the nucleus and to form a high angular momentum. Therefore one has no correlation between the shape of the nucleus and the high angular momentum. In this view a rotation around a prolate or oblate symmetry axis would be a mere accident due to special shell effects. (A "rotation" around a prolate symmetry axis is found in the Hf isotopes.)

(3) If one is below a doubly closed shell nucleus, one has only a few holes. To get a high angular momentum, one needs the single-particle angular momenta of all the holes. The holes have therefore to be concentrated rotating in the same direction around the equator. They take away nuclear matter near the equator and produce a slightly prolate deformed nucleus "rotating" around the symmetry axis.

Figure 20 shows an isotopic chart of the rare earth region for the even mass nuclei. The shaded area indicates nuclei for which we find yrast isomers rotating around an oblate symmetry axis. The circles indicate isomers rotating around a prolate symmetry axis. The dots indicate nuclei in which experiments at GSI [40] and Jülich [41] found yrast isomers. Figure 20 clearly indicates the qualitative description given by the MONA effect: before the rare earth nuclei we have mainly slightly oblate deformed shapes and after the rare earth nuclei we have mainly slightly prolate deformed shapes with the nuclei rotating around the symmetry axis.

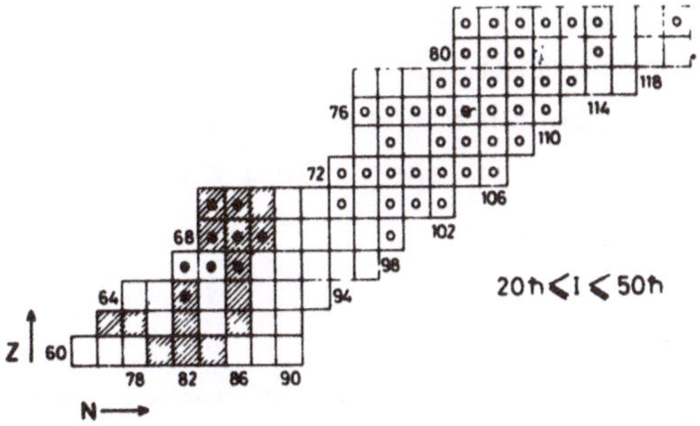

Fig. 20 Isotope chart for even mass nuclei plotted as a function of the charge number Z and the neutron number N. The dashed area indicates nuclei which rotate around an oblate symmetry axis. The circles give nuclei which rotate around a prolate symmetry axis. The black dots indicate nuclei in which yrast isomers have been found experimentally [40,41]

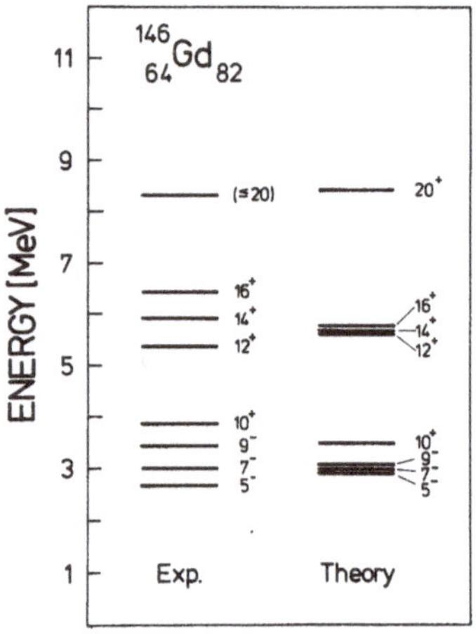

Fig. 21 Experimental data [41] compared with theoretical calculations for the doubly closed shell nucleus $^{146}_{64}Gd_{82}$. The 7^- and state with angular momentum (≤ 20) are yrast isomers

Figure 21 shows the experimental data from Jülich [41] compared with a theoretical calculation using the Hartree-Fock-Bogoliubov approach and pairing plus quadrupole force Hamiltonian according to a method developed in [42]. The lowest group of states is formed by two proton quasi-particle states where one particle mainly occupies the $h_{11/2}$ single-particle level. The second group is formed by four proton quasi-particle states. There is a good qualitative agreement between theory and experiment. The additional splitting of the levels found in experiment may be due to the short-range nature of the nucleon-nucleon interaction which is contained in the higher multipolarities of the interaction not included in the pairing plus quadrupole force Hamiltonian or it may be connected with the only average conservation of the total angular momentum.

4. Description of Transitional Nuclei

4.1 Introduction

The strong coupling model has been well known for more than 25 years. If one has an odd-mass nucleus, the strong coupling picture describes the odd nucleon rotating around the deformed core. The motion of the odd nucleon is practically not disturbed due to the rotation of the whole system. Such a distortion could come from the Coriolis or the centrifugal force. They are due to the fact that the odd nucleon is moving in a rotating system. The condition that the strong coupling picture is valid is:

$$|\langle -\hbar\omega\beta r^2 Y_{20}\rangle| \gg |\langle -\frac{\hbar^2}{\Theta}\vec{I}\cdot\vec{j}\rangle| \qquad (4\text{-}14)$$

This means that the average absolute value of the matrix elements which describe the coupling of the odd particle to the deformation β of the core has to be much larger than the matrix element of the Coriolis force. If one goes now to nuclei with smaller deformations β, the value of the matrix element which describes the coupling of the odd nucleon to the quadrupole moment of the core, is reduced. On the other hand, the Coriolis force is increasing since the moment of inertia decreases. So, finally one may end up with the opposite situation of (4-14) in which the Coriolis force is primarily determining the motion of the odd nucleon. In this case, the nucleus likes to minimize the Coriolis energy which corresponds to maximizing the scalar product $\vec{I}\cdot\vec{j}$ between the total angular momentum I and the single-particle angular momentum j. Thus, in this situation the odd nucleon and the even mass core are rotating in the same plane but independent of each other. The motion of the nucleon and the core are "decoupled" [45-47]. The decoupling model for transitional odd-mass nuclei has been extended to study triaxial deformations by MEYER-TER-VEHN [46] and to include the competition between strong coupling and decoupling and also the variable moment of inertia with increasing spin by TOKI and FAESSLER [47].

For most transitional nuclei, it is not obvious whether they show decoupling or strong coupling. Often only the position of the odd nucleon in the Nilsson energy diagram decides about these properties. Fig. 22 shows that with the Fermi surface on the lowest Nilsson state (particle states) one expects for the prolate deformation decoupling and for the oblate deformation strong coupling. This is connected with the projection of the single-particle angular momentum to the symmetry

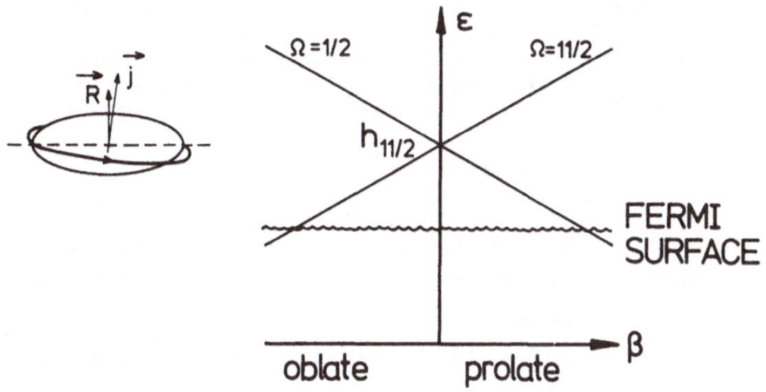

Fig. 22 Schematic Nilsson diagram of the $h_{11/2}$ level. Only the splitting of the $\Omega = 1/2$ and $\Omega = 11/2$ magnetic substates is indicated. A Fermi surface below the $h_{11/2}$ level indicates particle states and favors decoupling of the odd particle for positive and strong coupling for negative deformations. For hole states (Fermi surface on top of the $h_{11/2}$ level) the situation is reversed.

axis. For $\Omega = 1/2$ the single-particle angular momentum is perpendicular to the symmetry axis and prefers decoupling, while for $\Omega = j$ the single-particle angular momentum is pointing along the symmetry axis and prefers strong coupling. The situation is reversed when the Fermi surface lies above the spherical single-particle level (hole states).

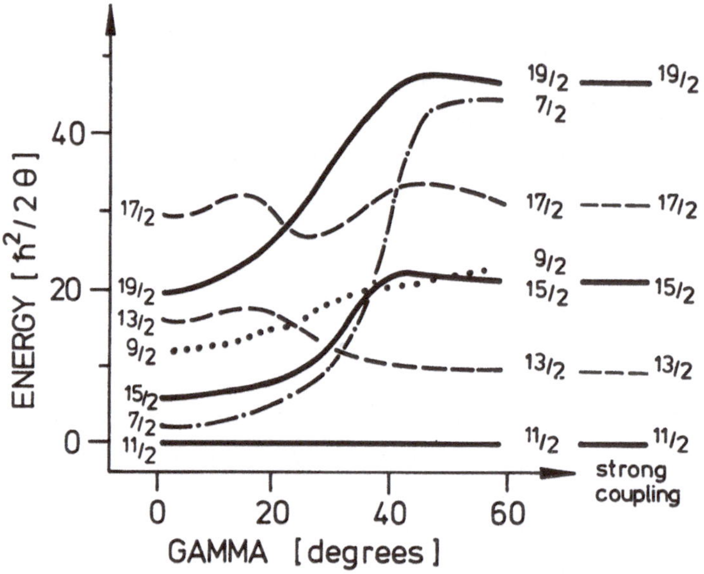

Fig. 23 The excitation energies of a triaxial core (deformation γ) with an odd $h_{11/2}$ nucleon are shown for the Fermi surface λ on the lowest Nilsson level. For $\gamma = 0^{o}$ one has decoupling and for $\gamma = 60^{o}$ strong coupling ($I = 11/2$, $13/2$, $15/2$, $17/2$, ...). On the decoupling side ($\gamma = 0^{o}$ and $I = 11/2$, $15/2$, $19/2$, ...) one recognizes also the unfavored band $I = 13/2$, $17/2$, ... for which the core angular momentum is tilted with respect to the single-particle angular momentum. The favored antiparallel case $I = 11/2$, $7/2$, ... has also an unfavored counterpart $I = 9/2$, $5/2$, ...

In Fig. 23 we show the spectrum of the excited states of an odd $h_{11/2}$ particle coupled to a triaxial core [46]. For $\gamma = 0$ one has decoupling ($I = 11/2$, $15/2$, $19/2$, ...) and for $\gamma = 60^{o}$ one finds strong coupling ($I = 11/2$, $13/2$, $15/2$, $17/2$, $19/2$, ...). On the decoupled side, one sees also the unfavored band ($I = 13/2$, $17/2$, ...) which stems from the coupling of the core angular momentum R and the single-particle angular momentum j not parallel to each other. In this way, the Coriolis force cannot gain so much energy and therefore this band lies higher than the favored band ($I = 11/2$, $15/2$, $19/2$, ...) in which core and single-particle angular momenta are parallel.

The properties of the odd-mass nuclei have been discussed extensively in [47] and especially in [48] where the qualitative explanations for the behavior of the transitional odd-mass nuclei are included also. In this lecture I will therefore concentrate on the transitional odd-odd mass nuclei and the description of the zero and two quasi-particle excitations in transitional even mass nuclei. But we shall see that the few hints about odd-mass nuclei given above and in Figs. 22 and 23 are very useful for understanding qualitatively the properties of the odd-odd and even mass nuclei.

If we extend the model to odd-odd mass nuclei and even mass nuclei with zero and two quasi-particle excitations, one can have the situation that both quasi particles which are coupled to the core are decoupling. In this situation we speak of the peaceful case. But one may also have the situation that one particle is decoupling and the other particle is coupling strongly to the triaxially deformed core. This situation we call the conflicting case. (If both particles are coupling strongly to the core, we have again a peaceful case. But this situation has already been studied in the past.)

Our description of the even mass nuclei does include collective and single-particle degrees of freedom. The latter may play even at low energies an important role and therefore this approach goes beyond the collective description with five quadrupole bosons. Already in 1974, JANSEN, DÖNAU and JOLOS [49] have shown that a special choice of the deformation energy surface in the five-dimensional quadrupole vibrational model is equivalent to the interacting boson model of ARIMA and IACHELLO [50]. This work is therefore in this sense also more general than the IBM. But at the moment we have still restricted the core to a rigid triaxial rotor. Here it is more restricted than IBM for the description of the purely collective states. But we are at the moment working on a description of the core by five-dimensional quadrupole bosons with general deformation energy surfaces [51]. After this extension our description of the core is more general than IBM.

4.2 Odd-Odd Mass Nuclei

The Hamiltonian for odd-odd mass nuclei has a first term which describes a triaxial rotor (DAVYDOV-Hamiltonian [52]).

$$H = \sum_{n=1}^{3} \frac{R_n^2}{2\Theta_n(\beta,\gamma)} + H_{Nil}(\pi) + H_{Nil}(\nu) + H_{pair} + V_{SDI}(\pi,\nu) \qquad (4\text{-}15)$$

The sum in this part of the Hamiltonian runs over $n = 1,2,3$, the three intrinsic axes of the rotor. R_n are the three projections to these axes of the core angular momentum and $\Theta_n(\beta,\gamma)$ are the three moments of inertia according to Bohr and Mottelson depending on the deformation parameters β and γ. The next terms are the Nilsson Hamiltonian with the triaxially deformed potential for the odd proton and the odd neutron. Furthermore, one has the pairing Hamiltonian which describes the pairing correlations between the nucleons. The last term is the residual interaction between the odd proton and the odd neutron. We shall normally take for this force the surface delta interaction. But we shall also use experimentally determined matrix elements.

If one wants to solve the Hamiltonian (4-15) one expands the solution into a complete set of basis states. One could think of the following two sets of basis states which span up the same Hilbert space and are therefore in principle equivalent.

1. Strong coupling basis:

$$|i_\pi, i_\nu; IMK\rangle = N\, D^I_{MK} \alpha^+_{i_\pi} \alpha^+_{i_\nu} |BCS\rangle \tag{4-16}$$

2. Quasi weak coupling basis:

$$|[(j_\pi j_\nu)J;\alpha R]IM\rangle = [(\sum_K A_K^{(\alpha)} D^R_{MK})(\beta^+_{j_\pi} \beta^+_{j_\nu})_J]_{IM}|BCS\rangle \tag{4-17}$$

The creation operators α^+_i describe the deformed odd quasi-particle states for the protons and/or the neutrons. The D-function D^I_{MK} describes the rotation of the whole system. N is a normalization constant. (The symmetrization under a rotation of π around the y-axis is not indicated in the formula to simplify this qualitative explanation.) The quasi weak coupling basis diagonalizes the triaxial rotor exactly. The solutions corresponding to a ground state band, γ-band and so on are characterized by the upper index α. The creation operators β^+_j represent spherical quasi-particle states. J is the total two quasi-particle angular momentum and I is the total angular momentum of the system. Since both basis sets, 1 and 2, span up the same Hilbert space, there exists a unitary transformation between them which can be calculated,

$$\langle (i_\pi i_\nu)K;I|(j_\pi j_\nu)J;\alpha R;I\rangle . \tag{4-18}$$

The transformation brackets (4-18) represent these unitary transformations. As stressed already above, both basis sets are in principle equivalent. But for a numerical calculation one has always to truncate the basis set and then it turns out that such a truncation is easier for the quasi-weak coupling basis. In this basis one needs only about 250 basis states (instead of about 1200 for the strong coupling basis) to get stable results for the low-lying states which do not change if one is increasing the basis. The reason is due to the physical quantum number α for the triaxial rotor. A restriction to the ground state band $\alpha = 0$ and the γ-band $\alpha = 1$ already yields good results.

First we would like to apply this description to the odd-odd mass Au isotopes. We shall see that they are an example for a peaceful case in which both odd nucleons like to decouple. In Fig. 24 the Nilsson diagram of the proton $h_{11/2}$ and neutron $i_{13/2}$ levels are schematically indicated. In the spherical case those levels are filled if the nucleus has 82 protons and 114 neutrons. The proton number for Au is $Z = 79$ and thus the Fermi surface lies on top of the proton $h_{11/2}$ level. For the neutrons the situation is not so clear. We shall assume for the moment that the Fermi surface for the neutrons also lies on top of most Nilsson levels coming from the $i_{13/2}$ state. If one takes

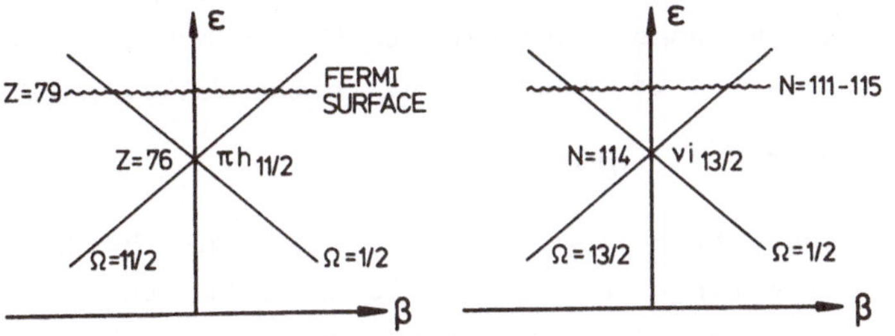

Fig. 24 Schematic Nilsson diagrams for the $h_{11/2}$ proton and the $i_{13/2}$ neutron which are mainly responsible for states in the odd-odd mass Au isotopes. For the protons the Fermi surface is above the $h_{11/2}$ level. For the neutrons we discuss first the case in which the Fermi surface is also above the $i_{13/2}$ level. From the neighboring odd mass nuclei we expect in the odd-odd mass Au isotopes a deformation $\gamma = 40^o$. This means that both particles have a tendency to decouple as discussed in Fig. 22.

into account that one finds from the odd Au isotopes ^{195}Au, ^{193}Au a γ-deformation of $\gamma = 35^o$ and from the odd mass ^{195}Hg a deformation $\gamma = 43^o$, one expects a deformation of about $\gamma = 40^o$. Thus, we have a peaceful case in which both particles, the $h_{11/2}$ proton and the $i_{13/2}$ neutron, are decoupling. The lowest angular momentum state should therefore be the two quasi-particle angular momentum $J = 12^-$ with a core angular momentum $R = 0$. On top of this state, the favored band is built up by rotating the core and one obtains the angular momenta $I = 12^-, 14^-, 16^-, 18^-, \dots$. If one adds the two quasi angular momentum $J = 12^-$ to the core angular momentum R in a tilted way, one gets the unfavored rotational band $I = 13^-, 15^-, 17^-, \dots$. In addition, one can now couple the two single-particle angular momenta $h_{11/2}$ and $i_{13/2}$ in a slightly tilted way to $J = 11^-$ or $J = 10^-$. In the situation where the Fermi surface is also for the neutrons on the topmost Nilsson level of the $i_{13/2}$ state, this coupling to $J = 11^-$ or 10^- is unfavored due to the Coriolis force, and the favored bands $I = 11^-$, 13^-, 15^- and the unfavored bands $I = 12^-$, 14^-, ... lie higher in energy than the corresponding favored and unfavored bands built on the fully aligned two quasi-particle state. The corresponding favored and unfavored bands built on the $J = 10^-$ state should lie even higher. One sees therefore that in the peaceful case the spectrum of an odd-odd mass nucleus is in some sense a superposition of spectra of several odd mass nuclei. The two quasi-particles are coupled together to the total quasi-particle angular momentum J and they act as one superparticle. The angular momentum of the two quasi-particles J is almost a good quantum number and this fact favors the above qualitative description. The quantum number J is mixed by the coupling of the two quasi particles to the quadrupole moment of the core. The matrix element is proportional to the following expression:

$$V \propto \langle \pi J|Q|\pi J'\rangle + (-)^{J-J'}\langle \nu J|Q|\nu J'\rangle \qquad (4\text{-}19)$$

The structure of the matrix elements of this part of the Hamiltonian is given in (4-19). The first matrix element describes the coupling of the proton to the quadrupole moment of the core, while the second matrix element describes this coupling for the neutron. Since the quadrupole moment for the core is the same for both nucleons, one gets roughly the same size for both matrix elements. The phase factor $(-)^{J-J'}$ now guarantees that for $\Delta J = 1$ the sum of these matrix elements which would describe the mixing of the quantum number J, is practically zero.

Whether two levels belong to the same two quasi-particle angular momentum J or they are distinguished by $\Delta J=1$, can be seen experimentally in the following way: for the same two quasi-particle angular momentum J one changes only the core and expects therefore strong collective E2 transitions. The E2 transitions between bands with $\Delta J = 1$ are reduced due to (4-19) below the single-particle limit. But, on the other hand, the magnetic dipole transitions are increased between $\Delta J = 1$ states. The corresponding proton and neutron matrix elements have opposite signs and with the phase factor they add up coherently. So, if one finds strong collective E2 transitions, the two states belong to the same quasi-particle angular momentum J, but if one finds a strong M1 transition, they belong to states with $\Delta J = 1$.

The odd-odd mass Au isotopes which have been measured [53] have neutron numbers between 111 and 115. This corresponds for negative deformations (or $\gamma = 40^o$) to a less than fully filled $i_{13/2}$ neutron shell. Therefore we have to lower the Fermi surface for the neutrons. This means that the full alignment of the $i_{13/2}$ neutron level is lost and we get projections of the neutron single-particle angular momentum to the intrinsic z-axis of 3/2, 5/2, 7/2 and so on. This means that now the favored and unfavored bands built on the total two quasi angular momenta $J = 11^-$, 10^- and so on are getting lowest states since it is energetically favored to have the $i_{13/2}$ single-particle angular momentum tilted against the $h_{11/2}$ proton momentum (see Fig. 25). In this figure we have plotted the excitation energy as a function of the Fermi surface of the neutrons. We see that for the full $i_{13/2}$ shell the lowest state is a 12^- state. With decreasing the Fermi surface the lowest state gets the $J = 11^-$ and then the $J = 10^-$ state, as discussed above.

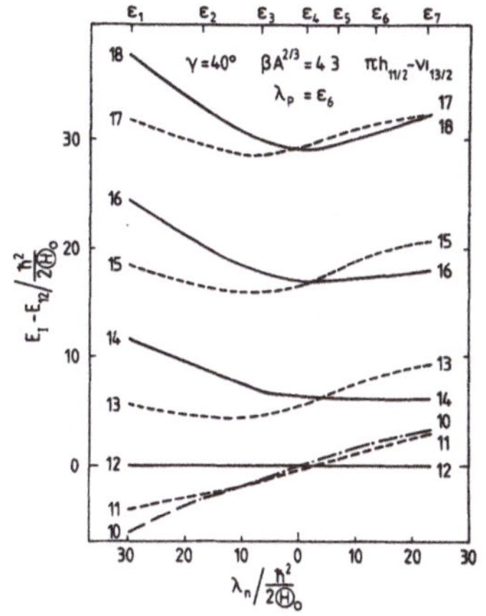

Fig. 25 Calculated excitation energy spectrum as a function of the neutron Fermi energy λ_n for the $(\pi h_{11/2}-\nu i_{13/2})$ configuration of the two odd quasi-particles. The proton Fermi energy is fixed and is equal to that of the highest $h_{11/2}$ Nilsson level. The other parameters are the same as described in Fig. 26. The results are obtained without any residual interaction. On the top scale the positions of $i_{13/2}$ Nilsson levels are indicated. All the energies are normalized to the $I = 12^-$ state. It is seen that the level spacings for the values of λ_n between $\lambda_n \simeq \varepsilon_2$ and $\lambda_n \simeq \varepsilon_4$ have trends similar to those observed experimentally for the three Au isotopes shown in Fig. 26. For the above values of λ_n the states with spin 12, 14, 16, ... are built with the intrinsic spin $J = 12$, and the states 11, 13, 15, ... are constituted by the $J = 11$ state. These are distinguished by solid and dashed lines, respectively.

Fig. 26 shows a comparison between the experimental data [53] and the theory with $(k = -0.125)$ and without $(k = 0)$ a residual interaction between the odd proton and the odd neutron. The comparison between theory and experiment is made by putting the neutron Fermi surface on the third lowest Nilsson level.

An example for a conflicting case in odd-odd mass nuclei is given by the Tl isotopes 198 and 196 which have recently been measured in Jülich and München [54]. The situation of these odd-odd mass nuclei in the Nilsson diagrams is represented in Fig. 27. Since the neighboring odd mass nuclei indicate a γ-deformation of $\gamma \sim 40^{\circ}$, one expects that the odd $h_{9/2}$ proton shows strong coupling while the odd $i_{13/2}$ neutron decouples from the motion of the even mass core. With the help of

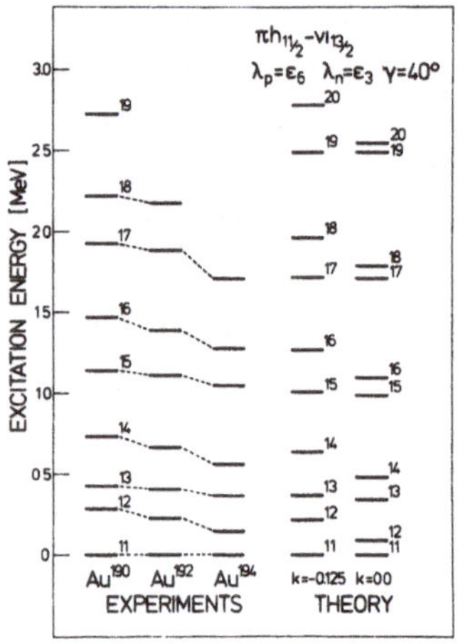

Fig. 26 Experimental data for the excitation of the Au isotopes (ref. 53) depicting the mass dependence. The states of the same spin for the three isotopes are connected by the dashed lines. Note that the energy spacings among 11^-, 13^-, 15^-, ... etc. and those among 12^-, 14^-, 16^-, ... etc. are similar to those in the rotational-like bands. The trends in the experimental energy spacing are compared with a theoretical calculation assuming the $(\pi h_{11/2}-\nu i_{13/2})$ configuration for the two odd quasi-particles. In this calculation the $\gamma = 40^o$ and the proton and neutron Fermi energies are given by the highest $h_{11/2}$ and the lowest $i_{13/2}$ Nilsson levels, respectively. The surface delta interaction is used as the residual interaction between the two quasi-particles. Both results obtained with (k = 0.125) and without (k = 0.0) the residual interaction are shown. The other parameters employed in the calculations are $\beta = 0.128$, and the gap energies $\Delta_P = \Delta_N = 12/A$ (A = 194).

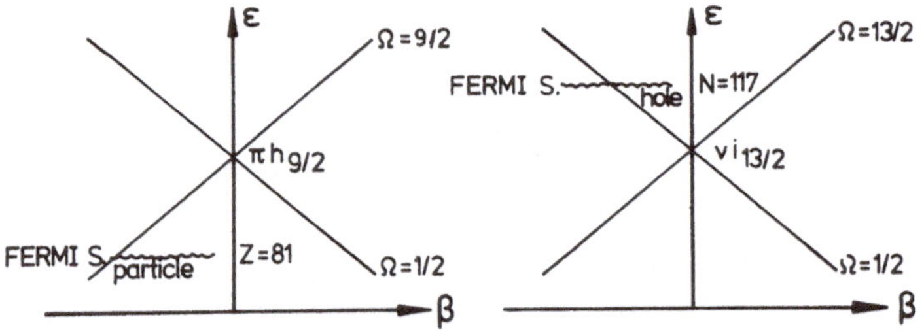

Fig. 27 Schematic Nilsson diagrams for ${}^{198}_{81}TL_{117}$. The neighboring odd-mass nuclei indicate a deformation $\gamma = 40^o$. One has a conflicting case with the $h_{9/2}$ proton coupled strongly to the core while the $i_{13/2}$ neutron decouples.

Pythagoras one calculates then easily $J_{min} = [(13/2)^2 + (9/2)^2]^{1/2} = 7.9 \sim 8$. If the core angular momentum R is increasing, the Coriolis force tries to align also the $h_{9/2}$ proton single-particle angular momentum. At higher angular momenta, one expects that the odd proton is aligned together with the core angular momentum and the single-particle neutron angular momentum. The two quasi-particle angular momentum J may then be 10^- or 11^-. The Coriolis force favors the $J = 11^-$ configuration. If one adds the even angular momentum of the core $I = R + J$, one expects that the odd angular momenta I lie lower than expected from the position of the even angular momenta. But the data show that it is just the other way around. This can be understood by the residual interaction. The particle-hole interaction is repulsive and of short range. If we take the zero-range surface delta interaction, it acts only between particle hole states of natural parity. Since in the fully aligned situation the spin coming from the $h_{9/2}$ proton and $i_{13/2}$ neutron level is zero, this means that we have only a repulsion for odd J and therefore also for odd I. This repulsion pushes the odd total angular momenta I up. Fig. 28 shows that this is in agreement with data.

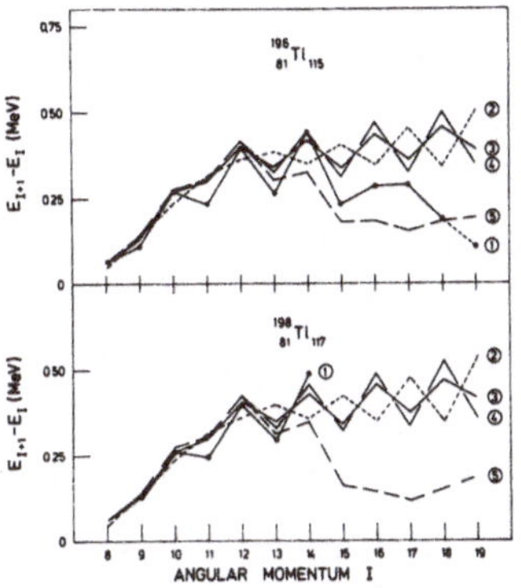

Fig. 28 Energy difference $E_{I+1} - E_I$ of the odd-even angular momentum in $^{196,198}Tl$ plotted as a function of the total angular momentum I. The experimental results marked 1 are compared with various calculations. Curve 2 is obtained without any residual interaction. Curve 3 results from including the surface delta force ($k = -0.3$) as the residual interaction. Curve 4 shows the results obtained with the empirical residual interaction from the splitting of the $\pi h_{9/2} - \nu i_{13/2}$ in ^{208}Bi. The curve 5 is obtained using the experimental energies for the even mass core. In the first three calculations listed above the extended VMI model has been employed to describe the even mass core states

4.3 Even-Mass Nuclei with Zero and Two Quasi-Particle Excitations

The Hamiltonian (4-15) can also describe even-mass nuclei if we understand the two Nilsson Hamiltonians as single-particle Hamiltonians for particles of the same charge, for example two protons or two neutrons and V_{SDI} as the residual interaction between them. In additiona, we have to allow in the basis states (4-16) and (4-17) that we also admix zero quasi-particle states. This allows one to describe the collective and two quasi-particle excitations in even-mass transitional nuclei. One has only one further difficulty: due to the admixture of the zero quasi-particle state, the basis (4-16) or (4-17) contains also the particle number spurious state due to particle number nonconservation. We construct explicitly this state and then Schmid-orthogonalize the basis to it. In this way the particle number spurious state is totally removed.

Figure 29 indicates schematically the position of the two quasi-particle states in the Nilsson diagram. Since we describe now even-mass nuclei, both quasi-particles have to be either protons or neutrons. But the total wavefunction can naturally contain a superposition of different two protons and two neutrons two quasi-particle states. Since neighboring odd-mass nuclei indicate a γ-deformation of roughly $\gamma = 40^{\circ}$, one expects that both quasi-particles like to decouple if they are

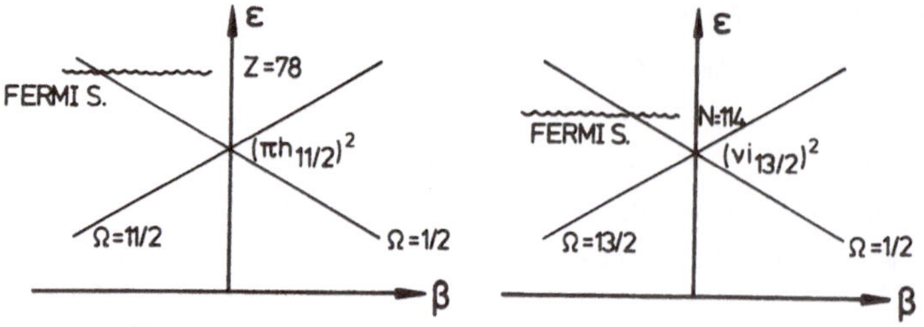

Fig. 29 Schematic Nilsson diagrams for the $^{192}_{78}Pt_{114}$ even mass nucleus. Neighboring nuclei indicate $\gamma \sim 40^{\circ}$. The nucleus prefers decoupling for the 2qp admixed states

admixed. We want now to apply this theory to describe anomalies found within the last few years in the Hg and Pt isotopes [55]. Fig. 30 shows the excitation energies of positive parity for ^{196}Hg and ^{198}Hg.

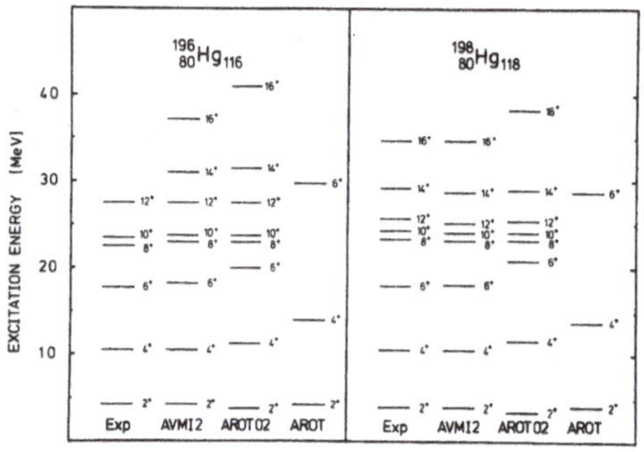

Fig. 30 Experimental and calculated results for the positive parity states in $^{196,198}Hg$. Experimental data (on the left) are compared with the asymmetric rotor model (AROT) in column 4, the results of the asymmetric rotor model with the 2qp mixing wherein the increase of the moment of inertia with growing angular momentum is described by the extended VMI model (AVMI2) in column 2. The parameters used in the calculations are $\beta = 0.115$, $\gamma = 60^{\circ}$, $\lambda_P = 42.20$ MeV, $\lambda_N = 48.05$ MeV, $\Delta_P = 0.90$ MeV and $\Delta_N = 1.24$ MeV for ^{196}Hg, and $\beta = 0.099$, $\gamma = 60^{\circ}$, $\lambda_P = 42.04$ MeV, $\lambda_N = 48.18$ MeV, $\Delta_P = 0.90$ MeV and $\Delta_N = 1.12$ MeV for ^{198}Hg

One sees that an anomaly exists for the 8^+ and 10^+ states in ^{196}Hg and for the 8^+, 10^+ and 12^+ states in ^{198}Hg. This anomaly and its modification between neutron number 116 and 118 corresponding to these two isotopes, are explained in the following way: in $^{196}_{80}Hg_{116}$, the 0^+, 2^+, 4^+ and 6^+ states are essentially members of the ground state (zero quasi-particle) band. Then the two quasi particle $(\pi h_{11/2})^2$ band intersects with the ground state band and forms the yrast states. The 8^+ state corresponds to the almost aligned configuration, while the 10^+ state is the fully aligned case of the $(\pi h_{11/2})^2$ 2qp states. Since the overlap between the two single-particle functions is increasing if full alignment is obtained, the energy difference between the 8^+ and 10^+

states is small. If one wants to increase the angular momentum even more, one has to rotate the core so that the energy difference between the 10^+ and the 12^+ states should correspond roughly to the energy difference between the 0^+ and 2^+ states in the ground state rotational band. These features are nicely reproduced with our calculation which is shown in the third column labelled with (0 + 2)qp. In ${}^{198}_{80}Hg_{118}$, the two neutron quasi-particle states $(i_{13/2})^2$ move down due to the increase of the neutron Fermi surface. This lowering of the neutron two quasi-particle configuration is mainly due to the fact that it is easier to align the two neutrons if we move the neutron Fermi surface closer to the $\Omega = 1/2$ or 3/2 states. Again the 0^+, 2^+, 4^+, 6^+ states are mainly members of the ground state rotational band. The 8^+, 10^+ and 12^+ states are the neutron two quasi-particle states with a partial up to a full alignment. To obtain the 14^+ state, one has to start to rotate the core and therefore the energy difference of the 12^+ to 14^+ state should correspond to the energy difference in the ground state rotational band between the 0^+ and the 2^+ states. The result of our calculation reproduces these features in almost quantitative agreement with the data. It is shown in the third column of the second part of Fig. 30 labelled (0 + 2)qp.

In the last column of Fig. 30 we also show the results of the asymmetric Davydov rotor without the Coriolis interaction between the zero and the two quasi-particle states. Comparing columns 3 and 4, one sees that this interaction depresses the 4^+ and 6^+ states considerably and describes the increase of the moment of inertia. We would like therefore to conclude that the variable moment of inertia (VMI) is due to the admixture of two quasi-particle states by the Coriolis force. This effect describes semi-phenomenologically the Coriolis antipairing effect. This explanation is supported by the result of a simple calculation shown in column 2. In this calculation, we put all the Coriolis matrix elements connecting the zero and two quasi-particle states equal to zero, and instead introduced the phenomenological variable moment of inertia model slightly extended to triaxial deformations. The restoring force constant C has been adjusted to the position of the 4^+ state. The result labelled VMI is almost in complete agreement with our more refined calculation.

In Fig. 31 the same results are shown for ${}^{190}Pt$. The interesting points here are the three 10^+ states which lie very close together and which are explained as coming from the 10^+ state with a $(\nu i_{13/2})^2$, a $(\pi h_{11/2})^2$ and an almost pure ground state configuration. The close lying position of these three 10^+ states suggests that they are not interacting too much with each other. This is also borne out by the

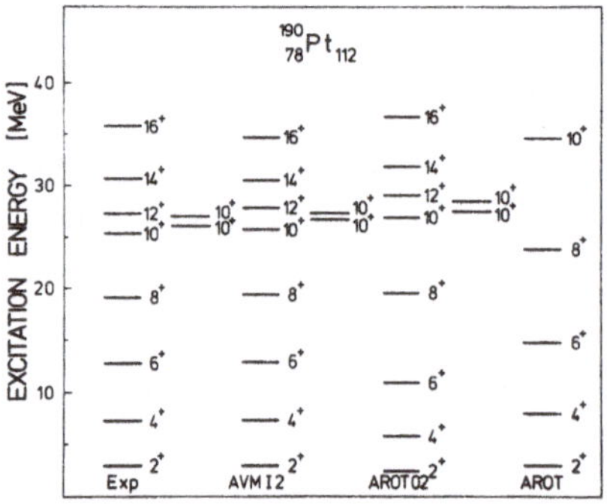

Fig. 31 Experimental and calculated results for the positive parity states in ^{190}Pt. For the details see Fig. 30. Here the second and third 10^+ states have been shown beside the yrast band in each case. The parameters employed in the calculations are $\beta = 0.155$, $\gamma = 30^o$, $\lambda_P = 43.30$ MeV, $\lambda_N = 48.76$ MeV, $\Delta_P = 0.80$ MeV, and $\Delta_N = 1.10$ MeV

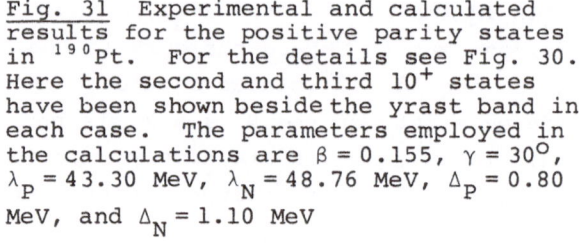

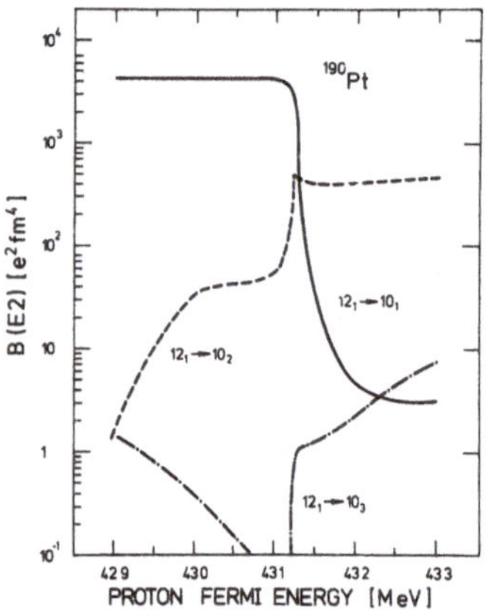

Fig. 32 Reduced transition probabilities from the 12^+ state to the three 10^+ states in ^{190}Pt as a function of the proton Fermi surface λ_P. The experimental values are [56-58] $B(E2,12^+\to10^+)$ $[e^2fm^4] = 950 \pm 60$; 520 ± 80 in ^{190}Pt, and 4040 ± 450; 1770 ± 260 and 10 ± 2 in ^{188}Pt in the sequence $12^+\to10^+_1$, $12^+\to10^+_2$, $12^+\to10^+_3$

calculation. But, on the other hand, one finds also that they admix relatively strongly many higher lying two quasi-particle configurations due to the Coriolis interaction.

Figure 32 gives the reduced E2-transition probabilities between the 12^+ and the three 10^+ states in ^{190}Pt. One sees that the transition probability depends sensitively on the proton Fermi surface. A similar plot could be obtained plotting those transition probabilities as a function of the neutron Fermi surface. The essential feature is the relative position of the proton against the neutron Fermi surface. Qualitatively the theoretical results indicate that we have a large transition probability and one which is roughly half as big and a very small one. This is also corroborated by the experimental data [56-58] given in the figure caption of Fig. 32. The transition probabilities between the purely collective states are not so strongly affected by the two quasi-particle admixtures, although one finds also appreciable changes. The transition between the 4^+ and 2^+ states of the γ-band in units of the 2^+ to 0^+ transition of the ground state band, is measured to be $0.40 \pm .1$ and $0.48 \pm .17$. Our result is 0.48, while the Davydov model yields 0.6. The transition probability between the 4^+ state of the γ-band and the 4^+ state in the ground band is in the same units 0.30 ± 0.05 measured by Coulomb excitation from the Berkeley group. Our value is 0.28, the value of the Davydov model is 0.27 and the value of the interacting boson model is 0.79.

4.4 Summary

In transitional nuclei, the Coriolis force plays a decisive role. It is the competition between the action of the Coriolis force on the odd particles and the coupling of the odd particles to the quadrupole moment of the core which decides if one has decoupling or strong coupling in these nuclei.

If one has an odd-odd mass nucleus or a two quasi-particle state in an even-mass nucleus, both quasi-particles can decouple, then we speak of a peaceful case, or one can decouple and the other can strongly couple to the quadrupole moment of the core depending on the position of the odd particles in the Nilsson diagram. In both cases one gets quite different single-particle spectra. In the peaceful case, the two odd quasi-particles couple to an almost good total quasi-particle angular momentum J and the two quasi-particles together behave like one super quasi-particle. Each coupling to a value J yields a spectrum corresponding to an odd-mass nucleus with a favored and an unfavored band. In this way the spectra of odd-odd mass nuclei consist in the peaceful case out of a superposition of spectra of several decoupled odd-mass nuclei.

As an example for a peaceful case we treated the Au isotopes. We saw that even if one empties the neutron $i_{13/2}$ shell half, one still has the chracteristics of the peaceful spectrum. The band-head position for the different total quasi-particle angular momenta J is only shifting relative to each other.

As an example for the conflicting case we treated ^{198}Tl and ^{196}Tl. We could explain the experimentally found staggering in these spectra by the residual particle-hole interaction.

In even-mass nuclei, we explain by the intersection of the ground state rotational band and the two quasi-particle bands anomalies found in the Pt and Hg isotopes. The Coriolis force admixes high-lying two quasi-particle states to the ground state rotational band. This is increasing the moment of inertia. In this semi-microscopic way, one can describe the variable moment of inertia which is due to the reduction of the pairing correlations by the Coriolis force (Coriolis antipairing effect). The transition probabilities between the collective states are not so strongly affected by the admixture of two quasi-particle states, a result which is expected. But we found at the example of the transition from the 12^+ to the three 10^+ states in ^{188}Pt and ^{190}Pt a strong dependence of the transitions on the position of the single-particle states.

4.5 Further Trends

Until now one has mainly studied even mass and odd mass nuclei. But I hope I have shown in Section 3 that, although it is more difficult to analyze the data from odd-odd mass nuclei, it should not be impossible to extract useful information from them. The spectra and effects found in odd-odd mass nuclei are richer than in odd mass nuclei. Therefore one should also dare to measure and analyze those spectra. The results will also be more rewarding than for odd mass nuclei.

In studying even-mass nuclei, one should not restrict the interest to the collective states only. As we have seen in Section 4, one gets interesting anomalies due to the interplay between the collective and the two quasi-particle degrees of freedom. The collective states alone can reasonably well be described by the five-dimensional quadrupole vibrations with a general deformation energy surface. The interacting boson model is equivalent to a specialization of this collective description. But we know from the data and also from theoretical considerations that already at relative low energies the two quasi-particle excitations play an important role in even-mass nuclei. The progress of the theoretical description of the interplay of collective and two quasi-particle excitations makes it worthwhile to study this interaction also experimentally.

The theoretical model which I presented here has one serious drawback. We assumed for the description of the core a rigid triaxial rotor. But we know that the quadrupole operators for the core do not commute with the Hamiltonian. Therefore the deformation of the triaxial core cannot be a rigid one. The nucleus has to β- and γ-vibrate. All microscopic calculations of the deformation energy surfaces and the corresponding masses yield a low-lying soft β- and γ-vibration which should affect the measured spectra. But the analysis of the data shows that until now we are able to understand the data to a large extent by a rigid triaxially deformed core. To study this effect, we are working to include for the description of the core the exact solution of the Bohr-Mottelson Hamiltonian (five-dimensional quadrupole vibrator) with a general deformation energy surface in the β-γ space. We hope that we shall be able in the near future to predict with the help of these calculations an experimentum crucis which may hopefully distinguish between a soft vibrator or an almost rigid rotor. Experimentally one should be able to study this difference in soft transitional nuclei. One possibility would be to investigate odd-mass nuclei in these regions. We know from theoretical calculations that a soft core is polarized by the odd particle in different ways depending on the single-particle state of the odd nucleon. Transitions between states built on different single-particle levels would have a reduction factor which should get information about the stiffness of the deformation energy surface. So, one should try to measure transition probabilities in odd-mass transitional nuclei just near the region where the shape of the nuclei tend to go over into large deformations as in the Ru and Ba isotopes. Especially interesting are transitions between states built on different single-particle levels. Hopefully a careful analysis of these transitions might give information on whether the nuclei are almost rigid triaxial rotors or whether they are soft β- and γ-vibrators around an equilibrium position as predicted by the microscopic calculations.

References

1. R.M. Lieder, H. Ryde, Phenomena in fast rotating heavy nuclei. Adv. in Nuclear Physics, Vol. 10 (1978) 1, edited by M. Baranger and E. Vogt.
2. A. Johnson, H. Ryde, J. Sztarkier, Phys. Lett. 34B (1971) 605.
3. R.M. Lieder, et al., Phys. Lett. 39B (1972) 196, and Phys. Lett. 40B (1972) 449.
4. P. Thieberger, et al., Phys. Rev. Lett. 28 (1972) 972.
5. M.A.J. Mariscotti, G. Scharff-Goldhaber, B. Buck, Phys. Rev. 178 (1969) 1864.
6. A. Faessler, W. Greiner, R.K. Sheline, Nucl. Phys. 62 (1965) 241.
7. B.R. Mottelson, J.G. Valatin, Phys. Rev. Lett. 5 (1960) 511.
8. A. Faessler, W. Greiner, R.K. Sheline, Nucl. Phys. 70 (1965) 33.
9. F.S. Stephens, R.S. Simon, Nucl. Phys. A183 (1972) 257.
10. A. Faessler, K.R. Sandhya Devi, F. Grümmer, K.W. Schmid, R.R. Hilton, Nucl. Phys. A256 (1976) 106.
11. A. Goodman, Nucl. Phys. A265 (1976) 113.
12. M. Baranger, K. Kumar, Nucl. Phys. 110 (1968) 490, 529.
13. F. Grümmer, K.W. Schmid, A. Faessler, Nucl. Phys. A239 (1975) 289.
14. M. Wakai, A. Faessler, Nucl. Phys. A295 (1978) 86.
15. A. Faessler, Fortschr. der Physik 16 (1968) 309.
16. A.B. Migdal, Nucl. Phys. 13 (1959) 655.
17. S.T. Beliaev, Nucl. Phys. 24 (1961) 322.
18. T.Y. Lee, et al., Phys. Rev. Lett. 38 (1977) 1454.
19. A. Faessler, M. Ploszajczak, Phys. Lett. 76B (1978) 1.
20. F.A. Beck, et al., Phys. Rev. Lett. 42 (1979) 493.
21. L.L. Riedinger, et al., to be published.

22. R. Bengtsson, I. Hamamoto, B. Mottelson, Phys. Lett. 73B (1978) 259.

23. F. Grümmer, K.W. Schmid, A. Faessler, Nucl. Phys. A308 (1978) 77.

24. D.L. Hendrie, et al., Phys. Lett. 26B (1968) 127.

25. S.G. Nilsson, et al., Nucl. Phys. A131 (1969) 1.

26. R.R. Chasman, Argonne National Laboratory, preprint 1979.

27. M. Wakai, A. Faessler, Nucl. Phys. A307 (1978) 349.

28. R.J. Liotta, R.A. Sorensen, Nucl. Phys. A297 (1978) 136.

29. J.O. Newton, et al., Proc. Int. Conf. on Nucl. Structure, Tokyo 1977, ed. by T. Marumori, Suppl. J. Phys. Soc. Japan 44 (1978).

30. S. Cohen, F. Plasil, W.J. Swiatecki, Ann. Phys. 82 (1974) 557.

31. A. Bohr, B.R. Mottelson, Physica 10A (1974) 13.

32. D.G. Sarantites, et al., Nucl. Phys. A93 (1967) 545; A180 (1972) 177.

33. G.B. Hagemann, et al., Nucl. Phys. A245 (1975) 166; R. Broda, et al., Nucl. Phys. A248 (1975) 356.

34. M. Ploszajczak, H. Toki, A. Faessler, Z. Physik A287 (1978) 103.

35. G. Anderson, et al., Nucl. Phys. A268 (1976) 205; Nucl. Phys. A309 (1978) 141.

36. K. Neergard, H. Toki, M. Ploszajczak, A. Faessler, Nucl. Phys. A287 (1977) 48; M. Ploszajczak, H. Toki, A. Faessler, J. Phys. G: Nucl. Phys. 4 (1978) 743.

37. M. Ploszajczak, A. Faessler, G. Leander, S.G. Nilsson, Nucl. Phys. A301 (1978) 477.

38. K. Matsuyanagi, T. Døssing, K. Neergard, Nucl. Phys. A307 (1978) 253.

39. A. Faessler, M. Ploszajczak, K.R. Sandhya Devi, Phys. Rev. Lett. 36 (1976) 1028.

40. J. Pedersen, et al., Phys. Rev. Lett. 39 (1977) 990.

41. R. Broda, M. Ogawa, S. Lunardi, M.R. Meier, P.J. Daly, P. Kleinheinz, Z. Physik A285 (1978) 423.

42. A. Faessler, K.R. Sandhya Devi, F. Grümmer, K.W. Schmid, R.R. Hilton, Nucl. Phys. A256 (1976) 106; A. Faessler, K.R. Sandhya Devi, A. Barroso, Nucl. Phys. A286 (1977) 101.

43. H. Emling, P. Fuchs, E. Grosse, D. Husar, D. Schwalm, R.S. Simon, H.J. Wollersheim, D. Pelte, GSI-preprint 1979.

44. O. Civitarese, A. Faessler, M. Wakai, Jülich-preprint 1979.

45. F.S. Stephens, R.M. Diamond, J.R. Leigh, T. Kammuri, K. Nakai, Phys. Rev. Lett. 29 (1972) 438.

46. J. Meyer-ter-Vehn, Nucl. Phys. A249 (1975) 111 and 141.

47. H. Toki, A. Faessler, Nucl. Phys. A253 (1975) 231 and Z. Physik A276 (1976) 35.

48. A. Faessler, Proc. Int. School on Nuclear Physics, Predial, Romania, 1976.

49. D. Jansen, R.V. Jolos, F. Dönau, Nucl. Phys. A224 (1974) 93.

50. A. Arima, F. Iachello, Ann. Phys. 99 (1976) 253, 111 (1978) 201; O. Scholten, F. Iachello, A. Arima, preprint 1978.

51. A. Faessler, C.L. de Lima, H. Müther, work in progress.

52. A.S. Davydov, G.F. Filippov, Nucl. Phys. 8 (1958) 237; A.S. Davydov, Nucl. Phys. 24 (1961) 682.

53. A. Neskakis, R.M. Lieder, H. Beuscher, Y. Gono, D.R. Haenni, M. Müller-Veggian, C. Mayer-Böricke, preprint, KFA Jülich, 1978.

54. A.J. Kreiner, M. Fenzl, S. Lunardi, M.A.J. Mariscotti, Nucl. Phys. A282 (1977) 243; A.J. Kreiner, M. Fenzl, W. Kutschera, R.M. Lieder, H. Beuscher, Y. Gono, D.R. Haenni, M. Müller-Veggian, preprint, KFA Jülich and TU München, 1978.

55. C. Günther, et al., Phys. Rev. 15 (1977) 1298; D. Proetel, et al., Proc. Int. Conf. on Reactions between Complex Nuclei, Nashville, Tenn., 1974, ed. R.L. Robinson, et al. (North-Holland, Amsterdam, 1974) p. 162; H. Beuscher, et al., Phys. Rev. Lett. 32 (1974) 843; M. Piiparinen, et al., Phys. Rev. Lett. 34 (1975) 1110; S.A. Hjorth, et al., Nucl. Phys. A262 (1976) 328.

56. L. Richter, et al., preprint, TH Darmstadt, 1978.

57. J.C. Cunnane, et al., Phys. Rev. C13 (1976) 2197.

58. S.A. Hjorth, et al., Nucl. Phys. A262 (1976) 328.

The work presented in these lectures results mainly from collaborations at Jülich with several people. For the results presented in the second section, I especially have to thank Marek Ploszajczak, K.R. Sandhya Devi, F. Grümmer, K.W. Schmid and M. Wakai. The results in Section 3 were obtained in collaboration with M. Faber, M. Ploszajczak, O. Civitarese and H. Toki. For the results in the last section, I have to thank H. Toki and H.L. Yadav.

Chapter V

GROUP THEORY AND NUCLEAR SPECTROSCOPY

F. Iachello
Physics Department
Yale University
New Haven, CT 06520
and
Kernfysisch Versneller Instituut
University of Groningen
The Netherlands

1. Introduction

My lectures will apply the methods of group theory to a problem of current interest in nuclear physics, namely the study of collective quadrupole states in nuclei. The lectures are organized in the following way. Lectures 1 and 2 contain a brief review of group theoretical methods and techniques. For further explication of these basic notions, there are several excellent texts [1-3]. The group theoretic methods are applied to the study of collective states in nuclei in lectures 3 and 4. The first two lectures are quite general and may provide the framework for other applications of group theory to nuclear physics, in particular Wigner supermultiplet theory (which makes use of the group SU(4)), the Flowers scheme (which makes use of the symplectic groups Sp(n)), and the Elliott model (which makes use of the group SU(3)). The last two lectures instead discuss the group theory of the interacting boson model of the nucleus, which provides a framework for the description of collective states in nuclei. As it will be indicated in the notes, several possible realizations of the same Lie algebra are possible for example in terms of differential operators acting on the space of functions, or in terms of creation and annihilation operators acting on a Fock space. In these lectures, I will make use only of the Fock realization, although the other, differential realization, is of course equally possible and gives rise to the description of collective states in nuclei in terms of shape variables [4]. This geometrical approach, developed by Bohr and Mottelson, will not be discussed here. These lectures are excerpts of the lecture notes of the Nuclear Physics course I have given at the University of Groningen (1977-78) and at Yale University (1978-79).

2.1 Groups, Definitions and Examples (Lecture 1)

I begin with a brief review of group theory. In this review, wherever possible, I will take as an illustration of each abstract definition the ordinary rotation group for which most results can be obtained also by elementary methods, without the use of group theory.

An abstract group G is a set of elements a,b,c,..., for which a law of composition or "multiplication" is given so that the "product" ab of two elements is well defined, and which satisfies the following conditions:

(i) if a and b are elements of the set, then so is $c = ab$;

(ii) the multiplication law is associative; that is, $a(bc) = (ab)c$;

(iii) the set contains an element e called the identity such that $ae = ea = a$ for every element a of the set;

(iv) for every a in the set there exists an element a^{-1} such that $aa^{-1} = a^{-1}a = e$ and which is also in the set; a^{-1} is called the inverse of a.

Groups can be discrete or continuous depending on whether or not a definition of continuity is imposed on the elements of the group manifold. Discrete groups find most of their applications in solid state physics. In nuclear physics one makes almost exclusively use of continuous groups and thus I will restrict myself in these lectures to continuous groups. An r-parameter continuous group has its elements labelled by r continuously varying real parameters $a_1,...,a_r$. It is implied that the r parameters are all essential. Thus, in counting the number of parameters some care must be taken. Consider, for example, the set of orthogonal transformations in 2 dimensions, O(2). By writing these transformations as

$$\begin{cases} x' = a_{11}x + a_{12}y \\ y' = a_{21}x + a_{22}y, \end{cases} \tag{5-1}$$

it would appear as if one has 4 parameters. However the orthogonality condition, $(x')^2 + (y')^2 = x^2 + y^2$, (invariance of $x^2 + y^2$) gives rise to the three constraints

$$a_{11}^2 + a_{21}^2 = 1, \quad 2a_{11}a_{12} + 2a_{21}a_{22} = 0, \quad a_{22}^2 + a_{12}^2 = 1, \tag{5-2}$$

thus leaving only one real parameter. The group O(2) is then a 1 parameter group and the transformation (5-1) can be written as

$$\begin{cases} x' = \cos\phi\, x - \sin\phi\, y \\ y' = \sin\phi\, x + \cos\phi\, y, \end{cases} \qquad 0 \leq \phi \leq 2\pi . \tag{5-3}$$

This is the group of rotations around the z-axis. The group elements are $R(\phi)$, where ϕ is the angle of rotation. Similarly the group of orthogonal transformations in 3 dimensions, O(3), has 9 parameters, but the invariance of $x^2 + y^2 + z^2$ imposes six conditions, leaving a 3 parameter group. This is the ordinary rotation group. The group elements may be labelled by $R(\alpha,\beta,\gamma)$ where α,β,γ are the Euler angles. In general, for the rotation group in n dimensions, O(n), there are $\frac{1}{2}n(n-1)$ essential parameters.

All groups mentioned above can be written, in matrix notation,

$$x' = Ax, \quad \det A \neq 0, \tag{5-4}$$

and they are examples of Lie groups.

2.2 Lie Algebras

Associated with an r-parameter Lie group, there are r generators $X_a,\ldots,X_r$ satisfying the commutation relations $(X_aX_b - X_bX_a \equiv [X_a,X_b])$

$$[X_a,X_b] = \sum_c c^c_{ab} X_c . \tag{5-5}$$

These relations define a Lie algebra, and the c's are called Lie structure constants. If

$$c^c_{ab} = 0, \text{ for any } a,b,c, \rightarrow [X_a,X_b] = 0, \text{ for any } a,b \tag{5-6}$$

and the algebra is called Abelian.

For each algebra there exist one (or more) operators C such that

$$[C,X_a] = 0, \text{ any } a. \tag{5-7}$$

The operators C are called Casimir operators. The number of independent Casimir operators is called the rank of the algebra. As an example, consider the Lie algebra of the 3 dimensional rotation group, O(3). This is a 3 parameter group with 3 generators satisfying

$$[X_1,X_2] = X_3, \quad [X_2,X_3] = X_1, \quad [X_3,X_1] = X_2 . \tag{5-8}$$

The 3 generators here can be identified as the 3 components of the angular momentum operator L_x,L_yL_z. By replacing $L_x = iX_1$, $L_y = iX_2$, $L_z = iX_3$, the Lie algebra (8) becomes

$$[L_x,L_y] = iL_z, \quad [L_y,L_z] = iL_x, \quad [L_z,L_x] = iL_y, \tag{5-9}$$

which is the algebra of the angular momentum operators. Similarly, one can show that the group O(3) is a rank 1 group, with only 1 Casimir operator which can be written in the form

$$C = -(X_1^2 + X_2^2 + X_3^2). \tag{5-10}$$

With the replacement given above (10) becomes

$$C = L_x^2 + L_y^2 + L_z^2 = \vec{L}^2, \tag{5-11}$$

the square of the angular momentum.

An important remark here is that the abstract Lie algebra (5) may be realized in many ways. For example, a possible realization of the Lie algebra of O(3), (8), is in terms of differential operators

$$L_x = \frac{1}{i}(y \frac{\partial}{\partial z} - z \frac{\partial}{\partial y}), \quad L_y = \frac{1}{i}(z \frac{\partial}{\partial x} - x\frac{\partial}{\partial z}), \quad L_z = \frac{1}{i}(x \frac{\partial}{\partial y} - y \frac{\partial}{\partial x}) \qquad (5\text{-}12)$$

acting on the space of functions $\psi(x,y,z)$. However, the same algebra can be realized in other ways. For example, introduce fermion creation and annihilation operators in a two dimensional space (as the proton-neutron space in nuclei) a_p^+, a_n^+, a_p, a_n. Consider the operators

$$T_+ = a_p^+ a_n, \quad T_- = a_n^+ a_p, \quad T_0 = \frac{1}{2}(a_p^+ a_p - a_n^+ a_n) \qquad (5\text{-}13)$$

and from these form

$$T_x = \frac{1}{2}(T_+ + T_-), \quad T_y = \frac{1}{2i}(T_+ - T_-), \quad T_z = T_0 . \qquad (5\text{-}14)$$

Using the basic anticommutation relations of the a's, one can show that the operators (14) satisfy the same algebra (9). However these operators do not act now in the space of functions $\psi(x,y,z)$, but in a Fock space, $a_p^{+k} a_n^{+i} |0\rangle$. In this particular example, the generators (14) can be identified as the isospin operators.

2.3 Isomorphism and Homomorphism

We have seen that some relations may exist between groups. However, in order to fully understand these relations it is not sufficient to study the Lie algebras alone. Rather we must return to the Lie groups of Sect. 2.1. Here there are two kinds of relations which may arise.

First, it may happen that all elements of a group G can be put into a one-to-one correspondence with the elements of another group G'. In this case the two groups are said to be isomorphic ($G \approx G'$) and one does not need to study the properties of both groups G and G'. An example is $SO(6) \approx SU(4)$.

However, in some cases, a different correspondence may arise. In order to illustrate this point, consider the simple (but important) example of the group of complex transformations in two dimensions which leave $|u|^2 + |v|^2$ invariant, SU(2),

$$\begin{cases} u' = a_{11} u + a_{12} v \\ v' = a_{21} u + a_{22} v . \end{cases} \qquad (5\text{-}15)$$

This looks like an 8 parameter Lie group. However, introducing

$$A = \begin{pmatrix} a_{11} & a_{12} \\ a_{21} & a_{22} \end{pmatrix}, \quad A^+ = \begin{pmatrix} a^*_{11} & a^*_{21} \\ a^*_{12} & a^*_{22} \end{pmatrix} \qquad (5\text{-}16)$$

and insisting that the transformation (11) be unitary

$$A^{\dagger}A = 1, \tag{5-17}$$

one obtains four conditions

$$\begin{aligned} a^*_{11}a_{11} + a^*_{21}a_{21} &= 1, \quad a^*_{11}a_{12} + a^*_{21}a_{22} = 0 \\ a^*_{12}a_{11} + a^*_{22}a_{21} &= 0, \quad a^*_{12}a_{12} + a^*_{22}a_{22} = 1 . \end{aligned} \tag{5-18}$$

The group of unitary transformations in 2 dimensions, U(2), is thus a 4 parameter group. In general, the group of unitary transformations in n dimensions, U(n), is a n^2 parameter group. We may also insist that

$$\det A = 1, \tag{5-19}$$

which imposes the additional condition

$$a_{11}a_{22} - a_{12}a_{21} = 1, \tag{5-20}$$

leaving only 3 essential parameters. The corresponding group is called SU(2), where S stands for special (or unimodular). In general, the group SU(n) is a $n^2 - 1$ parameter group. We have also seen that the group of real orthogonal transformations in three dimensions, O(3), is a three parameter group, and so is the group of real orthogonal transformations in 3 dimensions with $\det A = 1$, SO(3). We may now inquire what is the relation between SU(2) and SO(3). This is easily seen by rewriting (15) as

$$\begin{cases} u' = a_{11}u + a_{12}v \\ v' = -a^*_{12}u + a^*_{11}v \end{cases}, \quad (a_{11}a^*_{11} + a_{12}a^*_{12} = 1), \tag{5-21}$$

and introducing the quantitites

$$x = \frac{u^2 - v^2}{2}, \quad y = \frac{u^2 + v^2}{2i}, \quad z = uv . \tag{5-22}$$

These three quantities transform as

$$\begin{cases} x' = \frac{1}{2}(a^2_{11} - a^{*2}_{12} - a^2_{12} + a^{*2}_{11})x + \frac{i}{2}(a^2_{11} - a^{*2}_{12} + a^2_{12} - a^{*2}_{11})y + (a_{11}a_{12} + a^*_{11}a^*_{12})z \\ y' = -\frac{i}{2}(a^2_{11} + a^{*2}_{12} - a^2_{12} - a^{*2}_{11})x + \frac{1}{2}(a^2_{11} + a^{*2}_{12} + a^2_{12} + a^{*2}_{11})y - i(a_{11}a_{12} - a^*_{11}a^*_{12})z \\ z' = -(a^*_{11}a_{12} + a_{11}a^*_{12})x + i(a^*_{11}a_{12} - a_{11}a^*_{12})y + (a_{11}a^*_{11} - a_{12}a^*_{12})z. \end{cases} \tag{5-23}$$

This is a real, orthogonal transformation in 3 variables. Thus, to each unitary transformation in two dimensions, there is associated an orthogonal transformation in three dimensions. Let us now explore this correspondence further. Consider the transformation

$$a_{11} = e^{i\alpha/2}, \quad a_{12} = 0 . \tag{5-24}$$

Using (23), we find

$$\begin{cases} x' = x\cos\alpha - y\sin\alpha \\ y' = x\sin\alpha + y\cos\alpha \\ z' = z \end{cases} \tag{5-25}$$

This is a rotation of an angle α around the z-axis, $R(\alpha,0,0)$.

$$\underset{SU(2)}{\begin{pmatrix} e^{i\alpha/2} & 0 \\ 0 & e^{-i\alpha/2} \end{pmatrix}} \longrightarrow \underset{SO(3)}{\begin{pmatrix} \cos\alpha & -\sin\alpha & 0 \\ \sin\alpha & \cos\alpha & 0 \\ 0 & 0 & 1 \end{pmatrix}} . \tag{5-26}$$

Similarly, we can associate with the general rotation $R(\alpha,\beta,\gamma)$, the unitary transformation

$$\underset{SU(2)}{\begin{pmatrix} \cos\frac{\beta}{2}\, e^{i/2(\alpha+\gamma)} & \sin\frac{\beta}{2}\, e^{i/2(\gamma-\alpha)} \\ -\sin\frac{\beta}{2}\, e^{i/2(\alpha-\gamma)} & \cos\frac{\beta}{2}\, e^{-i/2(\alpha+\gamma)} \end{pmatrix}} \longrightarrow \underset{SO(3)}{R(\alpha,\beta,\gamma)} \tag{5-27}$$

We now see that the two rotations $R(0,0,0)$

$$\underset{SU(2)}{\begin{pmatrix} 1 & 0 \\ 0 & 1 \end{pmatrix}} \rightarrow \underset{SO(3)}{1} \tag{5-28}$$

and $R(2\pi,0,0)$

$$\underset{SU(2)}{\begin{pmatrix} -1 & 0 \\ 0 & -1 \end{pmatrix}} \rightarrow \underset{SO(3)}{1} \tag{5-29}$$

both correspond to the same unit matrix 1 of SO(3). There is thus a two-to-one correspondence between SU(2) and SO(3). This correspondence is called homomorphism.

The group SU(2) plays a fundamental role in practically all branches of physics. Some examples of operators satisfying the Lie algebra of SU(2) are:

(i) the spin operators

$$\vec{S} = \frac{1}{2}\vec{\sigma} \tag{5-30}$$

whose algebra is usually realized in terms of matrices

$$\sigma_x = \begin{pmatrix} 0 & 1 \\ 1 & 0 \end{pmatrix}, \quad \sigma_y = \begin{pmatrix} 0 & -i \\ i & 0 \end{pmatrix}, \quad \sigma_z = \begin{pmatrix} 1 & 0 \\ 0 & -1 \end{pmatrix}; \tag{5-31}$$

(ii) the isospin operators of Sect. 2.2;

(iii) the quasi-spin operators S_+, S_- and S_z, defined in terms of creation (a^+_{jm}) and annihilation (a_{jm}) operators for fermions in a single j-shell

$$\left\{ \begin{aligned} S_+ &= \frac{1}{2}\sum_m (-)^{j-m}\, a^+_{jm} a^+_{j,-m} \\ S_- &= \frac{1}{2}\sum_m (-)^{j-m}\, a_{j,-m} a_{jm} \\ S_z &= \frac{1}{4}\sum_m (a^+_{jm} a_{jm} - a_{jm} a^+_{jm}) . \end{aligned} \right. \tag{5-32}$$

2.4 Cartan Classification of Lie Groups

All possible Lie groups have been classified by Cartan. They may be divided into four classes:

A_n - Unitary groups, U(n);
B_n - Orthogonal groups, O(n), n = odd;
C_n - Symplectic groups, Sp(n);
D_n - Orthogonal groups, O(n), n = even.

In addition, there are 5 exceptional groups, called E_6, E_7, E_8, F_4 and G_2. Most applications to physical problems make use of the classes A_n and B_n only. In the discussion of collective states in nuclei we will encounter also groups belonging to D_n. The Flowers scheme, not discussed here, makes use of the groups C_n. There are very few examples of use of the exceptional groups. The most interesting one is the use of the group G_2 made by Racah in its study of atomic f^n-electron configurations.

3.1 Tensor Representations of the Lie Groups A_n, B_n, C_n and D_n (Lecture 2)

For applications to physical problems we need to represent the abstract group G in some linear vector space. This representation will provide the quantum numbers which are necessary to classify the states. The most commonly used representations of the ordinary Lie groups A_n, B_n, C_n and D_n are in terms of irreducible tensors of rank r. I list here a few results:

Unitary Groups, U(n)

The irreducible representations of U(n) are characterized by the partition of r into n integers such that

$$\lambda_1 + \lambda_2 + \ldots + \lambda_n = r, \text{ with } \lambda_1 \geq \lambda_2 \geq \ldots \geq \lambda_n \geq 0. \tag{5-33}$$

To each partition there corresponds a graph (or tableau), called Young tableau

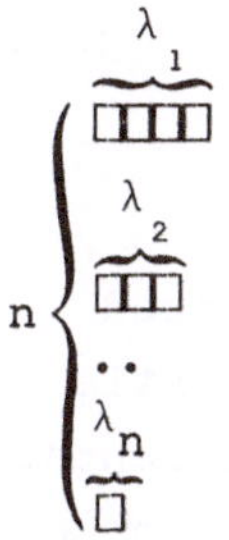

$$\equiv [4\ 3\ ..1]. \tag{5-34}$$

The tensor is symmetric under interchange in the rows and antisymmetric under interchange in the columns. The Young tableau $[\lambda_1\lambda_2\ldots\lambda_n]$ labels the irreducible representations of U(n).

Unimodular Unitary Groups, SU(n)

If we go to the unimodular groups SU(n) [det A = 1] the representations corresponding to the patterns $[\lambda_1\lambda_2\ldots\lambda_n]$ and $[\lambda_1 + s,\ \lambda_2 + s,\ \ldots,\ \lambda_n + s]$, s any integer, become equivalent. Thus, for SU(n) we need to consider only patterns with one row less

$$[\lambda_1, \lambda_2, \ldots, \lambda_n] \equiv [\lambda_1 - \lambda_n, \lambda_2 - \lambda_n, \ldots, \lambda_{n-1} - \lambda_n]. \tag{5-35}$$

For example

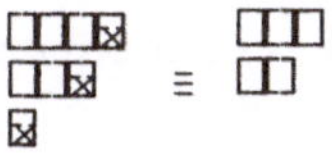

$$[4\ 3\ 1] \equiv [3\ 2]. \tag{5-36}$$

The Young tableau $[\lambda_1 - \lambda_n,\ \lambda_2 - \lambda_n,\ \ldots, \lambda_{n-1} - \lambda_n]$ labels the irreducible representations of SU(n). For example,

for SU(2) only 1 number is needed: j;

for SU(3) only 2 numbers are needed: λ, μ.

Orthogonal groups, O(n)

When we go from U(n) to O(n) the representations in terms of tensors of a given symmetry are no longer irreducible. The reason is that there is a new operation of contraction which commutes with the orthogonal transformations. Accordingly, a modification of the previous rules occurs, as follows: the irreducible representations of the orthogonal groups O(n) are characterized by the partition of r into ν integers such that

$$\mu_1 + \mu_2 + \ldots + \mu_\nu = r, \text{ with } \mu_1 \geq \mu_2 \geq \ldots \geq \mu_\nu \geq 0, \tag{5-37}$$

where

$$\nu = \frac{n}{2}, \quad n = \text{even}$$
$$\nu = \frac{n-1}{2}, \quad n = \text{odd} . \tag{5-38}$$

Proper Orthogonal Groups

For n = odd, the irreducible representations can still be described uniquely by the symbol $(\mu_1, \mu_2, \ldots, \mu_\nu)$. For n = even, some subtle problems arise from the nonequivalence of the irreducible representations corresponding to self-associate patterns. For the purposes of these lectures, these subtle problems will be neglected. For example

for SO(3), n = 3, $\nu = 1$, only one number is needed, ℓ;

for SO(5), n = 5, $\nu = 2$, two numbers are needed, μ_1, μ_2.

Symplectic Groups, Sp(n)

These are the groups of linear transformations which leave invariant a skew-symmetric bilinear form. For this reason the symplectic group can only be defined in an even dimensional space, $n = 2\nu$, ν = integer. The irreducible representations of Sp(n) are characterized by the partition of r into integers

$$\mu_1 + \mu_2 + \ldots + \mu_\nu = r, \quad \mu_1 \geq \mu_2 \geq \ldots \geq \mu_\nu \geq 0, \quad \nu = \frac{n}{2} . \tag{5-39}$$

The symplectic transformations are unimodular, det A = 1, and therefore there is no need to distinguish between proper and improper transformations, as for O(n).

3.2 The Classification Problem. Group Chains

In the solution of physical problems we often encounter the problem of constructing an appropriate complete basis. This may serve, for example, in order to diagonalize the Hamiltonian, H, or in order to discuss possible dynamical symmetries of H. The construction of a complete basis is simply illustrated with an example. Consider the rotation group SO(3). Its representations are characterized by the angular momentum, ℓ. However, this quantum number alone is not sufficient to describe uniquely the states. In addition to ℓ we need another quantum number, m, such that $-\ell \leq m \leq \ell$. It is not difficult to see that the number m is associated with the irreducible representations of SO(2). From the group theoretical point of view, the construction of a complete basis amounts to the construction of a complete chain of groups G ⊃ G' ⊃ ... . The irreducible representations of G, G', ... then label completely the states. The complete chain of groups here is SO(3) ⊃ SO(2). As we have seen in Sect. 3.1, there is only one quantum number associated with SO(3), ℓ, and one with SO(2), m. Thus the states are completely classified by $|\ell m\rangle$. In this case, the same result may be obtained by elementary methods (for example by solving the differential equation for the spherical harmonics). However, in general, elementary methods may be difficult to apply and one must resort to group theoretical methods in order to construct a complete classification scheme.

In constructing a classification scheme for complex systems, one often encounters the problem of finding the representations of G' contained in a representation of G. For the groups SO(3) and SO(2) the solution is simple. The values of m contained in a representation of SO(3) are $-\ell \leq m \leq \ell$, m = integer. Again, in general, this problem is rather difficult and in fact it is one of the most important problems of representation theory. It is usually solved by a building-up process. Before discussing this process, one must describe how to multiply representations. I give here only some rules for it. In order to find the components of the outer product of a representation

by a representation

draw the pattern for the first factor

In the pattern for the second factor, assign the symbol a to all boxes. Now apply a to the first pattern and enlarge it in all possible ways, subject to the rule that no two a's appear in the same column.

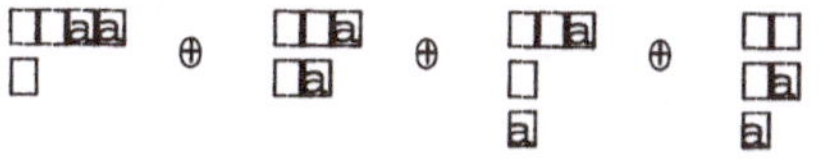

(5-40)

This may be rewritten as

$$[21] \otimes [2] = [41] \oplus [32] \oplus [311] \oplus [221]. \qquad (5\text{-}41)$$

As an example of the building up process consider the group reduction SU(3) ⊃ SO(3). One first considers the Young diagram [1]. This provides the irreducible representation [1] of SU(3) and since $2\ell + 1 = 3$ it also provides the representation $\ell = 1$ of SO(3). Thus

$$\begin{array}{ll} SU(3) & SO(3) \\ \square \equiv [1] & L=1 \end{array} \tag{5-42}$$

Next, consider the product

$$\square \otimes \square = \square\square \oplus \begin{array}{c}\square\\ \square\end{array} \tag{5-43}$$

$$L=1 \otimes L=1 \to L=0,1,2 \;.$$

Now, by virtue of the equivalence mentioned below in (5-49), the representation [11] is equivalent to [1], which contains $L = 1$. The representation [2] must thus contain the remaining angular momenta $L = 0,2$,

$$\begin{array}{ll} SU(3) & SO(3) \\ \square\square \equiv [2] & L=0,2 \;, \\ \begin{array}{c}\square\\ \square\end{array} \equiv [11] & L=1 \;. \end{array} \tag{5-44}$$

Next we consider

$$\begin{array}{c}\square\\ \square\end{array} \otimes \square = \begin{array}{l}\square\square\\ \square\end{array} \oplus \begin{array}{c}\square\\ \square\\ \square\end{array} \tag{5-45}$$

$$L=1 \otimes L=1 \to L=0,1,2$$

But

$$\begin{array}{c}\square\\ \square\\ \square\end{array} \equiv [111]$$

is equivalent to [0]. Thus

$$\begin{array}{ll} SU(3) & SO(3) \\ \begin{array}{c}\square\\ \square\\ \square\end{array} \equiv [111] & L = 0 \\ \begin{array}{l}\square\square\\ \square\end{array} \equiv [21] & L=1,2 \end{array} \tag{5-46}$$

Next we consider

$$\square\square \otimes \square = \square\square\square \oplus \begin{array}{l}\square\square\\ \square\end{array} \tag{5-47}$$

$$L=0,2 \otimes L=1 \to L=1,1,2,3$$

Thus

$$\begin{array}{ll} SU(3) & SO(3) \\ \square\square\square \equiv [3] & L=1,3 \end{array} \qquad (5\text{-}48)$$

and so on. We can therefore construct the following table.

Table 5-1 Angular momentum analysis of the representations of SU(3)

r	[f]	L
0	[0]	0
1	[1]	1
2	[2]	0,2
	[11] ≡ [1]	1
3	[3]	1,3
	[21]	1,2
	[111] ≡ [0]	0
4	[4]	0,2,4
	[31]	1,2,3
	[22] ≡ [2]	0,2
	[211] ≡ [1]	1

In constructing the table the following equivalences have been used, which hold for any group SU(n),

$$\begin{aligned} &[\lambda_1,\lambda_2,\ldots,\lambda_n] \equiv [\lambda_1-\lambda_n,\lambda_2-\lambda_n,\ldots,\lambda_{n-1}-\lambda_n,0], \\ &[\lambda_1,\lambda_2,\ldots,\lambda_n] \equiv [\lambda_1-\lambda_n,\lambda_1-\lambda_{n-1},\ldots,\lambda_1-\lambda_2,0]. \end{aligned} \qquad (5\text{-}49)$$

It is interesting to note that in some cases more than one representation of G' is contained in a given representation of G. In this case, the group G is said not to be fully decomposable with respect to G'. For example, the representation [42] of SU(3) contains $L=0,2^2,3,4$ of SO(3). Thus SU(3) is not fully decomposable with respect to SO(3) and we need an additional quantum number to characterize uniquely the states. The identification and use of this quantum number is one of the most difficult points of group representation. Another example of not fully decomposable groups which we will encounter later on is $SO(5) \supset SO(3)$.

3.3 Dynamical Symmetries. Eigenvalues of the Casimir Operators

Group theory is useful in providing a basis for the solution of physical problems. However, it becomes even more useful whenever the Hamiltonian H, which describes the system, has a dynamical symmetry. This arises when H (or a functional of H) can be written in terms only of the Casimir operators of the complete chain of groups $G \supset G' \supset \ldots$ which completely specifies the states. In that case, group theory provides an elegant and straightforward solution to the problem at hand, since the Hamiltonian

H is then diagonal in the basis $G \supset G' \supset \ldots$. To find the solution one has only to construct the eigenvalues of the various Casimir operators in the given representation.

As a simple example of this procedure, consider the problem of n identical fermions in a single j shell, interacting via a paring interaction. Introducing the quasi-spin operators (5-32) we can write the corresponding Hamiltonian as

$$H = -2GS_+S_-, \tag{5-50}$$

where G is the strength of the interaction. The basis states are characterized by the quantum numbers $|S,S_z\rangle$ corresponding to the group chain $SU(2) \supset SO(2)$. The expectation value of H in the state $|S, S_z\rangle$ can be obtained by rewriting S_+S_- as

$$S_+S_- = \vec{S}^2 - S_z(S_z - 1), \tag{5-51}$$

where

$$\vec{S}^2 = S_x^2 + S_y^2 + S_z^2 \tag{5-52}$$

is the quadratic Casimir operator of SU(2) and S_z is the Casimir operator of SO(2). [SO(2) is an Abelian, one parameter group.] The expectation value of H is thus

$$\langle S,S_z|H|S,S_z\rangle = 2G[S(S+1) - S_z(S_z - 1)]. \tag{5-53}$$

Instead of the pair of variables S, S_z it is customary to use two other variables, n (the number of particles) and v (the seniority). These are related to S and S_z by

$$S_z = \frac{n}{2} - \frac{\Omega}{2}; \quad S = \frac{\Omega}{2} - \frac{v}{2}; \quad \Omega = j + \frac{1}{2}. \tag{5-54}$$

In terms of these variables

$$\langle n,v|H|n,v\rangle = -\frac{1}{2}\, G(n - v)(2j + 3 - n - v). \tag{5-55}$$

The corresponding excitation spectrum looks as in Fig. 5-1.

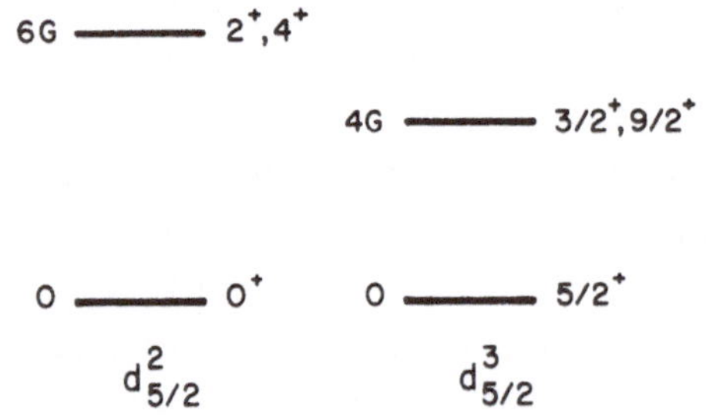

Fig. 1 Spectrum of two and three identical nucleons in a shell j=5/2 with a pairing interaction.

The eigenvalue of the Casimir operator of SU(2) (and SO(3)) can be obtained by elementary methods and it is $S(S+1)$. For larger groups, the derivation of the eigenvalues of the Casimir operators is more complicated. However, its full solution is known and I quote here the results [5,6].

Unitary Groups, U(n) and SU(n)

Denote by C_p the Casimir operator of order p. For linear operators $p = 1$, for quadratic operators $p = 2$, etc.. Construct the quantities

$$S_k = \sum_{i=1}^{n} (\lambda_i^k - \rho_i^k), \quad \rho_i = n - i, \quad \lambda_i = m_i + n - i$$

$$m_i = \begin{cases} f_i & \text{for } U(n) \\ f_i - \frac{f}{n} & \text{for } SU(n), \; f = f_1 + f_2 + \ldots + f_n . \end{cases} \tag{5-56}$$

Construct the function

$$\phi(z) = \sum_{k=2}^{\infty} a_k z^k, \quad a_k = \sum_{\ell=1}^{k-1} \frac{(k-1)!}{\ell!(k-\ell)!} S_\ell . \tag{5-57}$$

Define the quantities B_p by

$$\exp\{-\phi(z)\} = 1 - \sum_{p=0}^{\infty} B_p z^{p+1}, \quad B_0 = 0. \tag{5-58}$$

Then, the expectation value of C_p in the representation $[f_1, f_2, \ldots, f_n]$ is

$$\langle C_p \rangle = B_p - nB_{p-1} . \tag{5-59}$$

The expectation values of the first few Casimir operators are then given by

U(n)

$$\begin{aligned} \langle C_1 \rangle &= S_1 \\ \langle C_2 \rangle &= S_2 - (n-1) S_1 \\ \langle C_3 \rangle &= S_3 - (n - \tfrac{3}{2}) S_2 - \tfrac{1}{2} S_1^2 - (n-1) S_1 \end{aligned} \tag{5-60}$$

...

SU(n)

$$\begin{aligned} \langle C_1 \rangle &= 0 \\ \langle C_2 \rangle &= S_2 \\ \langle C_3 \rangle &= S_3 - (n - \tfrac{3}{2}) S_2 \\ &\dots \end{aligned} \tag{5-61}$$

Examples. For SU(3) this procedure gives

$$\langle C_2 \rangle = S_2 = \frac{6}{9} [f_1^2 + f_2^2 - f_1 f_2 + 3f_1]. \tag{5-62}$$

Instead of f_1 and f_2, it is customary to use Elliott quantum numbers $\lambda = f_1 - f_2$ and $\mu = f_2$. Then

$$\langle C_2 \rangle = \frac{6}{9}[\lambda^2 + \mu^2 + \lambda\mu + 3(\lambda + \mu)]. \tag{5-63}$$

For SU(5) the procedure gives

$$\langle C_2 \rangle = S_2 = f_1^2 + f_2^2 + f_3^2 + f_4^2 - \frac{f^2}{5} + 8f_1 + 6f_2 + 4f_3 + 2f_4 - 4f. \tag{5-64}$$

In the totally symmetric representation $f_1 = n_d$, $f_2 = f_3 = f_4 = 0$

$$\langle C_2 \rangle = \frac{4}{5} n_d (n_d + 5), \tag{5-65}$$

and so on.

Orthogonal Groups, O(2n + 1)

Construct the quantities

$$S_k = \sum_{i=-n}^{+n} (\lambda_i^k - \rho_i^k), \quad \rho_i = \lambda_i - f_i, \tag{5-66}$$

$$\begin{cases} \lambda_i = f_i + n + i - 1 & (i \neq 0) \\ \lambda_{-i} = -\lambda_i + 2n - 1 & (i \neq 0) \\ \lambda_0 = n, \end{cases} \tag{5-67}$$

$$f_{-i} = f_i, \quad f_0 = 0, \tag{5-68}$$

$$S_0 = S_1 = 0. \tag{5-69}$$

Construct the function

$$\phi(z) = \sum_{k=3}^{\infty} a_k z^k, \quad a_k = \sum_{\ell=2}^{k-1} \frac{(k-1)!}{\ell!(k-\ell)!} S_\ell . \tag{5-70}$$

Define the quantities B_p by

$$\exp\{-\phi(z)\} = 1 - \sum_{p=2}^{\infty} B_p z^{p+1}, \quad B_0 = B_1 = 0. \tag{5-71}$$

Then, the expectation value of C_p in the representation $[f_n, f_{n-1}, \ldots, f_1]$ is

$$\langle C_p \rangle = (2n+1)\delta_{p0} + B_p - (n+\tfrac{1}{2})B_{p-1} - \sum_{q=1}^{p-1} [B_q - (n+\tfrac{1}{2})B_{q-1}]n^{p-q}. \tag{5-72}$$

The first few Casimir operators have then expectation values given by

$$\begin{aligned} \langle C_0 \rangle &= 2n+1 \\ \langle C_1 \rangle &= 0 \\ \langle C_2 \rangle &= S_2 \end{aligned} \tag{5-73}$$

Examples. For O(3)

$$\langle C_2 \rangle = 2f_1(f_1+1) \tag{5-74}$$

In the representation L, $\langle C_2 \rangle = 2L(L+1)$. The Casimir operator defined here is thus twice $\vec{L}^2$.

For O(5)

$$\langle C_2 \rangle = 2f_2(f_2+3) + 2f_1(f_1+1). \tag{5-75}$$

In the representation (v,o),

$$\langle C_2 \rangle = 2v(v+3). \tag{5-76}$$

Symplectic Groups, Sp(2n)

Construct the quantities

$$S_k = \sum_{\substack{i=-n \\ i\neq 0}}^{+n} (\lambda_i^k - \rho_i^k), \quad \rho_i = \lambda_i - f_i \tag{5-77}$$

$$\left\{ \begin{aligned} \lambda_i &= f_i + n + i \\ \lambda_{-i} &= -\lambda_i + 2n \end{aligned} \right. \tag{5-78}$$

$$f_{-i} = -f_i \tag{5-79}$$

$$S_0 = S_1 = 0 \tag{5-80}$$

and the functions $\phi(z)$ as before. Then,

$$\langle C_p \rangle = 2n\delta_{p0} + (B_p - nB_{p-1}) - \sum_{q=1}^{p-1} (B_q - nB_{q-1})(n + \tfrac{1}{2})^{p-q}. \tag{5-81}$$

For the first few Casimir operators, this gives

$$\begin{aligned} \langle C_0 \rangle &= 2n \\ \langle C_1 \rangle &= 0 \\ \langle C_2 \rangle &= S_2 \\ &\cdots \end{aligned} \tag{5-82}$$

<u>Orthogonal Groups, O(2n)</u>

Construct the quantities

$$S_k = \sum_{\substack{i=-n \\ i\neq 0}}^{+n} (\lambda_i^k - \rho_i^k), \quad \rho_i = \lambda_i - f_i \tag{5-83}$$

$$\begin{cases} \lambda_i = f_i + n + i - 2 \\ \lambda_{-i} = -\lambda_i + 2n - 2 \end{cases} \tag{5-84}$$

$$f_{-i} = -f_i \tag{5-85}$$

$$S_0 = S_1 = 0 \tag{5-86}$$

and the function $\phi(z)$ as before. Then,

$$\langle C_p \rangle = 2n\delta_{p0} + (B_p - nB_{p-1}) - \sum_{q=1}^{p-1} (B_q - nB_{q-1})(n - \tfrac{1}{2})^{p-q} \tag{5-87}$$

For the first few Casimir operators, this gives

$$\begin{aligned} \langle C_0 \rangle &= 2n \\ \langle C_1 \rangle &= 0 \\ \langle C_2 \rangle &= S_2 \\ \langle C_3 \rangle &= (S_3 + \tfrac{3}{2} S_2) - (2n - \tfrac{1}{2}) S_2 \\ &\cdots \end{aligned} \tag{5-88}$$

Examples. For O(2)

$$\langle C_2 \rangle = 2f_1^2 . \tag{5-89}$$

For O(4)

$$\langle C_2 \rangle = 2f_2(f_2+2) + 2f_1^2 . \tag{5-90}$$

For O(6)

$$\langle C_2 \rangle = 2f_3(f_3+4) + 2f_2(f_2+2) + 2f_1^2 . \tag{5-91}$$

In the totally symmetric representation $(\sigma 0\dot{0})$,

$$\langle C_2 \rangle = 2\sigma(\sigma+4) . \tag{5-92}$$

4.1 Group Theory of the Interacting Boson Model (Lecture 3)

In the last few years it has become evident that a complete description of collective states in medium mass and heavy nuclei may be given in terms of the interacting boson model [7-11]. In this model an even-even nucleus is treated as a system of interacting bosons. The bosons represent correlated nucleon pairs and thus there are two kinds of bosons, proton bosons and neutron bosons. For the purposes of these lectures I will neglect altogether the difference between proton and neutron bosons and discuss only the case of one kind of bosons. These bosons, of which there are N in number (where N is the number of valence pairs), may occupy two levels, one with $L=0$ (called s-boson) and one with $L=2$ (called d-boson), as shown in Fig. 2.

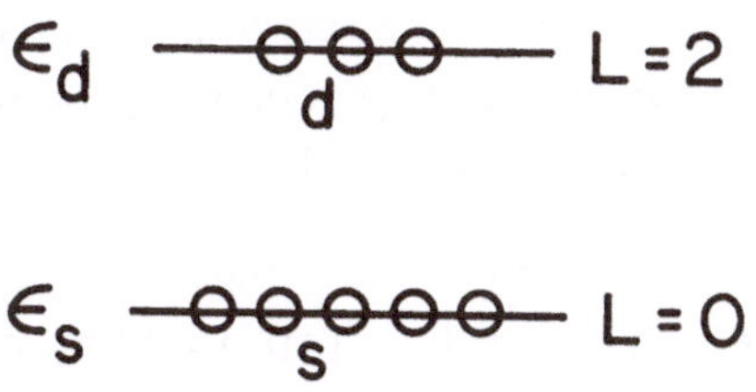

Fig. 2 The configuration s^5d^3 in the interacting boson model.

Introducing boson creation (s^+, d_μ^+) and annihilation (s, d_μ) operators satisfying the commutation relations

$$\begin{aligned}
&[s,s^+]=1, && [s,s]=0, && [s^+,s^+]=0 \\
&[d_\mu,d_{\mu'}^+]=\delta_{\mu\mu'}, && [d_\mu,d_{\mu'}]=0, && [d_\mu^+,d_{\mu'}^+]=0 \\
&[s,d_\mu^+]=0 && [s^+,d_\mu^+]=0 && \\
&[s^+,d_\mu]=0 && [s^+,d_\mu]=0 &&
\end{aligned} \tag{5-93}$$

one can write the most general Hamiltonian which contains only one boson terms and two-body boson-boson interactions, as

$$H = \varepsilon_s(s^+\cdot s) + \varepsilon_d(d^+\cdot\tilde{d}) + \sum_{L=0,2,4} \frac{1}{2}(2L+1)^{1/2} c_L[[d^+ x d^+]^{(L)} x[\tilde{d} x \tilde{d}]^{(L)}]^{(0)}$$

$$+ \frac{1}{2^{\frac{1}{2}}} \tilde{v}_2[[d^+ x d^+]^{(2)} x[\tilde{d} x s]^{(2)} + [d^+ x s^+]^{(2)} x[\tilde{d} x \tilde{d}]^{(2)}]^{(0)}$$

$$+ \frac{1}{2} \tilde{v}_0[[d^+ x d^+]^{(0)} x[s x s]^{(0)} + [s^+ x s^+]^{(0)} x[\tilde{d} x \tilde{d}]^{(0)}]^{(0)} \tag{5-94}$$

$$+ u_2[[d^+ x s^+]^{(2)} x[\tilde{d} x s]^{(2)}]^{(0)} + \frac{1}{2} u_0[[s^+ x s^+]^{(0)} x[s x s]^{(0)}]^{(0)} .$$

This Hamiltonian contains 2 one-body terms, $(\varepsilon_s, \varepsilon_d)$, and 7 two-body interactions $(c_L(L=0,2,4),\ \tilde{v}_L(L=0,2),\ u_L(L=0,2))$. However, it turns out that for fixed boson number N, only 1 of the one-body terms and 5 of the two-body terms are independent, as it can be seen by noting that $N = n_s + n_d$. In (5-94) the operator $\tilde{d}_\mu = (-)^\mu d_{-\mu}$ has been used instead of d_μ since the latter is not a spherical tensor with respect to O(3). Moreover, the cross denotes tensor products with respect to O(3).

In order to find the eigenvalues of the system, one should take H and diagonalize it in an appropriate basis. It turns out that a good fraction of the properties of this system, and thus of collective states in nuclei, may be obtained by studying the group structure of (5-94) to which we now turn. Since we have introduced creation and annihilation operators, our study will be done in the Fock space realization of the corresponding Lie algebra. An equivalent treatment is possible by realizing the Lie algebra with differential operators and it leads to the Bohr-Mottelson description of collective states in nuclei [12].

4.2 Algebras and Subalgebras

In studying the group structure of the model, the first step is that of identifying the corresponding algebra. Since there are six boson operators, denoted altogether b_i^+, it is well known that the 36 bilinear products $b_i^+ b_j = G_{ij}$ generate the Lie algebra of U(6). However, this Cartesian form of the algebra is not convenient for our study, because the G_{ij} turn out to be reducible with respect to some subgroups of U(6). The most convenient form is obtained by introducing the operators

$$G_\kappa^{(k)}(\ell\ell') = [b_\ell^+ x \tilde{b}_{\ell'}]_\kappa^{(k)} = \sum_{\mu_1,\mu_2} \langle \ell\mu_1 \ell'\mu_2' | k\kappa\rangle b_{\ell\mu_1}^+ (-)^{\mu_2} b_{\ell'\mu_2},$$

$$\ell, \ell' = 0,2 \equiv s,d, \tag{5-95}$$

where $\quad [b_{\ell\alpha}, b_{\ell'\alpha'}^+] = \delta_{\ell\ell'}\delta_{\alpha\alpha'}.$ (5-96)

With some simple manipulations the commutation relations of the G's can be written as

$$[G_\kappa^{(k)}(\ell\ell'), G_{\kappa'}^{(k')}(\ell''\ell''')] = \sum_{k'',\kappa''} (2k+1)^{\frac{1}{2}}(2k'+1)^{\frac{1}{2}} \langle k\kappa k'\kappa' | k''\kappa''\rangle (-)^{k-k'} \times \tag{5-97}$$

$$\times \left[(-)^{k+k'k''} \begin{Bmatrix} k & k' & k'' \\ \ell''' & \ell & \ell' \end{Bmatrix} \delta_{\ell'\ell''} G_{\kappa''}^{(k'')}(\ell\ell''') - \begin{Bmatrix} k & k' & k'' \\ \ell'' & \ell' & \ell \end{Bmatrix} \delta_{\ell\ell'''} G_{\kappa''}^{(k'')}(\ell''\ell') \right]$$

This is the Racah form of the Lie algebra of U(6). The structure constants

$$[G_\kappa^{(k)}, G_\kappa^{(k')}] = \sum_{k'',\kappa''} c_{k\kappa,k'\kappa'}^{k''\kappa''} G_{\kappa''}^{(k'')}, \tag{5-98}$$

can be read off directly from (5-97). In this form the $36 = 6^2$ generators are given by

$$\begin{aligned}
G_0^{(0)}(ss) &= (s^+ \times s)_0^{(0)} && 1 \\
G_0^{(0)}(dd) &= (d^+ \times \tilde{d})_0^{(0)} && 1 \\
G_\kappa^{(1)}(dd) &= (d^+ \times \tilde{d})_\kappa^{(1)} && 3 \\
G_\kappa^{(2)}(dd) &= (d^+ \times \tilde{d})_\kappa^{(2)} && 5 \\
G_\kappa^{(3)}(dd) &= (d^+ \times \tilde{d})_\kappa^{(3)} && 7 \\
G_\kappa^{(4)}(dd) &= (d^+ \times \tilde{d})_\kappa^{(4)} && 9 \\
G_\kappa^{(2)}(ds) &= (d^+ \times s)_\kappa^{(2)} && 5 \\
G_\kappa^{(2)}(sd) &= (s^+ \times \tilde{d})_\kappa^{(2)} && \underline{5} \\
& && 36=6^2
\end{aligned} \tag{5-99}$$

The next step is that of identifying all possible subalgebras of the full algebra G. A subalgebra is a subset of elements of G which is closed under commutation. It turns out that in this case there are three possible chains of subalgebras.

Subalgebras. I

Delete from the 36 operators the 11 operators $G_0^{(0)}(ss)$, $G_\kappa^{(2)}(ds)$, $G_\kappa^{(2)}(sd)$. The remaining 25 operators close under the algebra of U(5).

A) U(5)

$$G^{(0)}_{0}(dd) = (d^{+} \times \tilde{d})^{(0)}_{0} \qquad 1$$

$$G^{(1)}_{\kappa}(dd) = (d^{+} \times \tilde{d})^{(1)}_{\kappa} \qquad 3$$

$$G^{(2)}_{\kappa}(dd) = (d^{+} \times \tilde{d})^{(2)}_{\kappa} \qquad 5 \qquad (5\text{-}100)$$

$$G^{(3)}_{\kappa}(dd) = (d^{+} \times \tilde{d})^{(3)}_{\kappa} \qquad 7$$

$$G^{(4)}_{\kappa}(dd) = (d^{+} \times \tilde{d})^{(4)}_{\kappa} \qquad \frac{9}{25=5^2}$$

Delete from the 25 operators the 15 operators $G^{(0)}_{0}(dd)$, $G^{(2)}_{\kappa}(dd)$ and $G^{(4)}_{\kappa}(dd)$. The remaining 10 operators close under the algebra of O(5).

B) O(5)

$$G^{(1)}_{\kappa}(dd) = (d^{+} \times \tilde{d})^{(1)}_{\kappa} \qquad 3$$

$$G^{(3)}_{\kappa}(dd) = (d^{+} \times \tilde{d})^{(3)}_{\kappa} \qquad \frac{7}{10 = \frac{5x4}{2}} \qquad (5\text{-}101)$$

Delete from the 10 operators, the 7 operators $G^{(3)}_{\kappa}(dd)$. The remaining 3 operators close under the algebra of O(3).

C) O(3)

$$G^{(1)}_{\kappa}(dd) = (d^{+} \times \tilde{d})^{(1)}_{\kappa} \qquad 3 \qquad (5\text{-}102)$$

Finally, delete from the 3 operators, the 2 operators $G^{(1)}_{+1}(dd)$, and $G^{(1)}_{-1}(dd)$. The remaining operator is the generator of O(2).

D) O(2)

$$G^{(1)}_{0}(dd) = (d^{+} \times \tilde{d})^{(1)}_{0} \qquad 1 \qquad (5\text{-}103)$$

Thus, one possible chain of subgroups is [8,12]

$$U(6) \supset U(5) \supset O(5) \supset O(3) \supset O(2). \qquad (I)$$

Subalgebras. II

A) U(3)

Consider the operators

$$G_0^{(0)}(ss) + \sqrt{5}\, G_0^{(0)}(dd) = (s^+ \times s)_0^{(0)} + \sqrt{5}(d^+ \times \tilde{d})_0^{(0)} \qquad 1$$

$$G_K^{(1)}(dd) = (d^+ \times \tilde{d})_K^{(1)} \qquad 3 \qquad (5\text{-}104)$$

$$G_K^{(2)}(ds) + G_K^{(2)}(sd) - \tfrac{1}{2}\sqrt{7}\, G_K^{(2)}(dd)$$

$$= (d^+ \times s + s^+ \times \tilde{d})_K^{(2)} - \tfrac{1}{2}\sqrt{7}(d^+ \times \tilde{d})_K^{(2)} \qquad \frac{5}{9 = 3^2}$$

These operators close under commutation and form the algebra of U(3). Obvious subalgebras are now

B) O(3)

$$G_K^{(1)}(dd) = (d^+ \times \tilde{d})_K^{(1)} \qquad 3 \qquad (5\text{-}105)$$

and

C) O(2)

$$G_0^{(1)}(dd) = (d^+ \times \tilde{d})_0^{(1)} \qquad 1 \qquad (5\text{-}106)$$

Thus, a second possible chain of subgroups is [7,9,12]

$$U(6) \supset U(3) \supset O(3) \supset O(2). \qquad (II)$$

Subalgebras. III

A) O(6)

Consider the operators

$$G_K^{(1)}(dd) = (d^+ \times \tilde{d})_K^{(1)} \qquad 3$$

$$G_K^{(3)}(dd) = (d^+ \times \tilde{d})_K^{(3)} \qquad 7 \qquad (5\text{-}107)$$

$$G_K^{(2)}(ds) + G_K^{(2)}(sd) = (d^+ \times s + s^+ \times \tilde{d})_K^{(2)} \qquad \frac{5}{15 = \frac{6\times 5}{2}}$$

These operators close under commutation, yielding the Lie algebra of O(6). Obvious subalgebras are now

B) O(5)

$$G_K^{(1)}(dd) = (d^+ \times \tilde{d})_K^{(1)} \qquad 3$$

$$(5\text{-}108)$$

$$G_K^{(3)}(dd) = (d^+ \times \tilde{d})_K^{(3)} \qquad \frac{7}{10}$$

C) O(3)

$$G^{(1)}_{\kappa}(dd) = (d^{+} \times \tilde{d})^{(1)}_{\kappa} \qquad 3 \qquad (5\text{-}109)$$

D) O(2)

$$G^{(1)}_{0}(dd) = (d^{+} \times \tilde{d})^{(1)}_{0} \qquad 1 \qquad (5\text{-}110)$$

Thus, a third possible chain is [10,12]

$$U(6) \supset O(6) \supset O(5) \supset O(3) \supset O(2). \qquad (III)$$

It is possible to show that these are the only possible chains of subgroups, if one insists that the angular momentum L be a good quantum number (i.e. O(3) must be contained in the chain). In fact, starting from U(6) we have considered U(5), U(3) and O(6). But O(6) is locally isomorphic to SU(4). Thus we have considered all possible subgroups of U(6), namely U(5), U(4), U(3). In conclusion, there are three and only three possible chains

$$U(6) \begin{array}{ll} \nearrow U(5) \supset O(5) \supset O(3) \supset O(2) & \text{I} \\ \rightarrow U(3) \supset O(3) \supset O(2) & \text{II} \\ \searrow O(6) \supset O(5) \supset O(3) \supset O(2) & \text{III} \end{array} \qquad (5\text{-}111)$$

4.3 Classification of States

Once a group chain has been identified, its first important use is in constructing a basis in which the Hamiltonian H can be diagonalized. For this we need to know the quantum numbers which are needed to classify uniquely the states, in turn related to the labels of the irreducible representations of the groups in question. The number of labels needed for the groups of interest is as follows

U(6)	$[f_1 f_2 f_3 f_4 f_5 f_6]$	
SU(6)	$[f'_1 f'_2 f'_3 f'_4 f'_5]$	
U(5)	$[f''_1 f''_2 f''_3 f''_4 f''_5]$	
SU(5)	$[f'''_1 f'''_2 f'''_3 f'''_4]$	
U(3)	$[f^{iv}_1 f^{iv}_2 f^{iv}_3]$	(5-112)
SU(3)	$[f^{v}_1 f^{v}_2]$	
O(6)	$(\omega_1 \omega_2 \omega_3)$	
O(5)	$(\omega'_1 \omega'_2)$	
O(3)	(ω''_1)	
O(2)	(ω'''_1)	

We now proceed to construct explicitly the classification schemes for the three group chains in (5-111).

Group Chain. I

The complete classification is

$$\begin{array}{lll} SU(6) & [N] \equiv \overbrace{\square\square\ldots\square}^{N \text{ times}} \equiv [N0000] \\ SU(5) & (n_d) \equiv \overbrace{\square\square\ldots\square}^{n_d \text{ times}} \equiv [n_d000]; \; n_d = 0,1,\ldots,N \\ O(5) & (v) \equiv (v0); \quad v = n_d, n_d - 2, \ldots 1 \text{ or } 0 \\ O(3) & L \\ O(2) & M \end{array}$$

The representations of SU(6), SU(5), O(5) are the totally symmetric ones because we are dealing with a system of bosons. The step from O(5) to O(3) is not fully decomposable. We need an additional quantum number which we call n_Δ and which can be identified with the number of d-boson triplets coupled to zero angular momentum.

Algorithm to find the values of L contained in each representation n_d of SU(5). First, partition n_d as

$$n_d = 2n_\beta + 3n_\Delta + \lambda, \tag{5-113}$$

where $n_\beta = (n_d - v)/2$; $n_\beta = 0,1,\ldots,\frac{n_d}{2}$ or $\frac{n_d - 1}{2}$. (5-114)

Then,

$$L = \lambda,\ \lambda + 1,\ \lambda + 2, \ldots, 2\lambda - 2, 2\lambda \tag{5-115}$$

[Note that $2\lambda - 1$ is missing!]. This gives the following table.

Table 5-2 Classification scheme for the group chain I

SU(5)	O(5)		O(3)
n_d	v	n_Δ	L
0	0	0	0
1	1	0	2
2	2	0	4,2
	0	0	0
3	3	0	6,4,3
		1	0
	1	0	2
4	4	0	8,6,5,4
		1	2
	2	0	4,2
	0	0	0

The complete classification scheme for chain I is thus

$|[N](n_d)v\ n_\Delta\ L\ M>$.

Group Chain. II

The complete classification scheme is

$$
\begin{array}{ll}
SU(6) & [N] \equiv \overbrace{\square\square\ldots\square}^{N\ \text{times}} \equiv [N0000] \\
SU(3) & (\lambda,\mu);\ \lambda = f_1 - f_2,\ \mu = f_2 \\
O(3) & L \\
O(2) & M
\end{array}
$$

The step from SU(3) to O(3) is not fully decomposable. The simplest choice of the additional quantum number needed to classify the states is due to Elliott. The corresponding number is called K.

Algorithm to find the values of (λ,μ) contained in [N] and those of L contained in (λ,μ), in Elliott basis. Values of (λ,μ) contained in [N]:

$$
\begin{aligned}
[N] = {} & (2N,0) \oplus (2N-4,2) \oplus (2N-8,4) \oplus \ldots \oplus \begin{Bmatrix} (0,N) \\ (2,N-1) \end{Bmatrix} \oplus \begin{Bmatrix} N = \text{even} \\ N = \text{odd} \end{Bmatrix} \\
& \oplus (2N-6,0) \oplus (2N-10,2) \oplus \ldots \oplus \begin{Bmatrix} (0,N-3) \\ (2,N-4) \end{Bmatrix} \oplus \begin{Bmatrix} N-3 = \text{even} \\ N-3 = \text{odd} \end{Bmatrix} \\
& \oplus (2N-12,0) \oplus (2N-16,2) \oplus \ldots \oplus \begin{Bmatrix} (0,N-6) \\ (2,N-7) \end{Bmatrix} \oplus \begin{Bmatrix} N-6 = \text{even} \\ N-6 = \text{odd} \end{Bmatrix} \\
& , \ldots
\end{aligned}
\tag{5-116}
$$

Values of L contained in (λ,μ):

$$L = K,\ K+1,\ K+2,\ \ldots,\ K+\max\{\lambda,\mu\}, \tag{5-117}$$

where $K = \text{integer} = \min\{\lambda,\mu\},\ \min\{\lambda,\mu\}-2,\ \ldots,\ 1 \text{ or } 0,$ (5-118)

with the exception of $K = 0$ for which

$$L = \max\{\lambda,\mu\},\ \max\{\lambda,\mu\}-2,\ \ldots,\ 1 \text{ or } 0. \tag{5-119}$$

The Elliott basis has the drawback of not being orthogonal. For this reason, it is convenient to introduce another basis, called Vergados basis, which can be constructed from Elliott basis in the following way. Let $K_1, K_2, \ldots, K_n$ be the Elliott quantum numbers which occur in a given representation (λ,μ) with $K_1 < K_2 < \ldots < K_n$. The new basis is labelled by the quantum numbers $\chi_1 < \chi_2 < \ldots < \chi_n$ and defined by

$$|(\lambda,\mu)\chi_1 LM\rangle = |(\lambda,\mu)K_1\lambda\mu\rangle_0 ,$$

$$|(\lambda,\mu)\chi_2 LM\rangle = x_{21}|(\lambda,\mu)K_1 LM\rangle_0 + x_{22}|(\lambda,\mu)K_2 LM\rangle_0 , \tag{5-120}$$

$$\ldots$$

$$|(\lambda,\mu)\chi_i LM\rangle = \sum_{j=1}^{i} x_{ij}|(\lambda,\mu)K_j LM\rangle_0 ,$$

where the states $|(\lambda,\mu)KLM\rangle_0$ are related to Elliott states $|(\lambda,\mu)KLM\rangle$ by the phase convention

$$|(\lambda,\mu)KLM\rangle_0 = i^{\lambda+2\mu}|(\lambda,\mu)KLM\rangle \tag{5-121}$$

and the coefficients x_{ij} are obtained by the requirement

$$\langle(\lambda,\mu)\chi_i LM|(\lambda,\mu)\chi_j LM\rangle = \delta_{ij}. \tag{5-122}$$

Thus the sequence of quantum numbers χ_1, χ_2, ..., χ_n is the same as K_1, K_2, ..., K_n but the values of L contained in each χ_i are different from those contained in K_i. In fact, from its definition, it is clear that if a given L occurs in a given representation only once, it belongs to the lowest possible χ. If it occurs twice, it belongs to the two lowest possible χ's, etc. The only exception is when $\chi = 0$ for which the allowed L values are restricted to be even or odd for λ even or odd, respectively. In the following, the Vergados basis will be used. This gives the following table.

Table 5-3 Classification scheme for the group chain II

SU(6)	SU(3)		O(3)
N	(λ,μ)	χ	L
0	(0,0)	0	0
1	(2,0)	0	2,0
2	(4,0)	0	4,2,0
	(0,2)	0	2,0
3	(6,0)	0	6,4,2,0
	(2,2)	0	4,2,0
		2	3,2
	(0,0)	0	0

The complete classification for chain II is $|[N](\lambda,\mu)\chi LM\rangle$.

Group Chain. III

The complete classification scheme here is

SU(6)	$[N] \equiv \overbrace{\square\square\ldots\square}^{N\text{ times}} \equiv [N0000]$
O(6)	$\sigma \equiv (\sigma 00)$
O(5)	$\tau \equiv (\tau 0)$
O(3)	L
O(2)	M

The step from O(5) to O(3) is not fully decomposable. We need an additional quantum number, which we call ν_Δ.

Algorithm to find the values of σ contained in [N], those of τ contained in σ and those of L contained in τ.
Values of σ contained in N:

$$\sigma = N, N-2, \ldots, 0 \text{ or } 1, \text{ for } N = \text{even or } N = \text{odd}. \tag{5-123}$$

Values of τ contained in σ:

$$\tau = \sigma, \sigma-1, \ldots, 0. \tag{5-124}$$

In order to find the values of L contained in τ, partition τ as

$$\tau = 3\nu_\Delta + \lambda, \quad \nu_\Delta = 0,1,\ldots \tag{5-125}$$

and take

$$L = 2\lambda, 2\lambda-2, \ldots, \lambda+1, \lambda \tag{5-126}$$

[Note that $2\lambda-1$ is missing!]. This gives the following table.

Table 5-4 Classification scheme for the group chain III

SU(6) N	O(6) σ	O(5) τ	ν_Δ	O(3) L
0	0	0	0	0
1	1	1	0	2
		0	0	0
2	2	2	0	4,2
		1	0	2
		0	0	0
	0	0	0	0
3	3	3	0	6,4;3
			1	0
		2	0	4,2
		1	0	2
		0	0	0
	1	1	0	2
		0	0	0

The complete classification scheme for chain III is
$|[N]\sigma\tau\nu_\Delta LM\rangle$.

5.1 Dynamical Symmetries. Solution of the Eigenvalue Problem (Lecture 4)

We now return to the problem of finding the eigenvalues of the Hamiltonian

(5-94). This Hamiltonian can be rewritten, after some manipulations in terms of the Casimir operators of all the groups appearing in (5-111). For fixed boson number N, we can disregard the operators of U(6) and consider only those of U(5), O(5), O(3), SU(3) and O(6). [The group O(2) does not play any role unless the nucleus is placed in an external magnetic field. It will be henceforth neglected.] Of these only U(5) has a linear Casimir operator. Thus, the most general Hamiltonian (5-94) can be written as

$$H = \varepsilon\, C_{1U5} + \alpha\, C_{2U5} + \beta\, C_{2O5} + \gamma\, C_{2O3} + \delta\, C_{2SU3} + \eta\, C_{2O6}. \tag{5-127}$$

As in (5-94), there is in (5-127) 1 independent one-body term and 5 two-body interactions. Here, C_{1U5} denotes the linear Casimir operator of U(5), C_{2U5} the quadratic Casimir operator of U(5), etc.. This Hamiltonian is not diagonal in any of the group chains I, II or III. For example, C_{2SU3} is not diagonal in chain I, etc.. In order to find its eigenvalues one must diagonalize it numerically. The diagonalization can be done in any of the three group chains I, II or III, since all three provide a complete basis. Most of the numerical calculations have been done in the basis provided by the chain I.

However, it is interesting to know that the eigenvalue problem can be solved in closed, analytic form in some special cases. These special cases occur whenever the Hamiltonian can be written in terms only of Casimir operators of a group chain, $G \supset G' \supset \ldots$. In that case, one says that the Hamiltonian H has a dynamical symmetry. As it is clear from (5-127) these dynamical symmetries correspond to the vanishing of some coefficient and since, in the present problem, there are three possible group chains, there are also three possible dynamical symmetries.

<u>Dynamical Symmetry. I</u>

The group chain here is

$$U(6) \supset U(5) \supset O(5) \supset O(3) \supset O(2). \tag{5-128}$$

This dynamical symmetry corresponds to the vanishing of δ and η in (5-127). The corresponding Hamiltonian is

$$H = \varepsilon\, C_{1U5} + \alpha\, C_{2U5} + \beta\, C_{2O5} + \gamma\, C_{2O3}. \tag{5-129}$$

The expectation value of H in the representation $|[N]n_d v n_\Delta LM\rangle$ is given by (see Sect. 3.3)

$$\langle H\rangle = \varepsilon n_d + \alpha n_d(n_d+4) + 2\beta v(v+3) + 2\gamma L(L+1). \tag{5-130}$$

The structure of the corresponding spectrum is shown in Fig. 3, for $\varepsilon, \alpha, \beta, \gamma > 0$.

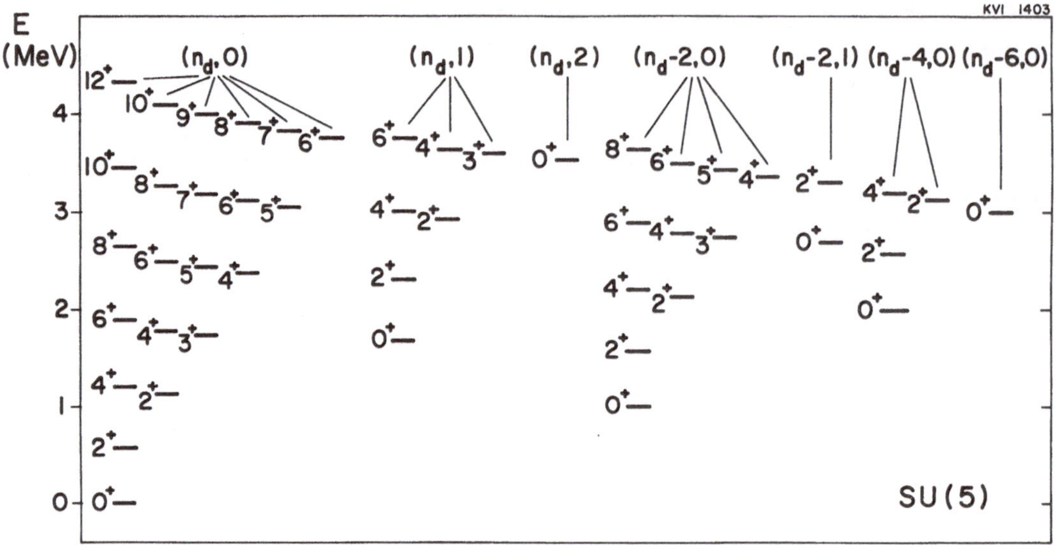

Fig. 3 A typical spectrum with U(5) symmetry and $N = 6$. In parenthesis are the values of v and n_{Δ}.

Dynamical Symmetry. II

The group chain here is

$$U(6) \supset SU(3) \supset O(3) \supset O(2). \tag{5-131}$$

This dynamical symmetry corresponds to the vanishing of ε, α, β and η in (5-127). The corresponding Hamiltonian is

$$H = \delta C_{2SU3} + \gamma C_{2O3}. \tag{5-132}$$

The expectation value of H in the representation $|[N](\lambda,\mu)\chi LM\rangle$ is given by

$$\langle H\rangle = \delta \frac{6}{9}[\lambda^2 + \mu^2 + \lambda\mu + 3(\lambda + \mu)] + \gamma 2L(L + 1). \tag{5-133}$$

The structure of the spectrum ($\delta < 0$, $\gamma > 0$) is shown in Fig. 4.

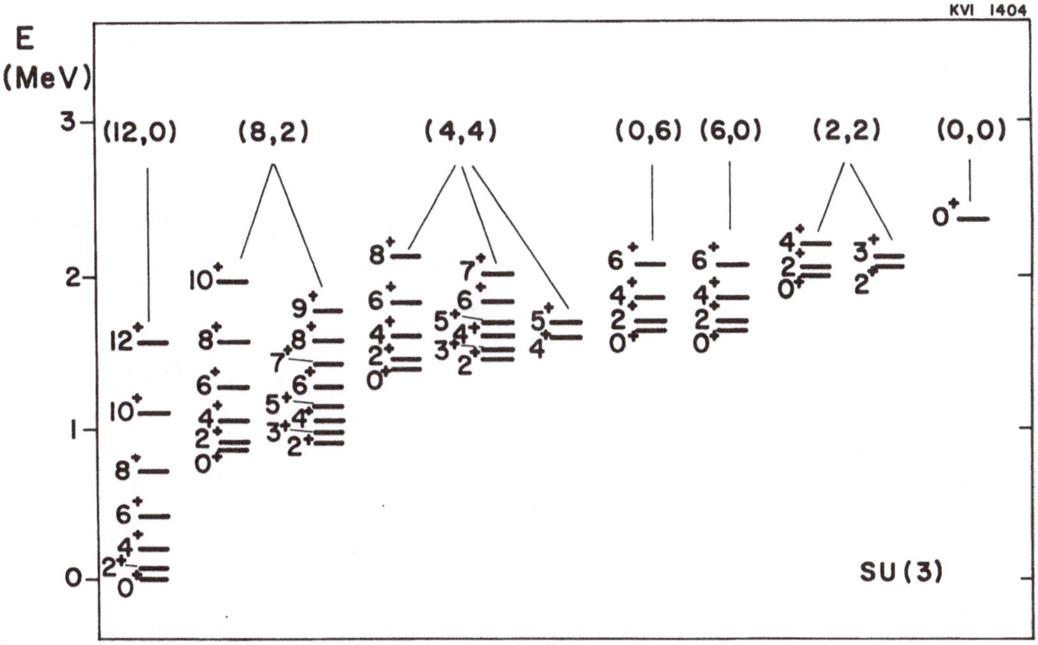

Fig. 4 A typical spectrum with SU(3) symmetry and N = 6. In parenthesis are the values of λ and μ

Dynamical Symmetry. III

The group chain is

$$U(6) \supset O(6) \supset O(5) \supset O(3) \supset O(2). \tag{5-134}$$

This symmetry corresponds to the vanishing of the coefficients ε, α, δ in (5-127). The corresponding Hamiltonian is

$$H = \beta C_{205} + \gamma C_{203} + \eta C_{206}. \tag{5-135}$$

The expectation value of H in the representation $|[N]\sigma\tau\nu_\Delta LM\rangle$ is

$$\langle H\rangle = \beta 2\tau(\tau+3) + \gamma 2L(L+1) + \eta 2\sigma(\sigma+4). \tag{5-136}$$

The corresponding structure of the spectrum is shown in Fig. 5.

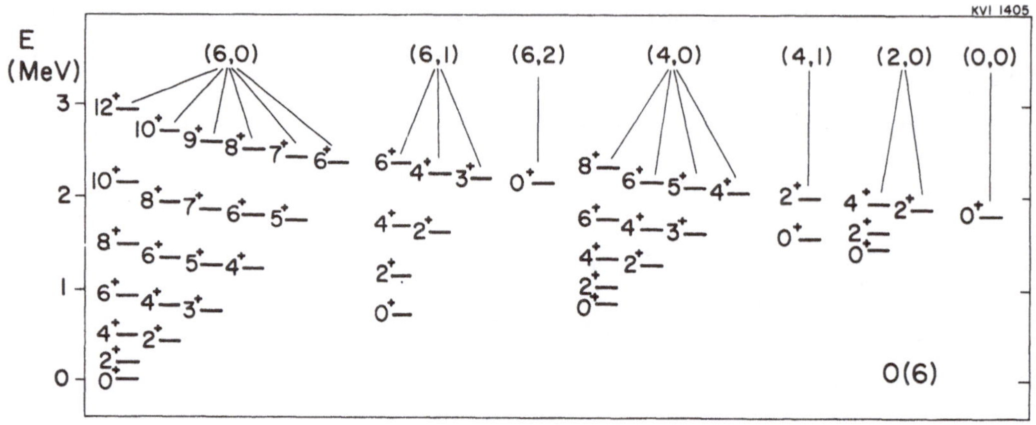

Fig. 5 A typical spectrum with O(6) symmetry and N = 6. In parenthesis are the values of σ and ν_{Δ}

5.2 Examples of Spectra with Dynamical Symmetries

There appear to be several nuclei whose spectrum can be well described by one of the limiting cases discussed in Sect. 5.1. Examples of these spectra are the following.

Dynamical Symmetry. I

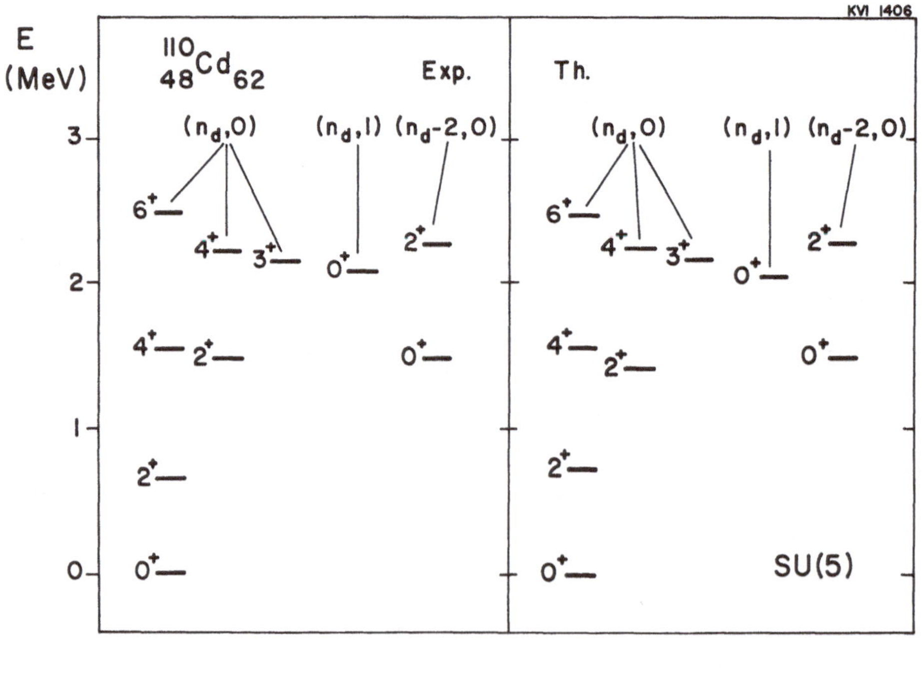

Fig. 6 An example of a spectrum with SU(5) symmetry: $^{110}_{48}Cd_{62}$.

Dynamical Symmetry. II

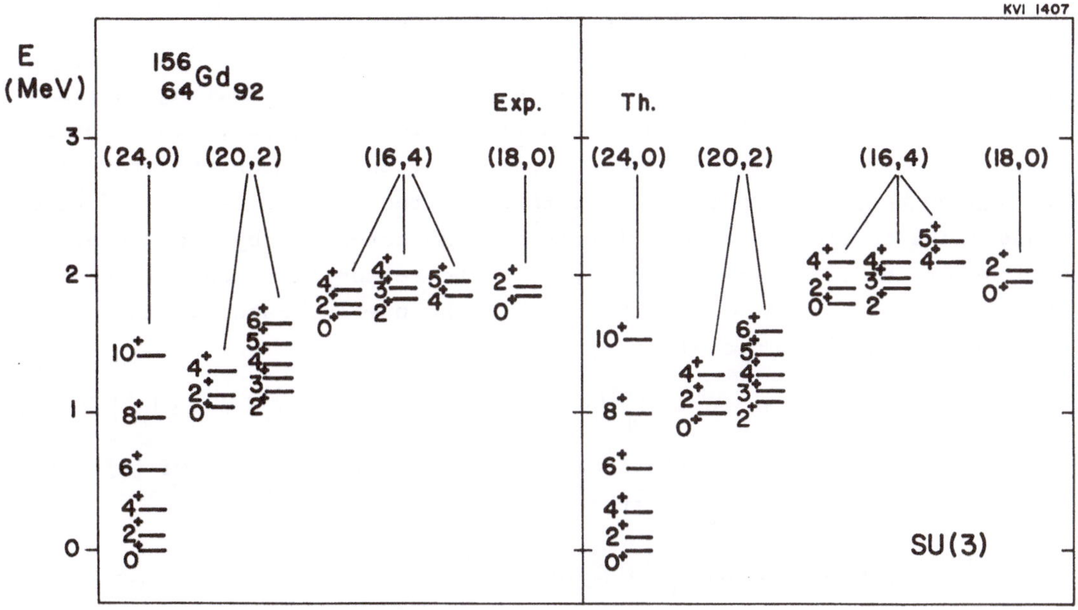

Fig. 7 An example of a spectrum with SU(3) symmetry: $^{156}_{64}Gd_{92}$.

Dynamical Symmetry. III

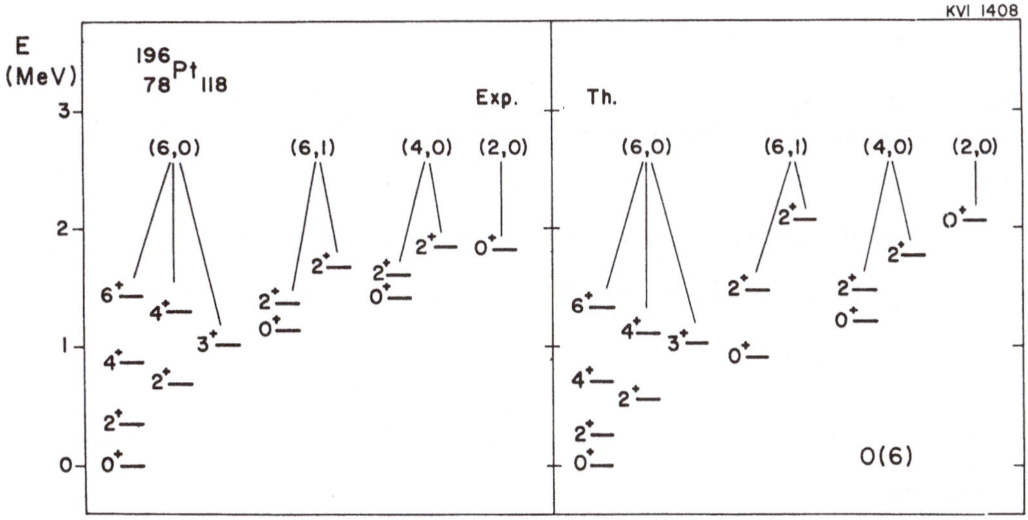

Fig. 8 An example of a spectrum with O(6) symmetry: $^{196}_{78}Pt_{118}$.

5.3 Selection Rules. Matrix Elements of Operators

We have discussed above the classification of states according to the irreducible representations of a chain of subgroups $G \supset G' \supset \ldots$. In addition to providing a basis in which the Hamiltonian H is diagonal, the chain of subgroups $G \supset G' \supset \ldots$ is also useful in evaluating matrix elements of operators T, such as the transition operators. The presence of a dynamical symmetry will make itself manifest through the presence of selection rules. An example of this, is the case of electric dipole transitions in atoms. If the levels are characterized by the representations of the rotation group $|L\rangle, |L'\rangle$, the selection rules of the dipole operator $\vec{D} = e\vec{x}$, are $\Delta L = \pm 1, 0$ $(0 \neq 0)$. In the present case, we must consider the transition operators

$$T_m^{(\ell)} = \alpha_2 \delta_{\ell 2} [d^+ \times s + s^+ \times \tilde{d}]_m^{(2)} + \beta_\ell [d^+ \times \tilde{d}]_m^{(\ell)} + \gamma_0 \delta_{\ell 0} \delta_{m 0} [s^+ \times s]_0^{(0)} . \quad (5\text{-}137)$$

These operators are all built from the generators of U(6). Of particular importance is the E2 operator

$$T_m^{(2)} = \alpha_2 [d^+ \times s + s^+ \times \tilde{d}]_m^{(2)} + \beta_2 [d^+ \times \tilde{d}]_m^{(2)} . \quad (5\text{-}138)$$

Matrix elements of this operator between the basis states of Sect. 4.3 can be evaluated using group theoretical methods. Complete expressions are given in the references. In the following, I will quote only their selection rules and some selected B(E2) values, which can be obtained from these matrix elements in the usual way

$$B(E2;\ J_i \to J_f) = \frac{1}{2J_i + 1} |\langle J_f \| T^{(2)} \| J_i \rangle|^2 . \quad (5\text{-}139)$$

Dynamical Symmetry. I

The selection rules of the operator (5-138) are

$$\Delta n_d = \pm 1,\ 0. \quad (5\text{-}140)$$

The B(E2) values along the ground state band are given by

$$B(E2;\ n_d + 1,\ v = n_d + 1,\ n_\Delta = 0,\ L' = 2n_d + 2 \to n_d,\ v = n_d,\ n_\Delta = 0,\ L = 2n_d)$$

$$= \alpha_2^2 \left(\frac{L+2}{2}\right) \left(\frac{2N - L}{2}\right) \quad (5\text{-}141)$$

Thus $B(E2;\ 2 \to 0) = \alpha_2^2 N$, in SU(5). (5-142)

Dynamical Symmetry. II

In this case it is more convenient to rewrite the E2 operator as

$$T_m^{(2)} = \alpha_2 Q_m^{(2)} + \alpha_2' Q_m'^{(2)} , \quad (5\text{-}143)$$

where

$$Q_m^{(2)} = (d^+ \times s + s^+ \times \tilde{d})_m^{(2)} - \tfrac{1}{2}\sqrt{7}(d^+ \times \tilde{d})_m^{(2)}, \qquad (5\text{-}144)$$

$$Q'^{(2)}_m = (d^+ \times \tilde{d})_m^{(2)} .$$

It turns out that the first term in (5-144) is by far the dominant term in the region where the symmetry II applies. The selection rules of this term are

$$\Delta\lambda = 0, \ \Delta\mu = 0, \qquad (5\text{-}145)$$

since $Q^{(2)}$ is a generator of SU(3) and thus cannot connect different SU(3) representations. The B(E2) values along the ground state are given by

$$B(E2;\ (\lambda = 2N,\ \mu = 0),\ \chi = 0,\ L' = L + 2 \to (\lambda = 2N,\ \mu = 0),\ \chi = 0,\ L) \qquad (5\text{-}146)$$

$$= \alpha_2^2\ \frac{3}{4}\ \frac{(L + 2)(L + 1)}{(2L+3)(2L+5)}\ (2N - L)(2N + L + 3).$$

Thus,

$$B(E2;\ 2 \to 0) = \alpha_2^2\ \tfrac{1}{5}N(2N + 3),\ \text{in SU(3)}. \qquad (5\text{-}147)$$

Comparing (5-142) with (5-147) one sees a change from a N to a N^2 dependence when going from SU(5) to SU(3). The N^2 dependence in SU(3) is responsible for the large B(E2) values observed in the middle of the shells, where the symmetry II applies.

Dynamical Symmetry. III

It turns out that the first term in (5-138) is the dominant term in the transition operators in the regions where the symmetry III applies. This term has the selection rules

$$\Delta\sigma = 0, \ \Delta\tau = \pm 1. \qquad (5\text{-}148)$$

The B(E2) values along the ground state band are given by

$$B(E2;\ \sigma = N,\ \tau + 1,\ \nu_\Delta = 0,\ L' = 2\tau + 2 \to \sigma = N,\ \tau,\ \nu_\Delta = 0,\ L = 2\tau)$$

$$= \alpha_2^2\ \frac{L + 2}{2(L+5)}\ \frac{1}{4}(2N - L)(2N + L + 8). \qquad (5\text{-}149)$$

Thus,

$$B(E2;\ 2 \to 0) = \alpha_2^2\ \frac{1}{5}\ N(N + 4). \qquad (5\text{-}150)$$

The calculated B(E2) values may be tested against experiment. An example for the symmetry III is shown in Fig. 9. Here the numbers in parenthesis are the O(6) quantum numbers $(\sigma\tau\nu_\Delta)$. The upper (lower) number on the transition arrow is the measured (predicted) relative

B(E2) value. The letter F indicates transitions which are forbidden in O(6) but whose branching becomes the dominant one whenever a small perturbation is added.

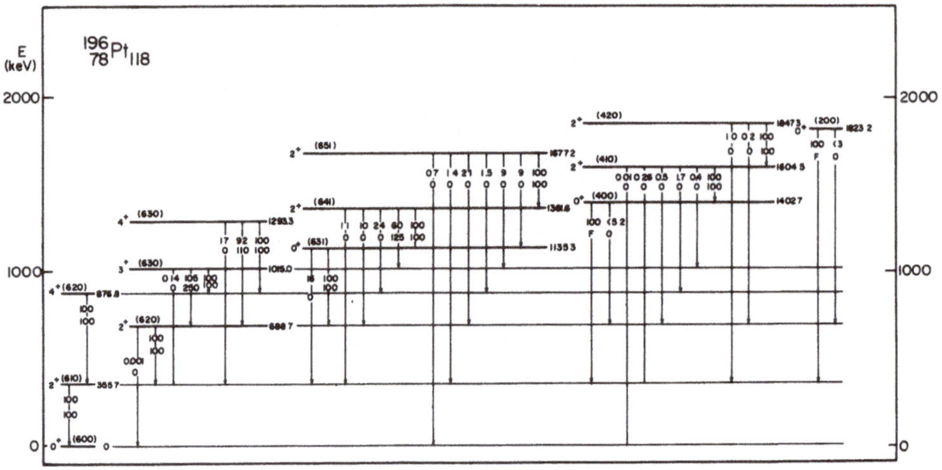

Fig. 9 Branching ratios for the decay of the positive-parity states in ^{196}Pt, from [13]

5.4 Group Lattices. Broken Symmetries and Phase Transitions

The dynamical symmetries discussed above are important because they provide simple, limiting cases to which the experimental results may be compared. In general, one must return to the Hamiltonian (5-94) and diagonalize it. However, here again, group theory may be useful in analyzing the salient features of the various transitional classes which may occur. Since the group structure is that given by (5-111) there are four possible classes of transitional nuclei: A) between I and II; B) between II and III; C) between III and I and D) a mixture of all three limits I, II and III. I will now discuss the main features of the transitional classes A and B.

Transitional Class A

This class can be studied by considering a mixture of Casimir operators of both groups I and II

$$H = \varepsilon C_{1U5} + \gamma C_{2O3} + \delta C_{2SU3} . \qquad (5\text{-}151)$$

The coefficients ε, γ and δ may then be expanded around some point $N = N_0$

$$\varepsilon(N) = \varepsilon(N_0) + \left.\frac{\partial\varepsilon}{\partial N}\right|_{N=N_0} (N - N_0) + \ldots,$$

$$\gamma(N) = \gamma(N_0) + \left.\frac{\partial\gamma}{\partial N}\right|_{N=N_0} (N - N_0) + \ldots, \qquad (5\text{-}152)$$

$$\delta(N) = \delta(N_0) + \left.\frac{\partial\delta}{\partial N}\right|_{N=N_0} (N - N_0) + \ldots .$$

when ε is large compared with γ and δ, the wavefunctions are those of the symmetry I, when ε is small the wavefunctions are those of the symmetry II. The simplest way to study this transitional class is to keep γ and δ constant and to let ε vary linearly with N

$$\varepsilon = \varepsilon_0 - \varepsilon_1 N. \qquad (5\text{-}153)$$

The corresponding spectra have the properties shown in Fig. 10 [11].

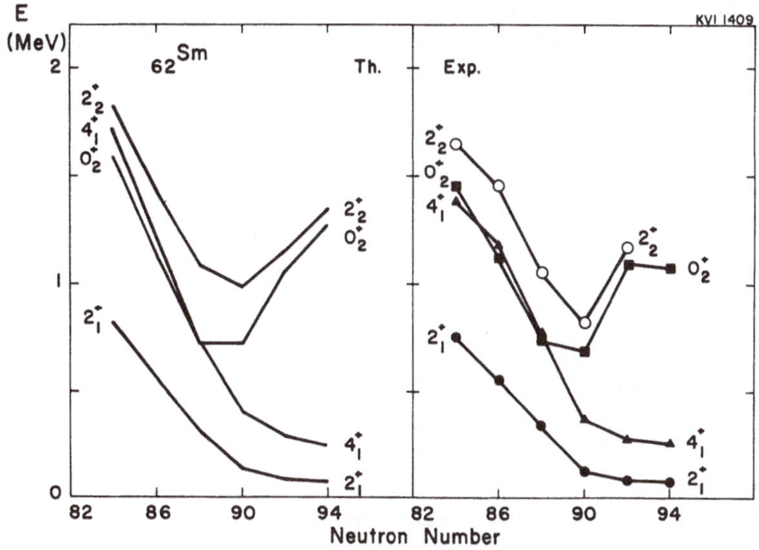

Fig. 10 Typical features of the transitional class A. Energies

Similarly, several typical electromagnetic transition rates change, as shown in Fig. 11.

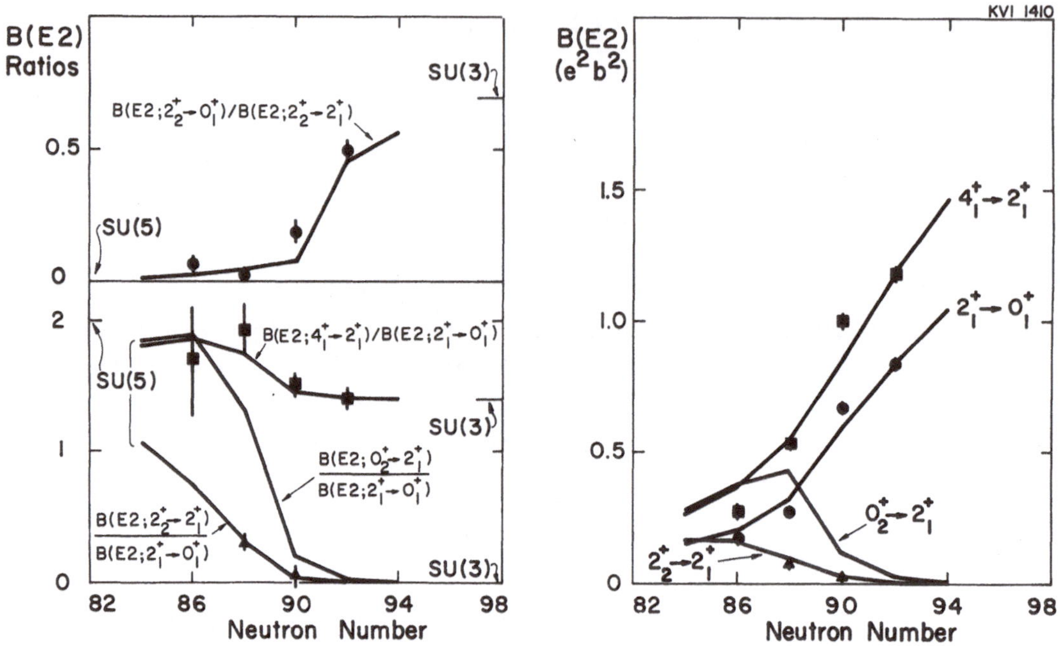

Fig. 11 Typical features of the transitional class A. Electromagnetic transition rates.

Of particular importance is the ratio

$$R = \frac{B(E2;\ 2^+_2\ 0^+_1)}{B(E2;\ 2^+_2\ 2^+_1)} \tag{5-154}$$

which is $R = 0$ in I,

$$R = \frac{7}{10} \text{ in II.} \tag{5-155}$$

Transitional Class B

This class can be studied by considering a mixture of Casimir operators of both groups II and III,

$$H = \beta C_{2O5} + \gamma C_{2O3} + \eta C_{2O6} + \delta C_{2SU3}. \tag{5-156}$$

Again, the coefficients β, γ, η and δ may be expanded around some point $N = N_0$

$$\beta(N) = \beta(N_0) + \left.\frac{\partial\beta}{\partial N}\right|_{N=N_0} (N - N_0) + \ldots,$$

$$\gamma(N) = \gamma(N_0) + \left.\frac{\partial\gamma}{\partial N}\right|_{N=N_0} (N - N_0) + \ldots,$$

$$\eta(N) = \eta(N_0) + \left.\frac{\partial\eta}{\partial N}\right|_{N=N_0} (N - N_0) + \ldots, \qquad (5\text{-}157)$$

$$\delta(N) = \delta(N_0) + \left.\frac{\partial\delta}{\partial N}\right|_{N=N_0} (N - N_0) + \ldots .$$

when δ is small the wavefunctions are those of the symmetry III, while when δ is large they are those of the symmetry type II. The simplest way to study this transitional class is to keep β, γ and η constant and to let δ vary linearly with N

$$\delta = \delta_0 + \delta_1 N. \qquad (5\text{-}158)$$

The resulting spectra have the properties shown in Fig. 12. The experimental spectra of nuclei in the Pt-Os region display this pattern [14].

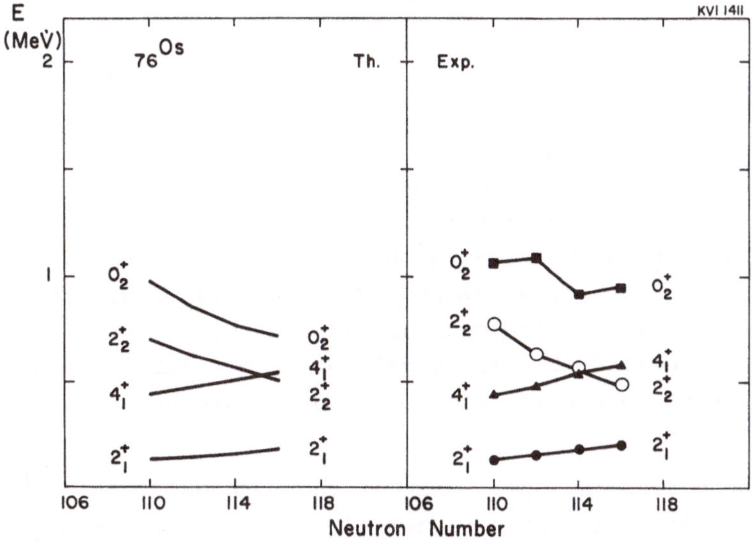

Fig. 12 Typical features of the transitional class B. Energies.

Similarly, electromagnetic transition rates change in a characteristic way from one limit to the other, as shown in Fig. 13.

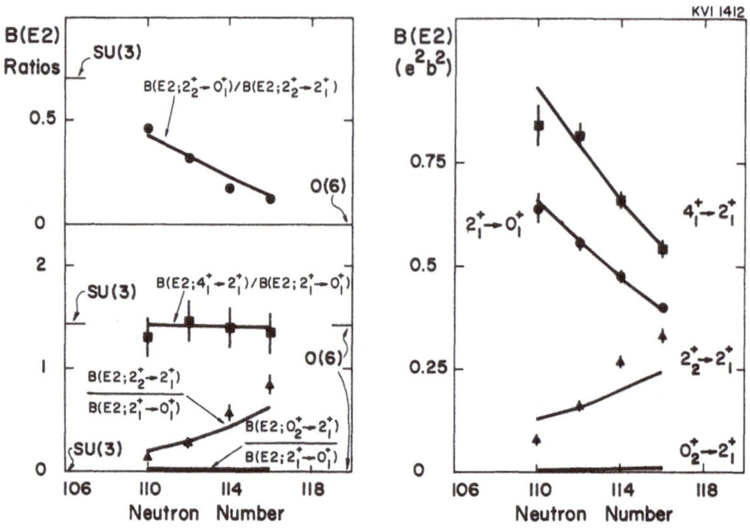

Fig. 13 Typical features of the transitional class B. Electromagnetic transition rates.

Of particular importance is the ratio R (see (5-154)) which is

$$R = 0 \text{ in III}, \qquad R = \frac{7}{10} \text{ in II.} \tag{5-159}$$

It is interesting to note that the change from one symmetry type to another is the equivalent, in finite systems, of the phase transitions which occur in infinite systems ($N \to \infty$). Phase transitions in finite systems described by a dynamical group SU(n) can be studied by means of the techniques recently developed by GILMORE [15,16]. The interacting boson model provides an example of application of these techniques [17]. The set of group chains (5-111) is called a group lattice and the transitions correspond to a shift from one side to another of the lattice.

6. Conclusions

I have presented here an application of the methods of group theory to the study of collective states in nuclei. Because of the large variety of observed spectra, with examples of both exact and broken symmetries, this application is one of the most complete examples encountered in physics so far, and thus it assumes particular importance both from the point of view of nuclear physics and of group theory. I hope that in the course of these lectures I have been able to give you at least a flavor of the underlying formalism and I would suggest the interested reader go back to the original references for a more detailed and complete exposition of this subject matter.

References

1. M. Hamermesh, "Group Theory", Addison-Wesley Publ. Co., Reading, Mass., 1962.
2. E.P. Wigner, "Group Theory and Its Application to the Quantum Mechanics of Atomic Spectra", Academic Press, New York, 1959.
3. H.J. Lipkin, "Lie groups for pedestrians", North-Holland Publ. Co., Amsterdam, 1966.
4. A. Bohr and B. Mottelson, Nuclear Structure Vol. II, (Benjamin, 1975) p. 677.
5. V.S. Popov and A.M. Perelomov, Sov. J. Nucl. Phys. 5, 489 (1967).
6. C.O. Nwachuku and M.A. Rashid, J. Math. Phys. 18, 1387 (1977).
7. A. Arima and F. Iachello, Phys. Rev. Lett. 35, 1069 (1975).
8. A. Arima and F. Iachello, Ann. Phys. (NY) 99, 253 (1976).
9. A. Arima and F. Iachello, Ann. Phys. (NY) 111, 201 (1978).
10. A. Arima and F. Iachello, Phys. Rev. Lett. 40, 385 (1978) and to be published in Ann. of Phys. (NY).
11. O. Scholten, F. Iachello and A. Arima, Ann. Phys. (NY) 115, 325 (1978).
12. O. Castaños, E. Chacon, A. Frank and M. Moshinsky, J. Math. Phys. 20, 35 (1979).
13. J.A. Cizewski, R.F. Casten, G.J. Smith, M.L. Stelts, W.R. Kane, H.G. Börner and W.F. Davidson, Phys. Rev. Lett. 40, 167 (1978).
14. R.F. Casten and J.A. Cizewski, Nucl. Phys. A309, 477 (1978).
15. R. Gilmore, J. Math. Phys. 20, 891 (1979).
16. R. Gilmore and D.H. Feng, Phys. Lett. 85B, 155 (1979).
17. A.E.L. Dieperink, R. Gilmore, D.H. Feng, F. Iachello and O. Scholten, to be published.

Chapter VI

STATISTICAL SPECTROSCOPY

J. B. French*
University of Rochester
Rochester, N.Y. 14627

1. Introduction and Preview

Statistical nuclear physics is by no means a new field. From the 1930's we have the BOHR [1] of the compound nucleus, BETHE's [2] theory of level density, and the beginnings of the statistical theory of nuclear reactions by WEISSKOPF and EWING [3]. Detailed studies of slow neutron reactions and associated fluctuations began in the 1950's and there was also the discovery [4] of optical-model giant resonances. More recently there are statistical theories of heavy-ion reactions. In all of these topics, and in many others, statistical considerations, either explicit or implicit, have played a major role. There is nothing surprising about that because the complexities expected and found at the high excitations which obtain in these processes seem to call for statistical analysis. During the past several years however it has gradually become clear that "statistical" behavior extends even into the ground-state domain, that there are major connections with symmetries, and that much more detailed statistical analysis is appropriate. This is what we shall discuss, excluding however processes in which nuclear reaction mechanisms play a large part; in other words we restrict ourselves to the domain of *statistical spectroscopy*. We shall stress the principles involved, give some idea of the formal techniques used, and briefly discuss some applications.

Dealing with spectroscopy via matrix diagonalization amounts to exactly solving the equations of motion in the model space; RPA, Hartree-Fock and other such methods correspond to approximate solutions. But we can give up entirely the notion of solving equations of motion, and apply standard methods of statistical mechanics, adapted to the finite-dimensional direct-product (shell-model) spaces encountered in spectroscopy (the finite dimensionality may in fact represent no real restriction at all). In place of the partition function, $Z(\beta)$, it is better to calculate, as a function of the system parameters, the eigenvalue density, $I(E)$, which is its inverse Laplace transform and carries

*Supported in part by the U. S. Department of Energy.

therefore the same information; it is moreover directly measureable for parameter values relevant to the system. We can expect then, just as in conventional statistical mechanics, that the quantities of interest (transition strengths, expectation values, etc.) will follow via parametric differentiation on I(E) and will emerge moreover as explicit functions of the parameters (matrix elements, etc.) of the system. To the extent that that is true we shall avoid then the high-order nonlinearities which are generated by the process of matrix diagonalization (which in most cases of interest is not feasible anyway).

We shall begin with a separation of the density into a fluctuation-free ("locally smoothed") part $\langle I(E)\rangle$ and a fluctuation part. $I_{f\ell}(E)$, which represents the deviations from the smoothed part, may for some purposes be regarded as describing the "noise" in which the physical information of most interest is embedded.

$$I(E) = \langle I(E)\rangle + I_{f\ell}(E) \tag{6-1}$$

from the first part of which (the usual density, which we shall usually write simply as I(E)) we shall expect to derive binding energies, low-lying spectra, locally-averaged expectation values and strengths, and most of the other quantities of interest. This will turn out to be feasible because, in a many-particle asymptotic limit, through the action of a *central-limit theorem* (CLT) the smoothed (fluctuation-free) density for all realistic Hamiltonians takes on a characteristic form defined by a few moments, traces of low powers of H, in particular the centroid and variance. Secular deviations from the characteristic form can be calculated similarly. For this part of the density we shall of course be dealing with a specified Hamiltonian; the averaging, which is characteristic of statistical mechanics, will be *spectral* (or *energy*) averaging, along the spectrum of H.

Shell-model spaces admit many symmetries, the most important of which correspond to subgroups of the basic group U(N) of unitary transformations in the (N-dimensional) single-particle space. A partitioning of the model space according to irreducible representations of a U(N) subgroup, or chain of subgroups, gives both increased accuracy in the calculations and methods for studying the goodness of the symmetries. Partitioning of the model space gives rise to partitioning of the moment traces and hence to a representation, $I(E) = \sum_{\Gamma} I_{\Gamma}(E)$, of the density as a *linear* superposition of subspace densities (which will in general no longer correspond individually to eigenvalue densities). Linearity here does *not* imply that the subspaces are treated independently, which would be a very bad approximation; the interaction between subspaces shows up in lowest order in the fact that the variance for one distribution has contributions from excitations which connect that one with any other. We find indeed a hierarchical classification of the interactions, the maximum complexity depending on the order of the moments which are needed to fix the distributions.

The smoothed level density, and its decomposition by symmetries (configuration, isospin, angular momentum and others) are of major interest in themselves. Approximate binding energies and low-lying spectra derive from the density via correspondence between the smoothed distribution function and the "staircase" function which represents the exact distribution. Many calculations of these kinds have been made.

As we have said, other quantities come by parametric differentiation and related operations on the density function. For the expectation value of an operator[1] G

$$\langle E|G|E\rangle = \rho^{-1}(E)\langle G\delta(H-E)\rangle^{m} = -\rho^{-1}(E)\left(\frac{\partial F_{\alpha}(E)}{\partial\alpha}\right)_{\alpha=0} \qquad (6\text{-}2)$$

which derives from the response of the system under $H \rightarrow H + \alpha G$. Here F_α is the distribution function for the Hamiltonian $(H + \alpha G)$. A natural expansion of the intermediate form here in terms of the orthonormal polynomials $P_\mu(E)$, defined by the density as weight function so that $\int P_\mu(E)P_\nu(E)\rho(E)\,dE = \delta_{\mu\nu}$, gives a series which is strongly convergent "to within fluctuations",

$$\langle E|G|E\rangle = \sum_{\mu} \langle GP_{\mu}(H)\rangle^{m} P_{\mu}(E) \xrightarrow{\text{CLT}} \langle G\rangle^{m} + \langle G(H-\mathcal{E})\rangle^{m}\,\frac{(E-\mathcal{E})}{\sigma^{2}} \qquad (6\text{-}3)$$

In the last form $\mathcal{E} \equiv \langle H\rangle^{m}$ and $\sigma^{2} \equiv \langle (H-\mathcal{E})^{2}\rangle^{m}$ are the spectral centroid and variance and we have taken the CLT limit, which then gives linearity in the energy as the characteristic form for smoothed expectation values. One finds similarly that transition strengths are asymptotically bilinear in the energies (linear both in the starting and final energies). These results have been used in the calculation of occupancies, spin-cut-off factors and β-decay strengths, and in sum rules for electromagnetic transitions; modifications of them are being used for studying effective interactions. Strength and expectation-value fluctuations are treated via ensemble averaging which leads to

---

[1] Notation: For model spaces of finite dimensional d(m), we shall write $I(E) = d(m) \times \rho(E)$, so that $\int \rho(x)dx = 1$. The distribution function is then $F(x) = \int^{x} \rho(z)dz$. For an arbitrary operator G we shall write $Tr(G) = \langle\langle G\rangle\rangle^{m} = d(m) \times \langle G\rangle^{m}$.

the Porter-Thomas distribution and various extensions of it. Finite-response results for the density are also of interest, while infinitesimal response for the ensemble average of the product of two densities leads very directly to a theory for the two-point fluctuations.

The variance of the eigenvalue distribution defines a norm (for traceless operators) and hence a unitary geometry in which operator magnitudes are expressed in terms of the standard unitary norm $||G||$ given by

$$||G||^2 = d^{-1} T_r^{(m)} (G^+G) = d^{-1} \sum \lambda_i^2 \tag{6-4}$$

where λ_i are the eigenvalues of G if $G = G^+$, and in any case the λ_i^2 are the eigenvalues of G^+G. The effectiveness of this geometry which gives precise definitions of orthogonality, projection (e.g. how much $Q \cdot Q$ is contained in H acting in a given space) and so forth, is then guaranteed by the CLT. The linear form in (6-3) then simply transcribes into the energy domain a model-space scalar product, or correlation coefficient, of the two operators G and (H-E). Many of the other forms encountered in statistical spectroscopy have similar direct geometrical interpretations.

Since all of the relevant information in the spectroscopic space is expressed in terms of many-particle traces, the question arises of how such traces may be calculated. For the simplest possible case, that of a k-body operator in an unpartitioned m-particle space, there is an elementary result that $\langle G(k) \rangle^m = \binom{m}{k} \langle G(k) \rangle^k$, so that the same combination of defining matrix elements which gives the eigenvalue centroid in the defining or "input" space does the same for all particle number m; in other words these traces "propagate". Trace evaluation in general then requires the extension of this result to operators of mixed particle rank, such as $(G(k))^p$, and to partitioned spaces. More or less straightforward methods are available for low powers (say $p \leq 4$), and similar products, of two-body operators, and with rather elegant methods one can go a little further than that. As long as the subspaces are defined in terms of group representations a factoring analogous to that in the elementary result obtains and propagates the input information to the parts of the model space where it is needed. The propagators themselves are traces of operators which, by construction, are scalars with respect to the subgroup (or the smallest subgroup in the case of a subgroup chain) so that the problem of explicitly constructing them reduces to a much studied group-theoretical problem, that of cataloging and constructing the polynomial invariants.

In some cases this is a simple operation and then there is a simple propagation of the traces; in other cases this is not so. But in all cases these appear to be physically significant approximations (for "dilute" systems for example).

It should be clear by now that our major interest will not be with *highly accurate* calculations of energy levels, transition strengths and so forth. We shall be more concerned with general questions, for example about the "information content" of spectra (which in principle at least sets a limit on the accuracy which it is worthwhile trying to achieve), about those features of the Hamiltonian which generate collectivities and other special phenomena, and with the goodness of symmetries. We shall however not ignore these specific questions which, after all, make the closest contact with experimental data.

We have stressed that we work with shell-model spaces. We do not explicitly consider continuum states though that should be feasible, and in any case we can deal with high excitations by implicitly invoking the Wigner-Eisenbud separation. With regard to conventional shell-model procedures we note first that: they are restricted to "small" model spaces which often (as with E2 collectivity) do not fully "contain" many of the interesting phenomena: that because of their highly non-linear nature they do not give the interesting results as explicit functions of the system parameters; that they are inelegant and, in a sense wrong in principle in their insistence on generating many-particle wave functions with all their many-particle correlations, even though the high-order correlations are not at all trustworthy and the things measured or needed for understanding involve only low-order density matrices. On the other hand shell-model calculations sometimes settle problems more quickly and more decisively than other methods and are correspondingly important both to theorists and to analysts of the data. It is a remarkable fact that statistical methods can be successfully applied in the ground-state domain (and that certain experimental results measured at high excitation and regarded as "of statistical origin" have their counterparts near the ground state); but it is not really certain that, with the statistical techniques available, we can really capture *all* of the significant information which resides in low-lying phenomena. Though we shall describe all phenomena in statistical terms it might eventually be feasible and useful to treat a part of the space, including the ground-state domain, in microscopic detail and the rest of it statistically. We would then have a division of the entire many-particle Hilbert space into three parts; for the theory of effective interactions (to which it appears that statistical procedures may be able to make significant contributions) already partitions it into two and gives us a permit for ignoring one of them.

The presence in (6-1) of the fluctuation density imposes a limit on the accuracy with which the smoothed quantities may be determined and, if for no other reason than that, we must study its properties. But beyond that, detailed spectra, spanning long runs of levels, are experimentally measureable by slow-neutron reactions in heavier nuclei [5], and by proton reactions in lighter ones [6] so that we can study the fluctuations in detail. $I_{f\ell}(E)$ cannot however be treated by any kind of spectral averaging, nor indeed calculated in non-trivial cases for a specified Hamiltonian (it is described in terms of high moments for whose calculation no spectral techniques are available). Instead we deal with it by introducing an ensemble of H's, WIGNER [7], and then averaging across the ensemble instead of along the spectrum.

In conventional statistical mechanics it is essential that the ensemble be ergodic so that the results of ensemble averaging can be taken as identical with the results of time averaging (along the phase-space orbit of the system). Similarly here the equivalence of the fluctuation results with those which would emerge from spectral averaging (if that could be done) requires an ergodic behavior of the ensemble. This behavior which has in the past been observed in Monte-Carlo calculations has recently been analytically derived [8] for the "standard" ensembles. The ergodicity which obtains is very strong, so that the ensemble results are relevant to averaging over only a small segment of the spectrum, exactly what is called for in applications. Besides that it turns out that when the spectra for a standard ensemble are mapped onto spectra with constant locally-averaged level density they are stationary, so that properly defined fluctuation measures, are the same all over the spectrum, a result for which there is some experimental evidence. It would be hasty however, to argue that there is nothing special about the ground-state region; for the standard ensembles pay no attention to model symmetries which are more liable to be good at low excitations and then have a large effect on the level repulsion which dominates all the fluctuation properties.

2. Some Simple Eigenvalue Distributions

We must first understand something about the original theory of level density, BETHE [2], which, though inadequate for our present purposes, has served as a basis for all but very recent work in that domain. Bethe ignored interactions between particles, as well as the Pauli blocking effects, and dealt therefore with an ideal gas of fermions for which everything is fixed by the single-particle spectrum. Let us start with a special case, that of equal unit spacings with no degeneracy. Suppose we have m active particles. Then, for the level density at energy E,

$$I_m(E) = \text{no. of solutions of } E = \sum_s s\, m_s \equiv \sum_s{}' s; \quad m = \sum_s m_s = \sum_s{}' 1$$

$= \text{no. of } \textit{unordered} \text{ partitions of E into m unequal parts}$
$= \text{no. of unordered partitions of } E^* = E - \binom{m+1}{2} \text{ into } (1,2,\ldots,m) \text{ parts}$
$= p_m(E^*)$ (6-5)

where we have written $\sum'$ as the sum over occupied states and, in the last step, have used a simple Eulerian theorem about partitions [9]. It will be seen that E* is the excitation energy. Then for $E^* \ll m$ we are dealing with a highly degenerate fermion gas, for which $p_m(E^*)$ is independent of m, so that

$$p_m(E^*) \to p(E^*) = \text{no. of unordered partitions of } E^*$$

$$\xrightarrow{E^* \gg 1} \frac{1}{\sqrt{48}\, E^*} \exp\left(\pi\sqrt{\frac{2E^*}{3}}\right) \qquad (6\text{-}6)$$

where the last step uses the asymptotic Hardy-Ramanujan formula for the number of partitions. A scale change to a single-particle density $g(\varepsilon)$ gives a more familiar result with the exponential as $\exp 2\sqrt{aE^*}$ and $a = \pi^2/6g$. More important than that, because most of the "activity" arises from s.p. states near the Fermi surface, the many-particle density is essentially proportional to the s.p. density in that region, so that, with more general s.p. densities, we still find the same form with $a = \frac{\pi^2}{6}\, g(\varepsilon_F)$.

In the same paper Bethe used the central limit theorem to decompose the level density according to angular momentum. Regarding the z-component of the i'th-particle angular momentum as a random variable whose distribution centers about zero with width $\sigma^2(1) = 1/3\ \overline{j(j+1)}$ (where the bar denotes an appropriate average), we have that J_z is a sum of similar random variables, which are independent if we can ignore Pauli blocking effects, as we assume we can. Then by the elementary CLT, J_z becomes a Gaussian variable when $m \gg 1$; its centroid will also be zero and its variance $m\sigma^2(1)$. Then J_z^2 is a $\chi^2(1)$ variable, and $J^2 = (J_x^2 + J_y^2 + J_z^2)$ is $\chi^2(3)$, which gives for J (or better for $(J+\frac{1}{2})$) essentially the Maxwellian distribution $\sim x^2 \exp\{-x^2/2\sigma^2(m)\}$ with $\sigma^2(m) = m\sigma^2(1)$; the latter result follows by considering $M_J = J$ and $(J-1)$ and subtracting.

The CLT used here combines the elementary fact that the density function for a sum of independent random variables is the convolution of the separate densities, with the theorem, whose genesis we see below, that the m-fold convolution of an integrable function $\rho_1(x)$ with itself approaches Gaussian[2] for large m. Specifically, defining $\rho_1 \otimes \rho_1[x] = \int dz\ \rho_1(z)\rho_1(x-z)$ and assuming that $\int \rho_1(x) = 1$, we have

$$\rho_m(x) = \rho_1 \otimes \rho_1 \otimes \ldots \otimes \rho_1[x] \xrightarrow{m\to\infty} (2\pi\sigma_m^2)^{-\frac{1}{2}} \exp[-(x-E_m)^2/2\sigma_m^2] \qquad (6\text{-}7)$$

with the centroid and variance given by $E_m = mE_1$, $\sigma_m^2 = m\sigma_1^2$. Bethe used (6-7) for the J_z distribution, but it was not applicable for the distribution in energy since his single-particle spectrum was unbounded; we see indeed that it gives a very different form. If however we truncate the s.p. spectrum, considering only $\varepsilon_1, \varepsilon_2, \ldots, \varepsilon_N$, we can apply the CLT directly to the energy spectrum also, at least as long as we can still ignore the blocking restriction; that would require that $m \ll N$. Then (6-7) gives the m-particle density[3] for large m, with

$$E_1 = N^{-1} \sum \varepsilon_i \ ; \qquad \sigma_1^2 = N^{-1} \sum (\varepsilon_i - E_1)^2 \qquad (6\text{-}8)$$

Before proceeding we must stress that, with a single-particle truncation, the resultant many-particle states for the most part (say all the states above 10 MeV excitation in $(ds)^{12}$, whose complete spectrum spans ∿100 MeV) can have only a tenuous relationship with the physical states, which necessarily involve-single-particle excitations outside the truncated s.p. spectrum. This feature, which of course

---

[2]The class of allowable functions $\rho_1(x)$ is extremely wide and includes the integrable functions which vanish outside a finite domain, which are all we shall encounter. It should be obvious from this that, for finite m, since $\rho_m(x)$ is then also similarly restricted, the deviations from Gaussian will be most apparent for large $|x|$. If $\rho_1(x)$ is discrete so also is ρ_m. There is a special class of "singular" bounded discrete spectra whose multiple convolution does not lead to Gaussian, as we shall see below. For future reference we remark also that the functions which we shall encounter are uniquely specified by their moments.

[3]Since $\rho_1(x)$ has unit integral, so also does $\rho_m(x)$. Because of the antisymmetry $d(m) = \binom{N}{m}$ instead of N^m. The blocking effects have usually only a small effect on the shape but they do affect the variance which becomes $m(N-m)(N-1)^{-1}\sigma^2(1)$.

is present in shell-model calculations as well, may be ignored for some purposes, but for others we shall deal with it by partitioning the space according to a symmetry, which at the same time enables us to extend the statistical methods to very large spaces and to study the symmetry involved as well. We shall see below a minor example of such a subspace distribution.

We have now to make major extensions of the above results, to take account of blocking effects, interactions between particles, and symmetries, to learn about the rate of convergence, and to give a good account of phenomena at low energies as well as high. The combinatorial methods, for unbounded s.p. spectra, are no longer applicable, nor is the method of convolutions (since the energies are not additive because of interactions, and the random variables are not independent because of blocking). But, since moment methods are quite feasible for interacting particles, we consider again our simple problem, m particles over N s.p. states, in that way.

In a single-particle basis, in which H is diagonal with ε_i the s.p. energy,

$$H = \sum_{i=1}^{N} \varepsilon_i n_i \tag{6-9}$$

Using the elementary result mentioned above we see that $\langle H\rangle^m = m\langle H\rangle^1 = \frac{m}{N}\sum \varepsilon_i$. Then, choosing the energy zero so that $\sum \varepsilon_i = 0$, we see that the m-particle centroid is also zero, and then the m-particle moments are central moments, $M_p(m) = \mathcal{M}_p(m)$. For the p'th power of H observe that:

(1) $n_i^2 = n_i = n_i^3 = n_i^4 \ldots$

(2) $n_i n_j$ ($i \neq j$) is a 2-body operator and $n_i n_j n_k \ldots n_t$ in which ℓ *different* operators appear is ℓ-body (e.g. $n_1 n_2 n_1 n_3 n_5 n_2 = n_1 n_2 n_3 n_5$ and is 4-body). Then, if ℓ is the number of different operators in the product

$$\langle n_i n_j n_k \ldots n_t\rangle^m = \binom{m}{\ell}\binom{N}{\ell}^{-1} \tag{6-10}$$

where the N dependence follows from the fact that $\langle n_i n_j n_k \ldots n\rangle^N = 1$, and the m-dependence from the elementary result.

Now we multiply out H^p and arrange the terms according to the (unordered) partitions of p into non-negative integers, $[p_1, p_2, \ldots, p_\ell]$ being associated with

$$\sum_{i_1 \neq i_2 \neq \ldots \neq i_\ell} \varepsilon_{i_1}^{p_1} \varepsilon_{i_2}^{p_2} \ldots \varepsilon_{i_\ell}^{p_\ell} \equiv N^\ell \langle \varepsilon_\wedge^{p_1} \varepsilon_\wedge^{p_2} \ldots {}_\wedge\varepsilon^{p_\ell}\rangle \tag{6-11}$$

where in the last form we have introduced a notation for the sum of correlated products. The number of ways of forming a partition $\pi(p) = [p] = [p_1 \dots p_\ell]$ is seen to be

$$d_\pi = \binom{p}{p_1}\binom{p-p_1}{p_2}\dots\binom{p-p_1-p_2\dots p_{\ell-1}}{p_\ell}[s_1!s_2!\dots s\ !]^{-1}$$

$$= \frac{p!}{\pi(p_i!)\pi(s_k!)} \tag{6-12}$$

where s_k is the number of the p_j's which are equal to k. For example with $p = 11$, $[p] = [332111]$, $s_1 = 3$, $s_2 = 1$, $s_3 = 2$, $s_{r>3} = 0$ and $d_\pi = 46{,}200$. We have now for the p'th moment

$$M_p(m) = \sum_{\pi(p)} d_\pi \langle \varepsilon^{p_1} \wedge \varepsilon^{p_2} \wedge \dots \wedge \varepsilon^{p_\ell} \rangle\ N^\ell \binom{m}{\ell}\binom{N}{\ell}^{-1} \tag{6-13}$$

an exact result since we have not yet ignored the Pauli effects. If we do so by taking the large-N limit[4] we see that in the summation of (6-11) the index restrictions may be ignored, so that the sum factors, whence

$$\langle \varepsilon^{p_1} \wedge \varepsilon^{p_2} \wedge \dots \wedge \varepsilon^{p_\ell} \rangle \longrightarrow \langle \varepsilon^{p_1} \rangle \langle \varepsilon^{p_2} \rangle \dots \langle \varepsilon^{p_\ell} \rangle + O(N^{-1}) \tag{6-14}$$

with $\langle \varepsilon^r \rangle = \langle H^r \rangle^1 = M_r(1)$, the r'th moment of the single-particle density. Since also $\binom{N}{\ell} \to N^\ell/\ell!$ we have that

$$M_p(m) \to \sum_{\pi(p)} d_\pi [\langle \varepsilon^{p_1} \rangle \langle \varepsilon^{p_2} \rangle \dots \langle \varepsilon^{p_\ell} \rangle] \ell! \binom{m}{\ell} \tag{6-15}$$

an alternative form of which follows by simple re-arrangement, or directly by using the theory of partition polynomials,

$$M_p(m) \to p! \sum_{\pi(p)} \left[\left\{\frac{\langle \varepsilon \rangle}{1!}\right\}^{s_1} \left\{\frac{\langle \varepsilon^2 \rangle}{2!}\right\}^{s_2} \dots \left\{\frac{\langle \varepsilon^p \rangle}{p!}\right\}^{s_p} \right] \frac{\ell}{\pi(s_i!)} \binom{m}{\ell} \tag{6-16}$$

Since $\ell!\binom{m}{\ell} \to m^\ell$ in the large-m limit (but always with $m \ll N$) the

---

[4] This is a loose manner of speaking since as N changes so must the single-particle spectrum and H itself. We are really referring to an expansion in inverse powers of N in which we keep only the leading term. The correlated products are the "augmented symmetric functions" of the statisticians; they may be expanded as homogeneous functions in the "power" sums $\langle \varepsilon^q \rangle$, the expansions being given up to order 6 by KENDALL and STUART [10]. Eq. (6-14) gives the first term in this expansion.

dominant contributions will come from the highest possible ℓ. But any partition containing 1's cannot contribute (since $\langle\varepsilon\rangle = 0$); thus for even p (=2ν) the dominant contribution for large m arises from the partition $[2^\nu]$ and gives $M_{2\nu}(m)=(2\nu-1)!!\,\{M_2(m)\}^\nu$. For the *reduced central moments*

$$\mu_{2\nu}(m) \xrightarrow{m\to\infty} (2\nu-1)!! \tag{6-17}$$

For odd $p = (2\nu+1)$, $[2^{\nu-1},3]$ is dominant for large m and gives

$$\mu_{2\nu+1}(m) \xrightarrow{m\to\infty} \frac{1}{3}\,\nu(2\nu+1)!!\,\mu_3(m)$$

$$= \frac{1}{3}\,\nu(2\nu+1)!!\langle\varepsilon^3\rangle/\{\langle\varepsilon^2\rangle\}^{3/2}\,m^{1/2} \tag{6-18}$$

We see that the odd moments disappear for large m and the distribution at the same time becomes Gaussian. Note the importance of *binary correlations* (as displayed by the dominant partitions of p in the moment structure).

More simply still we can argue that for large m (but[5] $p \ll m$) the separate H's in $M_p(m) = \langle H^p\rangle^m$ effectively act on different particles and all binary associations are permitted and contribute equally. Since the number of ways of making pairings among 2ν objects is (2ν-1)!!, $\mu_{2\nu} \to (2\nu-1)!!$ as above and we have Gaussian. The significance of the result is that, while we will not have convolutions for interacting particles, we may well have binary domination and hence Gaussian in that case also.

It is not enough to demonstrate Gaussian in the large-m limit. We must know about the rate and manner of convergence,which we can usually infer from the third and fourth moments. Besides that however, we must ask whether *every* (discrete) s.p. spectrum is asymptotically Gaussian and we seem to have found. Connected with the answer to this, which is in the negative, is a distinction between two classes of asymptotic ($d\to\infty$) spectra, and hence operators, between those which are *non-singular* (or *global*) and those which are *singular*. The statement about Gaussian is true for all non-singular operators but *not* for the

---

[5] For $p \gtrsim m$ the argument is obviously not correct; but such high moments are needed only for the discrete spectrum structure, whereas when we speak of Gaussian we are referring to the smoothed density. Note that $M_p(m)$ as given by (6-15) is the moment of a multiple convolution, as is easily verified, so that we have derived the elementary CLT used above and shown that it arises in that way. Eq. (6-13) in fact gives blocking corrections to it.

singular ones.

Implicitly we have taken for granted that the m-independent quantities $\{<\varepsilon^{p_1}><\varepsilon^{p_2}> .. <\varepsilon^{p_\ell}>\}$ are also d-independent[6], varying among themselves by factors which are small compared with d. This is equivalent to the assumption that the even-order reduced central moments[7] $\mu_{2\nu}$ are d-independent, which is satisfied for conventional spectra. The prototypical exception is a spectrum $\{1,0^{d-1}\}$ for which $\mu_{2\nu} \xrightarrow{d\to\infty} d^{\nu-1}$. More generally a spectrum is singular iff $\mu_{2\nu}(\nu \geqslant 2)$ is dominated by a few (<<d) states, which implies in general terms that a few states are removed from the great majority by an amount $\sim d \times \hat{\sigma}$ where $\hat{\sigma}$ is the width of the major group. For spectra of this kind it is easy to see that binary correlations are no longer dominant and spectra no longer Gaussian.

If we partition the s.p. states into "orbits" which we label as 1,2,...t, so that $N \Rightarrow \sum N_i$, $m \Rightarrow \sum m_i$, we decompose the density into a sum of configuration densities, each defined by a partition $\vec{m} = [m_1, m_2 ... m_t]$. For non-interacting particles the separate densities will (obviously!) also be Gaussian in the large m-limit. As an example Fig. 1 shows a (k-particle, k-hole) subset of the total density for

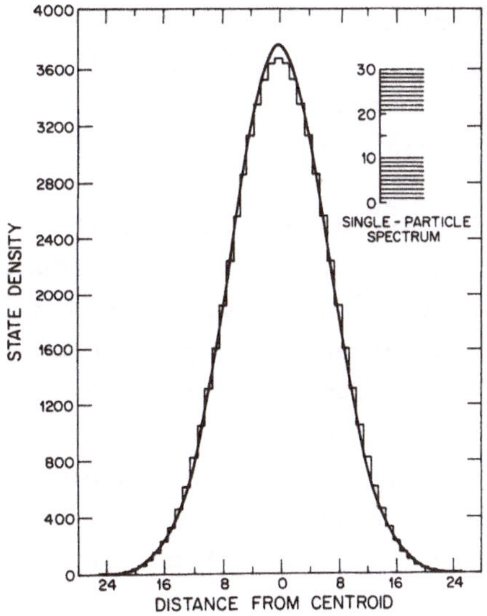

Fig.1 The exact 5 particle-5 hole state density, and its Gaussian approximation, for 10 non-interacting particles distributed over the single-particle spectrum indicated [11]. The single-particle degeneracies are unity, the total number of states is 63,504, σ=6.770, and the exact ground state comes at 25 units (=3.7σ) below the centroid.

---

[6]For the s.p. spectrum, d=N, but the argument works more generally than that.

[7]We encounter these because only the central moments are invariant under translations and, even for a singular spectrum, the odd central moments may vanish.

non-interacting particles. k=5 gives an excellent Gaussian, and this in fact is found also for the other subsets. Since, as we have said, the m-particle states form an irrep of a U(N) group and the subset does similarly for a direct-sum subgroup, we see here a simple case of a CLT acting within a sub-group space. Since the sum of all (kh-kp) states forms all the states, we have, in a sense, seen a decomposition of the CLT itself. In fact Bethe's CLT result for the J_z distribution derives in the same way since the angular momentum group involved is also a U(N) subgroup, and similarly and much more generally for other subgroups and subgroup structures. Obviously partitioning by irreps is going to be important for studying symmetries. It will be of major consequence too in increasing the accuracy of calculations and in clarifying the effects, e.g. on strength distributions, of various kinds of operator correlations.

We have stressed that convolution arguments don't work for interacting particles. But moment methods are valid in general, and for partitioned distributions as well as simpler ones. If the (smoothed) many-particle densities have a characteristic form described by a few parameters then we shall have to calculate only a few parameters and these will carry the significant information.

While all non-singular one-body operators have asymptotic Gaussian spectra this is not true at all for $(k > 1)$-body. For example $H=h^2$ where h is a non-singular traceless one-body operator will have a χ_1^2 asymptotic spectrum ($\sim x^{-1/2}e^{-x/2}$). The Hamiltonian $H = \sum_{\alpha=1}^{\nu} h_\alpha^2$, where every h_α is non-singular and traceless and $\langle h_\alpha h_\beta \rangle^1 = c\delta_{\alpha\beta}$, may be shown to have a χ_ν^2 spectrum for large m. It is a nice exercise to derive that result [12] which gives a χ_3^2 (Maxwellian) distribution for J^2, as found by Bethe, χ_5^2 for Q·Q, χ_8^2 for the SU(3) Casimir operator, etc. (the latter examples not being derivable, at least not easily, by Bethe's procedure). Since χ_ν^2 is the distribution for the sum of squares of ν independent identically distributed zero-centered Gaussian random variables, it follows by the CLT that for large ν, $\chi_\nu^2 \to$ Gaussian (in order to be indistinguishable by eye from Gaussian we need here $\nu \gtrsim \sim 30$) so that operators such as Q·Q (which are *not* however singular) have spectra considerably different from Gaussian.

It is nonetheless true, as illustrated in Fig. 2, that "standard" H's used in nuclear spectroscopy do generate close-to-Gaussian spectra. From our χ_ν^2 example we are tempted to argue that the essential feature of H for Gaussian is that it should decompose into a sum of a large number of "independent" terms of comparable magnitude; this is in fact the case. More formally, to see the origin of Gaussian behavior in

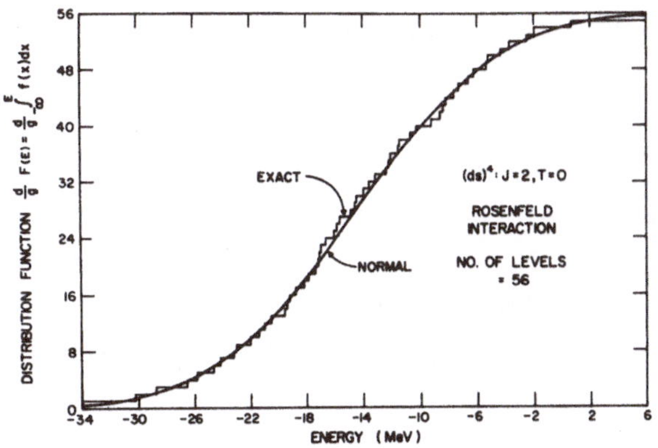

Fig.2 The exact distribution function and its Gaussian approximation for four interacting particles with J=2,T=0 in the ds shell [13]. The density here is written as f(x).

interacting-particle systems we know three ways to proceed:

(1) Evaluate low-order moments, in particular of order up to 4, and see how they fit with Gaussian. The skewness and excess (see ahead) are the usual parameters here. This is quite practical in many cases but the analytic forms are complicated.

(2) Evaluate the asymptotic spectral distribution, as done with the χ^2_ν cases above. This is not yet a general procedure.

(3) Introduce an ensemble [7] of H's and evaluate the ensemble-averaged spectrum, taking for granted that H is a characteristic member of an ensemble and that an ergodic property will give an essential equality between the ensemble spectrum and the spectrum of H.

The major use of the third method is in dealing with the fluctuating part, $I_{f\ell}(E)$, of the spectrum (6-1), but we shall use it here for the smoothed spectrum. We begin with the Gaussian Orthogonal Ensemble (GOE), the "logic" of which we discuss later. The GOE is an ensemble of d-dimensional real symmetric matrices with the distinct matrix elements chosen independently as zero-centered Gaussians with variances $\overline{W^2_{\alpha\beta}}=(1+\delta_{\alpha\beta})v^2$; we take $dv^2=1$ for normalization. We have now $\langle\bar{H}\rangle=M_1=0$ $=M_{2\nu+1}$, while $M_2=\langle\overline{H^2}\rangle=d^{-1}\sum\overline{W^2_{\alpha\beta}}=(1+d^{-1})\xrightarrow{d\to\infty}1$. For higher-order even moments consider $M_4=\langle\bar{H}^4\rangle=d^{-1}\sum\overline{W_{\alpha\beta}W_{\beta\gamma}W_{\gamma\delta}W_{\delta\alpha}}$, in which however: (1) only terms e.g. $W^2_{\alpha\beta}\ W^2_{\alpha\delta}$ which are fully paired can survive the ensemble average; (2) quartet correlations such as $\overline{W^4_{\alpha\beta}}$ can be ignored for large d, because their number is down by $\sim d^{-1}$ over the binary terms; (3) diagonal matrix elements can be ignored for the same reason. For $\langle H^4\rangle$ now there are three contributing pairings with structures AABB, ABBA, ABAB, the first two of which are equal while the third is small by

$\sim d^{-1}$ since under contraction of the A^2 pair about B only the diagonal m.e. of B could survive. Extending the argument we have then the result (always for asymptotic d) that

$$M_{2\nu} = \mu_{2\nu} = \text{(no. of pairings of } 2\nu \text{ objects with the restriction that any pair can only enclose a fully-paired subset)} \tag{6-19}$$

Thus M_2=1, M_4=2 (≡AABB, ABBA), M_6=5 (≡CCAABB, CABBAC, CAABBC, CCABBA, ABBACC). The general result follows from a simple recursion relationship which gives[8] [7]

$$\mu_{2\nu} = t_\nu = (\nu+1)^{-1}\binom{2\nu}{\nu} \Longrightarrow (2\pi)^{-1}(4-x^2)^{\frac{1}{2}} \tag{6-20}$$

which, as given in Table 1 below, are the *Catalan numbers*, the moments of Wigner's "semicircle" of radius 2.

A semicircle is far from Gaussian and so we seem to have proved the wrong thing. The trouble is that the GOE Hamiltonians, though non-singular, are "unreasonable" in that they involve simultaneous interactions between all particles. This is easy to see; since the dimensionality of a shell-model space grows rapidly, $d(m)=\binom{N}{m}$, with particle number, so do the number of matrix elements, $\sim(d(m))^2$, which must be fixed in order to define H. These m.e. are chosen independently; on the other hand a 2-body H is defined in the 2-particle space by a very much smaller number of matrix elements than $d^2(m)$ with large m. When operating in the larger space its matrix elements must then obey a huge number of constraints[9]: if these are ignored, as in the GOE, the result is an m-body operator.

If however a k-body GOE acts in an ($m \geq k$)-particle space (producing an *embedded GOE* or EGOE) the "matrix-element" constraints of the ordinary GOE gradually become ineffective as m increases, because the H's in a moment $M_p = \langle H^p \rangle^m$ gradually begin to act on different particles[10] (i.e. on particles in different s.p. states) and then [14]

---

[8]Notation: $A_p \Longrightarrow \rho(x)$ implies that the A_p are the moments of $\rho(x)$.

[9]In the simplest (direct-product) basis H(2-body) has a block form since in one step it cannot transfer more than 2-particles to other states. But these are by no means all of the constraints; satisfying them while ignoring the linear constraints on the non-vanishing matrix elements will still result in an m-body Hamiltonian.

[10]The argument fails for very high moments, i.e. when $2\nu k \gtrsim m$ (since then there aren't enough particles), as a result of which the CLT is unable to smooth away the level-to-level fluctuations. Note that we have already used this argument for k=1 to supplement the formal derivation of the Gaussian spectrum.

$$\mu_{2\nu} \xrightarrow{m>>k} \{\text{no. of pairings of } 2\nu \text{ objects without restriction}\}$$
$$= (2\nu-1)!! \qquad (6\text{-}21)$$

which generates a Gaussian behavior. The statistician's excess, the simplest measure of deviation from Gaussian (a value larger in magnitude than 0.3 is recognizable by eye), turns out to be $\binom{m-k}{k}\binom{m}{k}^{-1}$ $\xrightarrow[(N>>m)]{m\ >>\ k} \frac{-k^2}{m}$, so that a dozen particles would give a decent Gaussian for k=2. For spectroscopic cases, in which H is (1+2)-body and angular momentum is conserved, 5 or 6 is usually adequate. The transition from semicircular to Gaussian is shown in Fig.3.

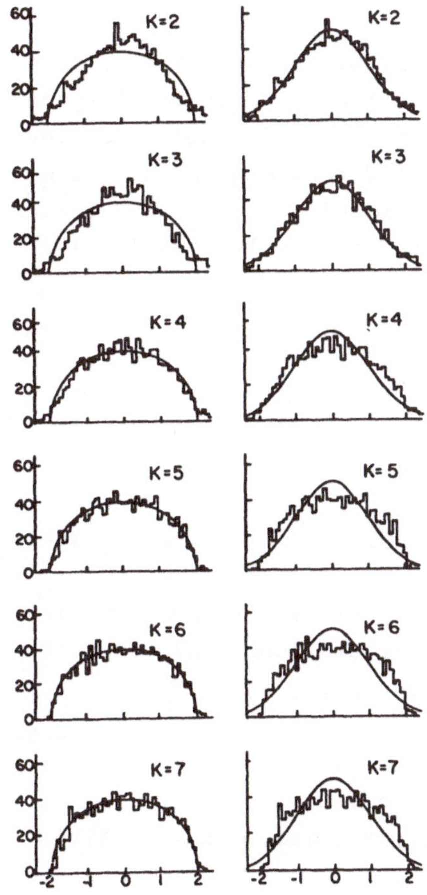

Fig.3 The $f^7(J=7/2)$ ensemble spectra for k-body J-scalar Hamiltonians with $2 \leq k \leq 7$. Each spectrum is given twice, compared on the left with a semicircular, and on the right with a Gaussian, spectrum of the proper centroid and width [15].

We can see now a close correspondence between the *ensemble* operations which lead to a Gaussian *ensemble* density for interacting particles and the *spectral* operations which lead to a Gaussian *spectral* density for non-interacting ones. In each case the moments are determined by a counting of allowed binary-correlation patterns, in the first case the correlations being defined by ensemble averaging, in the second by spectral averaging. Our real interest of course is in spectral results (for a given H), but, except for non-interacting particles, these are in general not accessible to calculation (except for very low-order moments). It is important then that there should be an ergodic behavior which relates the two; we come to that later. In the meantime note carefully that the ensemble-averaging results could not be valid for *all* H's in the GOE (since every H representable by a d-dimensional real-symmetric H is to be found there, and hence every spectrum) but only for *most* H's. Thus there is no conflict with the non-Gaussian many-particle spectra exhibited above.

3. Elementary Statistical Methods

It will be good to collect together a few elementary results, some of which we have already made use of. Suppose that, in our model space, $\underline{m}$, a Hamiltonian H and various other operators are defined. For any of these operators, G say, it is then implied that $G\psi_\alpha$ where $\psi_\alpha \in \underline{m}$ is itself a vector in $\underline{m}$. If the eigenvectors of H are Ψ_α, and corresponding eigenvalues W_α, we can write for the eigenvalue *density* (or *"frequency") function*

$$\rho_W(x) = d^{-1}\sum_{\alpha=1}^{d} \delta(x-W_\alpha) = d^{-1}\sum_\alpha \langle\alpha|\delta(x-H)|\alpha\rangle$$

$$= d^{-1}\mathrm{Tr}^{(m)}\{\delta(x-H)\} = \langle\delta(x-H)\rangle^m \qquad (6\text{-}22)$$

For the *distribution function* we have

$$F_W(x) = \int_{-\infty}^{x} \rho_W(z)\,dz \qquad (6\text{-}23)$$

and then $F_W(+\infty)=1$. Since the model space is finite the spectrum is necessarily discrete so that F(x) is a staircase function. The r'th (ordered) eigenvalue is defined by a discontinuity in F

$$F(W_{r_-}) = (r-1)/d \quad ; \quad F(W_{r_+}) = g_r r/d \qquad (6\text{-}24)$$

where g_r is the degeneracy of the level in question.

Let the moments of $\rho_W(x)$ be M_p and its Fourier transform (its "characteristic function") be $\phi_W(t)$; i.e.

$$M_p = \int \rho_W(x)x^p dx = d^{-1}\sum_r W_r^p = \langle H^p\rangle^m \qquad (6\text{-}25)$$

$$\phi_W(t) = \int e^{itx}\rho_W(x)\,dx = \sum_{p=0} \frac{(it)^p}{p!} M_p = \langle e^{itH}\rangle^m \qquad (6\text{-}26)$$

Note that by the normalization of $\rho_W(x)$ we necessarily have $\phi(0)=1$. Note also the way in which the moments determine $\phi(t)$ and hence $\rho(x)$.

M_0 is trivially unity and $M_1 \equiv E$ is the *centroid* of the distribution. The central moments[11] $\mathcal{M}_p (p \geq 2)$, which then carry no information about the *location* of the distribution, are $\langle (H-E)^p \rangle^m$, the moments of $(H-E)$. The set of $M_p (p \leq t)$ determines the $\mathcal{M}_p (p \leq t)$ and vice versa, by

$$\mathcal{M}_p = \sum_r (-1)^r \binom{p}{r} M_{p-r} (E)^r$$

$$M_p = \sum_r \binom{p}{r} \mathcal{M}_{p-r} (E)^r \qquad (6\text{-}27)$$

The variance of the distribution, $\mathcal{M}_2$, determines the *scale*. We shall write for the *width* of the distribution,

$$\sigma = \mathcal{M}_2^{1/2} = \{M_2 - M_1^2\}^{1/2} \qquad (6\text{-}28)$$

(For a Gaussian distribution 1.18σ is in fact the 1/2-width at 1/2-maximum.) It follows now that the higher-order central moments $\mathcal{M}_{p \geq 3}$, or better the *reduced central moments*, μ_p for $p \geq 3$, which are invariant under translations and scale changes, carry the information about the *shape* of the distribution.

It is often good to measure the random variable with respect to the centroid of its distribution and in units of its width. Then introducing $\chi = \chi(x) = (x-E)/\sigma$ we have

$$\rho(x) = \sigma^{-1} \hat{\rho}\left(\frac{x-E}{\sigma}\right) = \sigma^{-1} \hat{\rho}(\chi) = \rho(\sigma\chi + E) \qquad (6\text{-}29)$$

The implied separation of the centroid and width dependence from the shape dependence will be seen to be of major physical significance. The moments of $\hat{\rho}(\chi)$, in the re-scaled variable χ are the *reduced central moments* of $\rho(x)$

$$\int \hat{\rho}(\chi) \chi^p d\chi = \mathcal{M}_p / \sigma^p = \mu_p \qquad (6\text{-}30)$$

For $p=0,1,2$ the μ_p values then are 1,0,1, while for $p \geq 3$ they determine and are determined by (translation- and scale-invariant) shape parameters $S_{\nu \geq 3}$, as of course is the function $\hat{\rho}(\chi)$.

As examples we have for asymptotic $(d \to \infty)$ distributions:

---

[11] Notation: We encounter several kinds of moments: (1) ordinary moments, M_p; (2) central moments $\mathcal{M}_p$; (3) reduced central moments, $\mu_p = \mathcal{M}_p/(\mathcal{M}_2)^{p/2}$; (4) cumulants K_p; (5) reduced cumulants $k_p = K_p/(\mathcal{M}_2)^{p/2}$; (6) polynomial moments, Λ_p.

Table 1

Gaussian	$\mu_{2\nu}=(2\nu-1)!!\Longrightarrow\rho_G(x)=(2\pi)^{-\frac{1}{2}}\exp(-x^2/2)$
Semicircular	$\mu_{2\nu}=(\nu+1)^{-1}\binom{2\nu}{\nu}\Longrightarrow\rho_0(x)=(2\pi)^{-1}(4-x^2)^{\frac{1}{2}}$
Uniform	$\mu_{2\nu}=(2\nu+1)^{-1}3^{\nu}\Longrightarrow 1;\quad (\lvert x\rvert \leqslant \frac{1}{2})$
K_0	$\mu_{2\nu}=\{(2\nu-1)!!\}^2\Longrightarrow\pi^{-1}K_0(\lvert x\rvert)$
Single-State Pairing:	$\mu_{p\geqslant 2}=d^{p/2-1}\Longrightarrow d^{-1}\{\delta(x-1)+(d-1)\delta(x)\}$
Hermite-polynomial Excitation ($p-\zeta$=even)	$M_p=\zeta!(p-\zeta-1)!!\binom{p}{\zeta}\Longrightarrow\rho_G(x)He_\zeta(x)$
Chebyshev-polynomial Excitation ($p-\zeta$=even)	$M_p=\left(\begin{matrix}p\\ \frac{p-\zeta}{2}\end{matrix}\right)\Longrightarrow -\zeta^{-1}\frac{d}{dx}\{\rho_0(x)v_{\zeta-1}(x)\}$
χ^2_ℓ	$\sigma^2=2E^2/\nu$

The polynomial-excitation functions given here, which we shall need later, are oscillatory, and therefore not true probability densities (for these the boundary values of μ_p with $p=0,1,2$ do not apply). In the Chebyshev case $v_\zeta(x)=(-1)^\zeta(\sin\psi)^{-1}\sin(\zeta+1)\psi$, with ψ defined on the semicircle as the angle between the negative x axis and the ordinate $y(x)$, is the Chebyshev polynomial of the second kind, defined for the domain $(-2,2)$ and orthonormal with $\rho_0(x)$ as weight function. Recall the combinatorial definitions of $\mu_{2\nu}$ given above for the first two cases; we shall encounter extensions of them for the polynomial excitations.

A particular set of distribution parameters, the reduced cumulants $k_p=K_p/\sigma^p$, which are non-trivial shape parameters for $p\geqslant 3$ will be of major interest. Just as the moments enter in the Taylor expansion of $\phi(t)$ the cumulants K_p enter in the expansion of its logarithm:

$$\ell n\phi(t) = \sum_{p=0}^{\infty}\frac{(it)^p}{p!}K_p \;;\quad \phi(t) = \exp\{\sum K_p(it)^p/p!\} \tag{6-31}$$

Using the two expansion forms to evaluate the derivatives of $\phi(t)$ at $t=0$ we find for the lowest-order reduced cumulants expressed in reduced moments and conversely,

$$k_0=0;\ k_1=E;\ k_2=1;\ k_3=\mu_3\equiv\gamma_1;\ k_4=\mu_4-3\equiv\gamma_2 \tag{6-32}$$

Note that for $p\geqslant 2$ the (reduced) cumulants are homogeneous polynomials in the (reduced) central moments. For the Gaussian distribution and its characteristic function we have

$$\rho(x) = (2\pi\sigma^2)^{-\frac{1}{2}}\exp\{-(x-E)^2/2\sigma^2\}$$
$$\phi(t) = \exp\{itE - \tfrac{1}{2}\sigma^2t^2\}, \tag{6-33}$$

defined then by the vanishing of all cumulants with $p>2$. The shape parameter $k_3=\gamma_1$ is called the *skewness* and $k_4=\gamma_2$ the *excess*. Broadly speaking $k_3>0$ defines a distribution which extends more in the $(x>E)$-domain than in $(x<E)$, and $k_4>0$ one more sharply peaked than is the Gaussian. Eq.(6-31) is the beginning of the usual demonstration of the CLT convergence to Gaussian, which draws on the facts that Fourier transforms multiply under convolution (so that their logarithms are additive), while Gaussian is recognized by the vanishing of $k_{p>2}$.

We encounter now the question of how to use values for some of the higher moments in describing the actual density. To do this we expand the density in terms of the asymptotic density multiplied by a series of orthonormal polynomials defined by that density as weight function; described otherwise we represent the density in terms of "polynomial excitations" of the asymptotic density. For an asymptotic Gaussian density we have the Gram-Charlier series [16]

$$\rho(x) = \rho_G(x)\left\{1 + \sum_{\nu\geq 3} P_\mu(x)\int_{-\infty}^{\infty}\rho(z)P_\mu(z)dz\right\} \tag{6-34}$$

or equivalently

$$\rho(x) = \rho_G(x)\left\{1 + \sum_{\nu\geq 3}(\nu!)^{-1}R_\nu He_\nu\left(\frac{x-E}{\sigma}\right)\right\} \tag{6-35}$$

while for the distribution function

$$F(x) = F_G(x) - \sigma\rho_G(x)\sum_{\nu\geq 3}(\nu!)^{-1}R_\nu He_{\nu-1}\left(\frac{x-E}{\sigma}\right) \tag{6-36}$$

where the shape parameters R_ν are the Hermite-polynomial moments

$$R_\nu = \int\rho(z)He_\nu\left(\frac{z-E}{\sigma}\right)dz \tag{6-37}$$

being therefore calculable from the ordinary central moments M_p with $p=\nu,\nu-2,\ldots$. For $\nu\leq 5$ R_ν is simply the reduced cumulant; for $\nu\geq 6$ however the correspondence is not precise. If, as usually happens, the Gaussian is generated by an explicit or implicit convolution process the convergence to zero of the shape parameters is faster, the higher the order ζ, so that a truncation is appropriate; in fact the series would be of no value if this were not the case. There is however no guarantee that the resultant approximate density is non-negative definite; there are in fact no useful related expansions in which this criterion is satisfied. That the density should become negative is simply a sign of truncation errors, but we are liable to take a harsher view of such errors than of others which give numerical deviations rather than deviations of kind. This effect however, which shows up first in the tails of the distribution, is much lessened by partitioning, though it does not go away completely; we must learn to live with it. For our purposes we would seldom deal with $\zeta>4$; we restrict ourselves therefore to the skewness and excess corrections. An exception will be when we deal with fluctuations, for which incidentally there will be no problem in satisfying the $\rho(x)\geq 0$ restriction. We remark finally that there is a rearrangement [16,17] of the Gram-Charlier series, due to Edgeworth, each term of which converges uniformly in the order of convolution (or particle number).

We shall need also the (orthonormal) polynomials $P_\mu(x)$ defined by $\rho(x)$ itself, instead of its asymptotic form. These can be generated, by the Schmidt orthogonalization process from the orthonormality relationship, $\int P_\mu(x)P_\nu(x)\rho(x)dx=\delta_{\mu\nu}$. We find for the first three, with $\chi=(x-\mathcal{E})/\sigma$ as above, that

$$P_0(x)=1; \quad P_1(x)=\chi; \quad P_2(x)=\frac{(\chi^2-1)-\mu_3\chi}{[\mu_4-\mu_3^2-1]^{\frac{1}{2}}} \tag{6-38}$$

It is easy to see that the unnormalized polynomial of order ν requires density moments of orders up to $(2\nu-1)$, while the normalization requires also the moment of order 2ν. The general solution for the polynomials, the phase being chosen so that the coefficient of the highest power (x^μ in $P_\mu(x)$) is positive, is

$$(D_\mu D_{\mu-1})^{\frac{1}{2}}P_\mu(x) = \begin{vmatrix} 1 & M_1\cdots\cdots M_\mu \\ M_1 & M_2\cdots\cdots M_{\mu+1} \\ \vdots & \\ M_{\mu-1} & \cdots\cdots\cdots M_{2\mu-1} \\ 1 & x^\mu \end{vmatrix} \tag{6-39}$$

where D_μ is the determinant with the last row replaced by $M_\mu, M_{\mu+1}, \cdots M_{2\mu}$. Observe that $P_2(x)$ is a function of $(x-\mathcal{E})$ and of the *central* moments only; this feature derives from the translational invariance given by (6-29) and of course holds for polynomials of arbitrary order, giving rise to a simpler form of (6-39). Note also that if we think of the density as discrete the polynomials are defined on a line and there is a finite number (d) of them.

The fact that $M_2=\sigma^2+M_1^2\geq M_1^2$ reminds us that not every set of numbers M_p defines a density. The non-negative-definite nature of $\rho(x)$ imposes for example a set of determinantal constraints on the moments, *viz*

$D_\nu \geq 0$ for all ν, which with $\nu=1$ yields $M_2 \geq M_1^2$ and with $\nu=2$ a relationship between the excess and skewness, $\gamma_2 \geq \gamma_1^2 - 2$.

The division of the parameters which define $\rho(x)$ into *location*, *scale* and *shape* parameters will be seen to be quite fundamental. On the other hand, we shall meet distributions for which the second central moment does not really define the physical extent in a satisfactory way; this is particularly true for a bimodal distribution (one which contains two local maxima), or for a multi-modal one, in which small "pieces" of the density occur a great distance away from the dominant piece whose extent is really the quantity of interest, and make thereby large contributions to σ^2. We shall encounter such distributions particularly when the "chaos" which is generated by strong interaction, and which is responsible for the success of the statistical procedures, is modified by the presence of relatively weak interactions, e.g. between particles in orbits which are very far apart in energy. It will be important for us to recognize this feature when it is present (which we shall be able to do by evaluating the skewness) and make proper allowance for it. It will be recognized that we are here encountering examples of singular spectra. We shall see also that, for our spectra, which are discrete with fluctuations, it is better to describe the higher-order parameters (labeling those "excitations" of the reference spectrum, as for example in (6-34), which have wavelength $\sim$ spacing) as *fluctuation* parameters, and agree then that "shape" implies "smoothed to within fluctuations".

We conclude this section with comments about correlation coefficients and functions. Everything about the statistical behavior of two random variables, F and G, is determined by their *joint probability distribution*, $\rho_{F,G}(x,y)$, which for cases of interest to us is fixed by its moments

$$M_{p,q} = \int \rho_{F,G}(x,y)\, x^p y^q dxdy \equiv \langle F^p G^q \rangle_{av} \tag{6-40}$$

or

$$\mathcal{M}_{p,q} = \langle (F-M_{1,0})^p (G-M_{0,1})^q \rangle_{av} = (\mathcal{M}_{2,0})^{p/2} (\mathcal{M}_{0,2})^{q/2} \mu_{p,q} \tag{6-41}$$

in which $\langle \; \rangle_{av}$ denotes the averaging operation of interest. For obvious reasons the variables will be described as *independent* iff $\rho_{F,G}(x,y) = \rho_F(x)\rho_G(y)$, in which case we have a factoring of the moments and a vanishing of the lowest-order mixed moment $\mathcal{M}_{11}$, the *covariance*. The renormalized version of this moment, i.e. μ_{11} is ζ the *correlation coefficient*; its value lies in the interval $(-1,1)$ and it measures the simplest kind of correlation between F, G. If $\zeta=0$ the variables are uncorrelated, but not necessarily independent. If $\zeta = \pm 1$ the variables are completely correlated, either positively or negatively ("anticorrelated"), one variable behaving then like a multiple of the other.

As an example let f_i, g_i represent values of time-dependent variables F, G at time t_i (i=1...d). Then a significant correlation coefficient, or an estimate of it, is

$$\zeta = \frac{\sum (f_i - f_{av})(g_i - g_{av})}{\left\{\sum (f_i - f_{av})^2 \sum (g_j - g_{av})^2\right\}^{1/2}} \tag{6-42}$$

The same form applies if f_i, g_i represent the spectra of two (commuting) Hermitian operators F,G which act in the model space $\underline{m}$ and have the same eigenfunctions Ψ_i (which establishes then the association between the two sequences). Expressing things in terms of traces we see that

$$\zeta = (\sigma_F \sigma_G)^{-1} \langle (F - \mathcal{E}_F)(G - \mathcal{E}_G) \rangle^m \tag{6-43}$$

where of course $\mathcal{E}, \sigma$ are spectral centroids and variances.

Even if F, G do not commute we see that the form (6-43) is still a valid correlation coefficient, namely between the eigenvalues of one operator and the expectation values of the other in the corresponding eigenstates of the first, or between the matrix elements in any basis.

For a random variable G which takes on values along a line (increasing time for example, or, more interesting for us, along the spectrum of some Hermitian operator) great interest attaches to the correlation between its own values at different points. For this we introduce the *autocovariance function* (or *two-point* or *autocorrelation* function, though the latter term might well be reserved for renormalized variables so that it measures correlation coefficients directly)

$$S^G(x,y) = \langle G(x) G(y) \rangle_{av} - \langle G(x) \rangle_{av} \langle G(y) \rangle_{av} \tag{6-44}$$

The averaging operation here could be that of ensemble averaging, $\langle Q \rangle_{av} \to \overline{Q}$, or spectral averaging, $\langle Q \rangle_{av} \to \langle Q \rangle$, relevant for a stationary process, in which we would fix (y-x) and vary x along the spectrum. The ensemble average is often analytically tractable while the spectral average may be estimated from data.

In the case that G is defined along a discrete spectrum, S^G has a natural decomposition into a two-level part and a one-level part which we may subtract out to produce a "true" two-level function. In particular for the eigenvalue density itself, $\rho(x) = d^{-1} \sum \delta(x - E_i)$, we have

$$\begin{aligned} \overline{\rho(x)\rho(y)} &= \{\overline{\rho(x)\rho(y)}\}_2 + \{\overline{\rho(x)\rho(y)}\}_1 \\ &= \{\overline{\rho(x)\rho(y)}\}_2 + d^{-1}\delta(x-y)\overline{\rho}(x) \end{aligned} \tag{6-45}$$

and then, for the two-point cluster function Y_2, DYSON [18],

$$\{\bar{\rho}(x)\bar{\rho}(y)\}^{-1}S^{\rho}(x,y) = (\bar{\rho}(x)d)^{-1}\delta(x-y) - Y_2(r,X) \qquad ,$$
$$= \delta(r) - Y_2(r,X) \rightarrow \delta(r) - Y_2(r) \qquad (6\text{-}46)$$

in which $X=(x+y)/2$ defines the center of the interval while $r=(y-x)/D(x)$ measures the length of the interval in local spacing units $D(x)$. The final form is appropriate when the locally renormalized spectrum is stationary.

4. Expectation Values, Transition Strengths and Statistical Response

We are interested in expectation values of various operators K in the Hamiltonian eigenstates; we encounter them for example in evaluating electromagnetic moments, in studies of symmetries where K is a Casimir operator of some group, and in evaluating sum-rule quantities for excitation operators O (in fact we shall need also to deal with the strength distribution for these excitations). We write

$$\langle\Psi_E|K|\Psi_E\rangle \equiv K(E) \qquad (6\text{-}47)$$

where E is the eigenenergy, and we have suppressed other quantum numbers. In case there are (non-accidental) degeneracies we will understand that an average is taken over the degenerate states[12].

Suppose that our system has an essentially asymptotic (say Gaussian) spectrum. Let us test its response by adding to H a multiple of an operator K, $H \rightarrow H_\alpha = H + \alpha K$. Then, for a wide class of K operators which also have a Gaussian spectrum, we can expect that H_α is also Gaussian so that the smoothed $\rho(x)$ and $\rho_\alpha(x)$ can differ only in centroid and variance. For a very much wider range of operators K (including operators whose spectrum is by no means Gaussian, but not for example $K=H^2$) the same will be true to lowest order in α, which is all we need for our present purposes. But then we have immediately the CLT result, given earlier, that

$$K(E) \xrightarrow{\text{CLT}} \langle K\rangle^m + \langle K(H-\mathcal{E})\rangle^m \frac{(E-\mathcal{E})}{\sigma^2} \qquad (6\text{-}48)$$

which of course is valid only to within fluctuations. As we have mentioned earlier, and will return to below, (6-48) gives the first two terms of an exact expansion in the orthonormal polynomials (6-39) defined by $\rho(x)$. In contrast to the higher-order terms, which arise from shape deformations of the density under the action of K, these

[12] This is not essential, for in (6-53) below we could make a finer selection of states by making use of further δ-function projection operators.

terms are *uninhibited* by the CLT. In common practice we would calculate one or two correction terms in the series.

To understand the geometrical significance of (6-48), and for more general purposes, we return to the question of norms and geometries mentioned in the introduction. When an operator G acts on a vector ψ_α it generates a new vector $G\psi_\alpha$. We could determine a magnitude for G by comparing the magnitude of $G\psi_\alpha$ with that of ψ_α itself, using of course the usual *vector norm* for the states. Taking $||\psi_\alpha||^2 \equiv (\psi_\alpha,\psi_\alpha)=1$ we would have then $||G\psi_\alpha||^2 = \sum_\beta |G_{\beta\alpha}|^2 = \langle\alpha|G^+G|\alpha\rangle$, and thus, when acting on ψ_α, G is measured by $\{\langle\alpha|G^+G|\alpha\rangle\}^{1/2}$; since we shall want a measure which treats all states on a democratic basis (and which is calculable without constructing any of the states) we take the norm as given by (6-4), the measure of an operator being then simply its RMS average eigenvalue.

Many other norms are possible but this, the *unitary norm* (or *Euclidean norm* for Hermitian operators), is exactly what we need. It satisfies the four standard requirements for a norm, *viz:* that it is non-negative and vanishes only when G=0: that on multiplying G by a constant c its norm is multiplied by $|c|$; that the norm of a product is not greater than the product of the norms, and similarly for the sum, equality holding in that case only when one operator is a multiple of the other. Moreover $||G\psi|| \leq ||G|| \times ||\psi||$ so that the norm is "compatible" with the wave-function norm. These conditions protect us against the possibility that the product or sum of two small operators be a huge one, and similarly for the action of an operator on a state.

Moreover $||G||$, being a function of a trace, is invariant under unitary transformations of the space so that its value does not depend on the basis used for evaluation, an obvious requirement for a physically significant quantity and one which is essential for evaluation.

Every norm defines a geometry in the sense that it defines distances between points (vectors!) in the space and thus angles between vectors, projections of one vector along another, and so forth. We have then, in particular for a Hermitian operator, that its second moment defines a geometry. So also does its variance, and in fact a geometry which is usually more interesting. In the case of the Hamiltonian for example we feel free to choose the zero of energy, this because the eigenstates are left unchanged under $H \to H + c1$, and similarly for other operators. Let us therefore *center* G by $G = \langle G\rangle^m + \{G - \langle G\rangle^m\}$ and then

$$||G||^2 = ||\langle G\rangle^m||^2 + ||G - \langle G\rangle^m||^2 \qquad (6\text{-}49)$$

so that the decomposition of G is orthogonal and defines for G two natural magnitudes as indicated. In statistical terms centering is in correspondence to our passing from moments to central moments, whereas, in geometrical language, it amounts to projecting the operator into the plane normal to the unit operator in the operator space. For an example of a more general decomposition the "amount of G contained in H" is λG where we have an orthogonal decomposition,

$$H = (H - \lambda G) + \lambda G \quad ; \quad \lambda = (G\cdot H)/||G||^2 \tag{6-50}$$

There is an important general question about the usefulness of the geometry we have introduced. Is it clear that our democratic measure (6-4) for an operator is really appropriate? For we commonly work in large model spaces whose H eigenvalues span a large energy domain ($\sim$100 MeV in the case of $(ds)^{12}$) with most of the states concentrated in the central region. Our interest on the other hand might well be in phenomena involving the ground state or low-lying excited states (and even if it isn't we may need an accurate treatment of the low energy end in order to fix the ground state energy as a reference energy for the calculation); why then should our measure deal equally with all the states? The situation in fact is "worse" than that, for the standard shell-model spaces involve a restricted number of single-particle states whose energy span (5 MeV for ds) is only a small fraction of the multiparticle spectrum span; thus, as we have already stressed, the higher-lying states, which almost completely determine the measure are hopelessly inadequate for representing the physical states which therefore play a negligible role in the measure. Why then should our geometry be physically relevant at all? The answer, broadly speaking, is that the geometry is made "effective" through the action of the CLT which generates eigenvalue and related distributions which have a characteristic asymptotic shape, the density then being described essentially in terms of the centroid and variance, the same two parameters which fix the geometry.

Return now to the expectation value whose CLT-limit (6-48) we rewrite in geometrical terms as

$$k(W) = (\vec{k}\cdot\vec{h})\,\frac{(W-E)}{\sigma} = kh\cos\theta_{kh}\,\frac{(W-E)}{\sigma} = \zeta_{K-H}\,\frac{(W-E)}{\sigma} \tag{6-51}$$

Here k and h are the *unit centered* (traceless) versions of K and H

$$k = \frac{K-\langle K\rangle^m}{||K-\langle K\rangle^m||} = \frac{K-E_K}{\sigma_K} \quad ; \quad h = \frac{H-\langle H\rangle}{||H-\langle H\rangle^m||} = \frac{H-E}{\sigma_H} \tag{6-52}$$

while θ_{kh} is the angle between the two centered operators and ζ_{K-H} is the correlation coefficient (6-43) between them.

Observe that when the density undergoes no first-order shape deformations as $H \to H_\alpha$, so that (6-48) is valid, we need have no further concern, at least for expectation values, that the norm being used weights the high states too heavily. Note also an aspect of *collectivity*; when K and H have a strong negative correlation the states with large expectation values lie low. For $K = Q \cdot Q$ where Q is the electromagnetic quadrupole operator, one commonly finds correlation coefficients ~ -0.5 (a very strong correlation) and then, since the $Q \cdot Q$ expectation fixes the E2 sum rule, the low-lying states are those with large quadrupole strength.

The linear (CLT) result gives a remarkably simple picture for the behavior of expectation values but, if the CLT is not fully effective[13], we shall need "shape corrections" as well. A compact exact result is given above (6-2), but it is simpler to write [19]

$$K(W) = \rho^{-1}(W) \int \rho(x) K(x) \delta(x-W)\, dx = \rho^{-1}(W) \langle K\delta(H-W)\rangle^m = \sum_\nu \langle KP_\nu(H)\rangle^m P_\nu(W) \qquad (6\text{-}53)$$

where in the last step we have introduced the orthonormal polynomials (6-39) and used completeness. The first two terms give back the CLT result while the terms with $\nu \gtrsim 2$ define the shape corrections. For "most" operators, K, H the expansion, or its partitioning extension, is strongly convergent (once again "to within fluctuations"), its convergence deriving from the CLT convergence in the density.

The sum in (6-53) extends to $\nu = \infty$ or $\nu = d$ according as ρ is continuous or discrete (as it will be in the cases of interest to us). In either circumstance we shall restrict the summation very severely, the necessity for that arising from the complexity of the traces $\langle KP_\nu(H)\rangle^m$ for ν large, and the permit for it arising in part from the CLT and in part from our lack of interest in the detailed fluctuations in K(x). To see something about the smoothing implied in restricting the ν summation, we write

---

[13] Either because the particle number is too small, or because we are interested in expectation values near the ground state in spaces which are too large, two conditions which are to some extent opposed to each other. The second is relevant because deviations from the characteristic form set in first in the tail of the distribution. If the spaces are very large, partitioning is more effective than taking terms of higher and higher order in the polynomial expansion (6-53). In practice two terms beyond CLT should then always be adequate.

$$K(x) = \sum_{\nu=0}^{\bar{\nu}} \langle KP_\nu(H)\rangle^m P_\nu(x) + \tilde{K}^{(\bar{\nu})}(x) \tag{6-54}$$

and observe then that $\tilde{K}^{(\bar{\nu})}$ is *fluctuating* in the sense that it has vanishing moments of order $p \leq \bar{\nu}$; or alternatively, being orthogonal to polynomials of lower order, it involves no wave-lengths longer than those characteristic of $P_{\bar{\nu}}(x)$. The smoothing involved is seen to be not *local* but global. The ν-sum in (6-54) may be taken as defining the secular behavior of $K(x)$, while $\tilde{K}^{\bar{\nu}}(x)$ defines the deviations about the local mean. This separation is really only significant if it is independent of $\bar{\nu}$ over a wide range; it can be shown that this separation actually occurs in many-particle spectra, for which indeed there are only long-range secular excitations and very short-range ($\lambda \sim$spacing) fluctuations with no intermediate-wave-length excitations at all. The former may be incorporated into the density itself which then has only short-range fluctuations.

For the *microscopic strength* (spectroscopic factor for nucleon transfer, BE2 value for electric-quadrupole excitation, and so forth) we have, with R, R^+ the strengths respectively for excitations O, O^+,

$$R(W',W) = R^+(W,W') = |\langle W'|O|W\rangle|^2 = \{I'(W')I(W)\}^{-1} \langle\langle O^+\delta(H-W')O\delta(H-W)\rangle\rangle^m \tag{6-55}$$

where, as in (6-1),

$$I(W) = d \times \rho(W) \quad ; \quad I'(W') = d' \times \rho'(W') \tag{6-56}$$

The starting and final model subspaces may of course be the same and then $I'(W) = I(W)$. *Ceteris paribus*, the strength (6-55) determines the relative cross-section or transition rate for the excitation, and is in many cases measurable, at least when one of the states is a ground state. The last form of (6-55), analogous to the last form of (6-22) arises on representing the squared matrix element in terms of a double integral over W and W'. Just as in (6-22), we automatically sum over degenerate states but could, in many cases, separate them by appropriately decomposing the model space or by introducing further δ-function projection operators.

For given W, $R(W',W)$ is an *unnormalized* frequency function. Its moments, which are in principle determined by the sum rules (with and without energy weighting) are of course expectation values, and as such have an expansion (6-53)

$$\begin{aligned} M_p(W) &= \sum_{W'} (W')^p R(W',W) = \int I'(W')(W')^p R(W',W)\,dW' \\ &= \langle W|O^+H^pO|W\rangle = \langle W|M_p|W\rangle = I^{-1}(W)\langle\langle M_p\delta(H-W)\rangle\rangle^m = \sum_\nu \langle M_pP_\nu(H)\rangle^m P_\nu(W) \end{aligned} \tag{6-57}$$

Here we have written (and confusion must be avoided with the M_p of (6-27))

$$M_p = O^+ H^p O \tag{6-58}$$

a "sum-rule operator" whose expectation value in Ψ_W which by our convention we write as $M_p(W)$ sums the energy-weighted strengths which start with Ψ_W. It is these quantities which are determined by the non-energy-weighted (p=0) and energy-weighted sum rules (p>0). $M_0(W)$ is the total strength originating with Ψ_W, while $M_1(W)/M_0(W)$ gives its centroid and $M_2(W)/M_0(W)-\{M_1(W)/M_0(W)\}^2$ its variance. In practice, in the sum-rule analysis of experimental strengths, one cannot normally go beyond p=2, and usually indeed not beyond p=1, primarily because small, unobservable, parts of the strength arising at relatively high energies would give large contributions to higher moments; on the other hand high-moment contributions are essential for strength fluctuations.

Observe now that even when the M_0 and M_1 operators give rise to only centroid and width deformations, and have therefore expectation values linear in the energy, the strength centroid is itself not linear but is rather the ratio of two linear forms; and similarly for the higher strength moments. We might expect also that departures from linearity for the $M_p(W)$ will be larger the higher is the order p because the H^p operator for $p \gtrsim 1$ will naturally tend to introduce a nonlinearity.

We shall often need the strength function itself rather than its sum-rule quantities. At low excitations, especially when one of the levels is a ground state, R(W',W) for a given pair of levels may be measureable. Besides that in some processes, electromagnetic excitation for example (as opposed to excitation by nucleon transfer) only a range W'>W may be reached (where Ψ_W is the starting state). $M_p(W)$, in which all the strengths are summed over, whether they can be realized experimentally or not, does give an upper bound to the observed strength (appropriately energy weighted), but for a better comparison we should decompose $M_p(W)$ into its *exothermic* (W>W') and *endothermic* (W<W') parts. We shall be able to do this once we have results for the strength itself.

The polynomial expansion of the strength function itself follows immediately from the last form of (6-55) along with the completeness expansion applied separately to the states $\Psi_{W'}$ in model-space $\underline{m}'$ and to the Ψ_W in model space $\underline{m}$ which may or may not coincide with $\underline{m}'$. The structure involved here is that O acting on a vector in $\underline{m}$ generates one in $\underline{m}'$, and of course conversely for O^+. We have now two sets of polynomials, $P_\nu(W)$ defined by the Ψ_W density and $P'_\mu(W')$ by that for $\Psi_{W'}$, and then

$$R(W',W) \equiv R(W'm',Wm) = \sum_{\mu\nu} G_{\mu\nu}(m',m) P'_\mu(W') P_\nu(W)$$

$$G_{\mu\nu}(m',m) = (dd')^{-1} \langle\langle O^+ P'_\mu(H) O P_\nu(H) \rangle\rangle^m$$

$$= (dd')^{-1} \langle\langle O P_\nu(H) O^+ P'_\mu(H) \rangle\rangle^{m'}$$

$$= G^{(+)}_{\nu\mu}(m,m') \tag{6-59}$$

where, in deriving the $G_{\mu\nu}$ symmetry we have used the elementary result that if operator E acting on vectors in m generates only $m \to m'$ transitions we have, for arbitrary F, that $\langle\langle FE \rangle\rangle^m = \langle\langle EF \rangle\rangle^{m'}$; thus $G^{(+)}_{\nu\mu}(m,m')$ denotes a coefficient associated with the inverse transitions.

For the sum rules it is often better to use, the "polynomial moments", in which we write $\Lambda_\alpha = O^+ P'_\alpha(H) O$ (note that the set $\Lambda_{\alpha \leq t}$ carries the same information as the set $M_{p \leq t}$). Then

$$\Lambda_\alpha(W) = \sum_{W'} R(W',W) P'_\alpha(W') = \int R(z,W) P'_\alpha(z) I'(z) dz$$

$$= \langle W | O^+ P'_\alpha(H) O | W \rangle = d' \times \sum_\nu G_{\alpha\nu}(m',m) P_\nu(W) \tag{6-60}$$

and for the strength itself

$$d' \times R(W',W) = \sum_{\alpha\nu} P'_\alpha(W) P_\nu(W) \int \Lambda_\alpha(z) P_\nu(z) \rho(z) dz$$

$$= \sum_\nu \Lambda'_\nu(W') P_\nu(W) = \sum_\mu \Lambda_\mu(W) P'_\mu(W') \tag{6-61}$$

In the CLT limit in which expectation values are linear in the energy we see from the last form of (6-60) that $R(W',W)$ is *bilinear*, being linear in both W' and W. Thus we shall find, as the fundamental behavior, *normality*, *linearity* and *bilinearity* for densities, expectation values and strength functions. Of course secular deviations will be found in practical cases and besides that there will also be short-wave-length fluctuations.

This is a good place to say something about partitioning of model spaces, for the effects of partitioning show up in an elegant way in expectation values and strengths. Partitioning is almost essential if we wish to deal with transitions between states of fixed symmetries such as isospin; besides that, even when symmetries are not of interest, results without partitioning may not be sufficiently accurate if the spaces are very large, for then we may need several correction terms beyond the CLT which are difficult to calculate. We can obviously hope to simplify things by partitioning the space; in fact at the same time we will get an excellent understanding of the structure of the strength distribution, of the way in which correlations between the excitation operator and the Hamiltonian shift the strength up or down in the spectrum, and of the relationship with simple schematic theories which neglect correlations in the system which are often important.

If we partition the space $\underline{m}$ into subspaces $\underline{m} \rightarrow \sum \alpha_i$ we have for the Hamiltonian traces and moments

$$\langle\langle H^p \rangle\rangle^m = \sum_i \langle\langle H^p \rangle\rangle^{\alpha_i} \quad ; \quad M_p(m) = \sum_i \frac{d(\alpha_i)}{d(m)} M_p(\alpha_i) \tag{6-62}$$

For the centroid we have of course

$$E(m) = \sum_i \frac{d(\alpha_i)}{d(m)} E(\alpha_i) \tag{6-63}$$

and for the variance

$$\begin{aligned} d(m)[\sigma^2(m) + E^2(m)] &= \langle\langle H^2 \rangle\rangle^m = \sum_i \langle\langle H^2 \rangle\rangle^{\alpha_i} \\ &= \sum_i d(\alpha_i)\{\sigma^2(\alpha_i) + E^2(\alpha_i)\} \end{aligned} \tag{6-64}$$

But since $\sum_i d(\alpha_i) = d(m)$ and $\sum_i d(\alpha_i) E(\alpha_i) E(m) = d(m) E^2(m)$ we have finally

$$\sigma^2(m) = \sum_i \frac{d(\alpha_i)}{d(m)} \{\sigma^2(\alpha_i) + [E(\alpha_i) - E(m)]^2\} \tag{6-65}$$

The first term in $\sigma^2(m)$ represents the variances of the individual subspectra while the second arises when not all of the subspectra have the same centroid. We may think of the first as giving an average "spreading variance" and the second a "displacement" variance. Note carefully however that that separation depends on the subspace decomposition. In the extreme case, where the subspaces are one-dimensional and H-eigenfunctions, the first term vanishes while the $E(\alpha_i)$ become the eigenvalues. We can regard the eigenvalue problem as one of decomposing the model space into 1-dimensional subspaces in such a way that the displacement variance is maximized.

In the complete model space the distribution of interest is that of the H-eigenvalues, but in those subspaces which are not eigenspaces (though in exceptional cases they may contain some H-eigenvectors) there is no such distribution. The appropriate way to think of the subspace distribution in this case (and it is valid for eigenspaces too) is as a distribution of the partial intensity of the eigenstates versus the energy; more specifically as $I_r(\alpha_i)$ vs E_r, where $H\Psi_r=E_r\Psi_r$ and $I_r(\alpha_i)$ is the intensity of Ψ_r in the subspace α_i, or equivalently the square of the norm of the projection of Ψ_r into α_i. If $\phi_{\alpha_i}(s)$, with $s=1,2,\ldots d(\alpha_i)$ forms an orthonormal basis for α_i we have

$$I_r(\alpha_i) = \sum_{s=1}^{d(\alpha_i)} |\langle\phi_{\alpha_i}(s)|\Psi_r\rangle|^2$$
$$\sum_{\alpha_i} I_r(\alpha_i) = 1 \quad ; \quad \sum_r I_r(\alpha_i) = d(\alpha_i) \qquad (6\text{-}66)$$

The subspace distributions are of course discrete just as the eigenvalue distributions are.

It will be seen later that the only useful subspace decompositions will be defined by groups, usually U(N) subgroups. There are CLT extensions to these subspaces; moreover the symmetries defined by the subspaces give rise to methods for evaluating the moments. It will be important also that the individual subspace distributions not span the entire spectrum; they should supply at least a very rough division of the spectrum span, not a difficult requirement to meet. And besides that of course an interest in particular symmetries will dictate the kind of decompositions to use, since the subspace decomposition carries information about the symmetries.

It is important to understand that the linearity of the subspace decomposition and that of the moments (6-62) does not at all imply that we are dealing with non-interacting subspaces. For example the subspace variance $\langle(H-E(\alpha_i))^2\rangle^{\alpha_i}$ has contributions to it from intermediate states in other subspaces α_j, corresponding to $\alpha_i \xrightarrow{H} \alpha_j \xrightarrow{H} \alpha_i$. we have indeed a natural decomposition of $\sigma^2(\alpha_i)$ into an *internal* variance $(\alpha_i\to\alpha_i\to\alpha_i)$ and an *external* variance, which latter of course has a decomposition according to α_j $(j\neq i)$. Formally we have, with $P(\alpha_j)$ a projection operator, that

$$\sigma^2(\alpha_i) = \sum_j \sigma^2(\alpha_i;\alpha_j) = \sum_j \langle HP(\alpha_j)H\rangle^{\alpha_i}$$
$$d^{-1}(\alpha_j)\sigma^2(\alpha_i;\alpha_j) = d^{-1}(\alpha_i)\sigma^2(\alpha_j;\alpha_i) \qquad (6\text{-}67)$$

the quantities appearing in the last equation being the mean-squared H matrix element connecting the two subspaces; of course if, for fixed i, $\sigma^2(\alpha_i,\alpha_j)$ vanishes for all $j\neq i$, α_i is necessarily an H-eigenspace, and conversely.

To the extent then that we consider only centroids and variances we are taking account of those interactions between subspaces which are described by single excitations connecting pairs of them; if the description is accurate these are the only ones of consequence. In practice however we should be prepared to go to fourth order in moments, which involves multiple excitations and as many as four subspaces simultaneously. The same kind of concept is of course employed in constructing the complex well in optical-model scattering and many other situations.

For expectation values we see immediately that under the partitioning $\underline{m} = \sum \alpha_i$, the expansion (6-53 becomes,

$$K(W) = d^{-1}(m) \sum_i d(\alpha_i) \sum_\nu \langle KP_\nu(H)\rangle P_\nu(W) \tag{6-68}$$

But this is by no means the best we can do; for, though we have partitioned the space and hence the density, $\rho(x)=d^{-1}(m)\sum d(\alpha_i)\rho_{\alpha_i}(x)$, we have used $\rho(x)$ in determining the polynomials. It would be better to use, for each subspace α_i, its own density $\rho_{\alpha_i}(x)$ as the weight function to define its own set of polynomials. For we take for granted that the subspace densities do roughly partition the total spectrum, and therefore, with this *physically motivated* choice, each expansion is required to "cover" only a part of the spectrum. We can expect then that lower polynomial orders will be required than in the non-partitioned case. True, there may be far more of them but low-order moments are easy to evaluate while for higher-order moments the difficulty increases extremely rapidly with the order.

To replace (6-68) by its fully-partitioned form we go back to (6-53) which gives $K(x)\rho(x)=\langle K\delta(H-x)\rangle^m$, partition the trace first and then use the completeness relation just as we have done in deriving the last form of (6-53). Then we have the final exact form

$$K(W)=\sum_i \frac{I_{\alpha_i}(W)}{I_m(W)} \sum_\nu \langle KP_\nu^{(\alpha_i)}(H)\rangle^{\alpha_i} \times P_\nu^{(\alpha_i)}(W) = \sum_i \frac{I_{\alpha_i}(W)}{I_m(W)} K(W;\alpha_i) \tag{6-69}$$

in which the polynomials $P_\nu^{(\alpha_i)}(x)$ are orthonormal with respect to $\rho_{\alpha_i}(x)$, no orthogonality however being implied between polynomials defined by different densities.

For the strength (6-55) let us consider first [19] how we might produce a form for $R(W'W)$ by the most elementary intuitive procedure, one which is sometimes used in practise. Suppose that we partition the densities by configurations which arise by partitioning the single-particle space into "orbits" (which may or may not be the usual spherical ones) and the number of particles accordingly; thus $N \to \sum_{i=1}^{\ell} N_i$; $m \to \sum [m_1, m_2, \ldots m_\ell] = \sum \vec{m}$. Then a first approach to the problem gives

$$R(W',W) \to \sum_{\vec{m}',\vec{m}} M^2(\vec{m}',\vec{m})\,\delta(W'-E(\vec{m}'))\,\delta(W-E(\vec{m}))$$
$$= \sum_{\vec{m}',\vec{m}} d(\vec{m}')^{-1} \langle O^+O(\vec{m}'-\vec{m})\rangle^{m} \delta(W'-E(\vec{m}'))\,\delta(W-E(\vec{m})) \quad (6\text{-}70)$$

in which $M^2(\vec{m}',\vec{m})$ is the mean-squared matrix element connecting the two configurations and we have expressed it in terms of a trace by using a decomposition[14] of O into parts which, when acting on a state in one configuration, produces a state of the other; thus $O \to \sum O(\vec{q})$ where $O(\vec{q})$ on a state in $\vec{m}$ produces one in $\vec{m}+\vec{q}=[m_1+q_1, \ldots m_\ell+q_\ell]$.

This very crude theory already accomplishes something by spreading the strength according to the centroids but, since it ignores the spreading of the individual configurations, it does not distribute the strength widely enough. As a plausible extension we can assume that the strength "follows" the densities. Then, since the $\vec{m}$ states at W form a fraction $I_{\vec{m}}(W)/I_m(W)$ of the total, we find

$$R(W',W) \to \sum_{m',m} \frac{I_{\vec{m}}(W')\,I_{\vec{m}}(W)}{I_{m'}(W')\,I_m(W)}\, d(\vec{m}')^{-1} \langle O^+O(\vec{m}'-\vec{m})\rangle^{\vec{m}} \quad (6\text{-}71)$$

which now gives strength over the entire spectra. With this form we find that the energy-weighted sums are proportional to the corresponding density moment,

$$M_p(W) \quad \sum_{\vec{m}',\vec{m}} \frac{I_{\vec{m}}(W)}{I_m(W)}\, M_p(\vec{m}') \langle O^+O(\vec{m}'-\vec{m})\rangle^{\vec{m}} \quad (6\text{-}72)$$

a form which, in strong contrast to (6-69), or even (6-63), takes no account whatever of the correlations between O^+O and H which would lower or raise the states with expectation values different from the average, an effect which *is* taken account of in the simplest CLT form (6-48). There are in fact a whole hierarchy of interactions and

---

[14]This decomposition is immediate in the configuration case. For other subspace decompositions we could in principle introduce a projection operator $P(\alpha_j')$ into the traces. In practice however it is usually better to use Racah transformations (for the group which defines the decomposition) in order to isolate a single final subspace [19].

correlations which are ignored in these schematic treatments; they can be distinguished and catalogued in a proper theory, and their effects are especially large when the transitions give rise to collectivities.

For the proper theory we simply go back to (6-61), and, proceeding as we did with the expectation value, we find easily the result for general partitioning that

$$R(W',W) = \sum_{i,j} \frac{I_{\alpha'_j}(W') \times I_{\alpha_i}(W)}{I_{m'}(W') \times I_m(W)} R(W'\alpha'_j;W\alpha_i)$$

$$R(W'\alpha'_j;W\alpha_i) = [d(\alpha'_j) \times d(\alpha_i)]^{-1} \times \sum_{\mu,\nu} \langle\langle O^{+} \times P_{\mu}^{(\alpha'_j)}(H) \times O(\alpha_j-\alpha_i) \times P_{\nu}^{(\alpha_i)}(H)\rangle\rangle^{\alpha_i} \times P_{\mu}^{(\alpha'_j)}(W') \times P_{\nu}^{(\alpha_i)}(W) \qquad (6\text{-}73)$$

so that once again the form is precisely what we would expect by considering density superpositions, *but with all the correlations built in.* In applying (6-73) for configurations the partitioning is immediate, just as above. For other subspace decompositions we could in principle introduce a projection operator $P(\alpha_j)$ into each of the traces of (6-73). In practice however it is usually better to use Racah transformations, for the group which defines the partitioning, in order to isolate a single final subspace.

5. Trace Evaluation; Information and its Propagation

The matrix elements of a k-body operator in an m-particle space are linear functions of its matrix elements in the k-particle space in which the operator is defined. We may say then that the matrix elements "propagate." But this is not very helpful since the coefficients which enter (the "propagators") are in general immensely complicated; if we were obliged to use them, it would follow that methods we have been describing are of little *practical* value, even though they do make clear what kinds of structures are involved in various phenomena. On the other hand, since the CLT obviously implies that most of the "information" contained in the specification of the operator is filtered out, we might well expect much simpler propagation for the low-order traces which survive the filtering process. This is indeed the case, provided that any partitioning which is done divides the model spaces into irreps (or sums of equivalent irreps) of some group. The complete m-particle model space itself forms an irrep, the antisymmetric[15] one [m] of the group U(N) of unitary transformations in the

[15]For fermions it is more convenient to describe the Young shape by columns rather than rows.

single-particle space; partitioning by configurations, in which we define "orbits" by $N \to \sum N_i$, involves a direct-sum subgroup of U(N), while, for example, isospin partitioning involves a direct-product subgroup, $U(N/2) \times U(2)$.

Though the relationships between statistical behavior, symmetries and information which are touched upon here are extremely interesting (though by no means completely understood) we shall not do much with them but shall instead turn to some more elementary and practical aspects of trace evaluation.

We have mentioned and used the elementary result that

$$\langle F(k) \rangle^m = \binom{m}{k} \langle F(k) \rangle^k \qquad (6\text{-}74)$$

which shows that in the complete space $\underline{m}$ we do have direct propagation of traces. A prosaic but instructive derivation of (6-74) uses a direct-product basis (each m-particle state having m filled and (N-m) empty s.p. states). A simpler more formal proof argues that, since the trace of the m-particle density is a unitary scalar, invariant under U(N) transformations of the s.p. space, and similarly for $\langle F(k) \rangle^m$ itself, only the U(N)-scalar part of F(k) can contribute to $\langle F(k) \rangle^m$; but the only U(N) invariants which can be realized in a fermion space are the number operator n and functions of it. The only k-body invariant (which must vanish in spaces with m<k) is then $n(n-1)(n-2)\ldots(n-k+1) \sim \binom{n}{k}$; thus $F(k) \equiv \langle F(k) \rangle^k \binom{n}{k}$, the equivalence being for linear traces only and the coefficient being fixed by taking m=k, for which $\binom{n}{k} \to 1$.

As an immediate extension of this we see that, for an operator F of mixed particle ranks $\leq u$ (i.e. $F = \sum_{t=0}^{u} F(t)$), $\langle F \rangle^m$ is a polynomial of rank u in m, and therefore expressible in terms of its own values at any (u+1) values of m, either by inspection or, formally, via Lagrange polynomials [20,21]. This result contains that of (6-74) since F(k) vanishes for m=0,1,...(k-1). Let us use it to consider again the asymptotic spectrum of a one-body Hermitian operator, but this time taking account of blocking corrections and not ruling out the singular spectra mentioned in the second footnote. We can, without loss, take H(1) to be traceless, whence $\langle H(1) \rangle^m = 0$ and then we see from its explicit form, $\sum \varepsilon_{ij} a_i^+ a_j$, that it simply changes sign under a hole$\leftrightarrow$particle transformation, $(a_i \leftrightarrow a_i^+$. Alternatively we can agree to deal with central moments only and then these satisfy $M_p(m) = (-1)^p M_p(N-m)$. Then by inspection:

$$M_2(m)=m(N-m)(N-1)^{-1}M_2(1)$$
$$M_3(m)=m(N-m)(N-2m)(N-1)^{-1}(N-2)^{-1}M_3(1) \qquad (6\text{-}75)$$
$$M_4(m)=m(N-m)(N-3)^{-1}\{-(m-2)(N-m-2)(N-1)^{-1}M_4(1)+(m-1)(N-m-1)(2N-4)^{-1}M_4(2)\}$$

the even ones being special cases of

$$M_{2\nu}(m)=\binom{m}{\nu+1}\binom{N-m}{\nu+1}(\nu+1)!\sum_{s=1}^{\nu}\frac{(N-2s)(-1)^{\nu-s}}{s!(\nu-s)!}\frac{1}{(m-s)(N-m-s)}\binom{N-s}{\nu+1}^{-1}M_{2\nu}(s)$$

$$\xrightarrow{N\to\infty}\binom{m}{\nu+1}(\nu+1)\sum_{s=1}^{\nu}(-1)^{\nu-s}\binom{\nu}{s}\frac{1}{(m-s)}M_{2\nu}(s) \qquad (6\text{-}76)$$

The first equation of (6-75) gives $\sim(1-m/N)$ as the blocking correction to the variance; the second shows that $\gamma_1(m)\xrightarrow{N\to\infty}\gamma_1(1)/m^{1/2}$ so that distributions become symmetrical for large m; combining the third equation with the elementary result that

$$M_4(2)=2(N-8)(N-1)^{-1}M_4(1)+6N(N-1)^{-1}\{M_2(1)\}^2 \qquad (6\text{-}77)$$

gives for the excess[16]

$$\gamma_2(m)=\frac{(N-1)}{(N-2)(N-3)m(N-m)}\left\{[N(N+1)-6m(N-m)]\gamma_2(1)+6N\left[1-\frac{m(N-m)}{(N-1)}\right]\right\}$$

$$\xrightarrow[\text{fixed } m]{N\to\infty}\gamma_2(1)/m \qquad (6\text{-}78)$$

The asymptotic result for $\gamma_2(m)$, just as for $\gamma_1(m)$, would follow directly from the additivity of the cumulants under convolution which we have discussed earlier. For non-singular spectra $\gamma_2(1)=O(1)$ and then the excess vanishes for large particle number. For example with m particles in a large orbit, $N=2j+1$, we have

$$\langle J_z^2\rangle^1=N^{-1}\sum_{-j}^{j}s^2=\frac{1}{12}(N^2-1)\ ;\quad \langle J_z^4\rangle^1=\frac{1}{240}(N^2-1)(3N^2-7)$$
$$\langle J_z^4\rangle^2=\frac{1}{30}(N-2)(N+1)(2N^2-2N-7) \qquad (6\text{-}79)$$

and then for J_z

$$\gamma_2(m)\to-\frac{6}{5}\frac{N(N+1)-m(N-m)}{m(N-m)(N+1)}\xrightarrow{N\to\infty}-\frac{6}{5}\frac{1}{m}\xrightarrow{m\to\infty}0 \qquad (6\text{-}80)$$

its effects being unobservable by eye as long as there are a half-dozen particles or more.

The asymptotic vanishing of the excess does not of course prove that

---

[16] Products and powers of traces propagate by the same rules as single traces and thus these rules apply for example to cumulants.

the asymptotic distribution is Gaussian but a contrary circumstance would be most unusual. On the other hand, for singular one-particle spectra, $\gamma_2(1)=O(N)$, and then the Gaussian form is not guaranteed for large particle number. For the extreme case with spectrum $\{0^{N-1},1\}$ we have simply $\gamma_2(1)=N$. We might say then that the fermion system cannot admit enough particles to make $\gamma_2(m)<<1$; in fact with $m=N/2$, $\gamma_2(m)\xrightarrow{N\to\infty}-2$.

If the operator whose trace, over the complete m-particle space, which we are evaluating is given to us in normal or antinormal form (i.e. decomposed as a sum of operators of fixed particle or hole rank) propagation is immediate and the problem is solved. We can avoid the decomposition by taking, as input values, $t=0,1,\ldots u$, whence[17]

$$\langle F\rangle^m = \sum_{t=0}^{u} \binom{u-m}{u-t}\binom{m}{t}\langle F\rangle^t \tag{6-81}$$

For $F=H^2$, where H is a (0+1+2)-body operator, we would still need 3-particle and 4-particle traces as input, and for large N these might not be directly calculable. We can avoid that however by using $t=0,1,2,N,N-1$ (i.e. 0,1,2 particles +0,1 holes), for which we would use both the particle and hole representations of H; it is trivial to write (in terms of Lagrange polynomials) the propagation form for that or any other case [20].

Of particular interest is the norm of a Hermitian k-body operator F(k), which to within a sign is identical with its *hole-particle adjoint*; i.e. $F(k)\xrightarrow{a_i\leftrightarrow a_i^+}\tilde{F}(k)=(-1)^k F(k)$. As we have seen this property obtains in general when k=1 but this is not so when k>1. Since the operator vanishes in 2k spaces, those with t<k particles and t<k holes, we choose these input values along with t=k and then the squared operator propagates as a simple multiple of its k-particle trace:

$$\langle F^2(k)\rangle^m=\binom{m}{k}\binom{N-m}{k}\binom{N-k}{k}^{-1}\langle F^2(k)\rangle^k \xrightarrow{N\to\infty} \binom{m}{k}\langle F^2(k)\rangle^k \tag{6-82}$$

and similarly for the product of two operators of fixed particle rank, at least one of which has the specified behavior under $h\leftrightarrow p$.

Note incidentally that, in spaces with $m<<N$, the $(k,k+1,\ldots 2k)$-body operator propagates as if it were a pure k-body operator, as we see from the propagator $\binom{m}{k}$. Because of this effective reduction in particle rank one might guess that the cumulants would vanish for

---

[17] Note that $\binom{-a}{b} = (-1)^b\binom{a+b-1}{b}$

large m and thus that a Gaussian distribution would necessarily obtain for these restricted k-body operators. It will be found however by using (6-76) that this is not the case and that, for Gaussian behavior of the spectrum, one further condition must be satisfied by F(k). For k=1, as we have seen, the condition is that $\gamma_2(1)$ is not comparable with N. For k=2 the condition has been given [21] as a restriction on a "classical" fourth-order ring diagram associated with H^4; but we know by this time that it is an "unusual" H for which that condition is *not* satisfied so that Gaussian obtains almost always.

A k-body operator which vanishes in t<k particle spaces and hole spaces both is an irreducible U(N) tensor [22], transforming according to the Young (columnar) shape (N-k,k); we write such an operator as $F^k(k)$, a k-body operator which transforms as $[N-\nu,\nu]$ being then $F^\nu(k)=\binom{n-\nu}{k-\nu}F(\nu)$. As indicated, such an operator has a unitary-scalar factor which arises from (a,a^+) contractions, while $F(\nu)$ would vanish under any further contraction. The corresponding unitary decomposition of operators

$$F(k) = \sum_{\nu=0}^{k} F^\nu(k) = \sum_\nu \binom{n-\nu}{k-\nu} F(\nu) \tag{6-83}$$

which is orthogonal with respect to our unitary norm, is important in all trace evaluations, as we have already seen in (6-82); it gives for example a very simple way [21] of evaluating configuration variances, which would be very complicated indeed if we had to evaluate them via Lagrange-polynomials in ℓ variables $m_1,m_2,\ldots m_\ell$.

Except for an (n-1) factor the ν=1 part of a two-body Hamiltonian H(2) is representable, (6-83), as an "induced" one-body operator which thus renormalizes the primary one-body term H(1) by an amount dependent on the total number of particles. If $\xi_i(1)$ are the primary traceless single-particle energies and $W_{ijk\ell}$ the two-body matrix elements we find that

$$H^{\nu=1}=\sum_i \xi_i n_i=\sum\left[\xi_i(1)+(n-1)(N-2)^{-1}\left\{\sum_j W_{ijij}-N^{-1}\sum_{jk} W_{jkjk}\right\}\right]n_i \tag{6-84}$$

in which we have taken for granted that the induced Hamiltonian is diagonalizable in the same basis as the primary one-body term, the usual case when we deal with spherical orbits; otherwise a simple modification of (6-84) is called for.

We can show now how we can determine, via our statistical considerations, some specific properties of the interaction Hamiltonian by measuring an expectation value, e.g. a single-state occupancy $\langle n_s \rangle$ in the ground state, which is accessible to measurement by means of single-particle transfer experiments. Let us first evaluate the *correlation integral* between n_s and a one-body traceless Hamiltonian, $H(1)=\sum_i \xi_i n_i$ with $\sum \xi_i=0$. We have

$$\langle n_s H(1)\rangle^m = \sum_{i\neq s} \xi_i(1)\langle n_s n_i\rangle^m + \langle \xi_s(1)\rangle^m = \binom{m}{2}\binom{N}{2}^{-1} \sum_{i\neq s} \xi_i(1) + \frac{m}{N}\xi_s(1)$$

$$= m(N-m)N^{-1}(N-1)^{-1}\xi_s(1) \qquad (6\text{-}85)$$

where we have used (6-74), but could equally well have written the result at sight. The important result, that the correlation integral measures the traceless single-particle energy, extends directly to the case of interacting particles whose Hamiltonian contribute to the integral only via the (ν=1) part (higher order tensors not contributing because of orthogonality). The CLT result is then

$$n_s(W) = \langle n_s\rangle^m + \frac{\langle n_s H^{\nu=1}\rangle^m (W-E)}{\sigma^2} = \frac{m}{N}\left\{1+\frac{N-m}{(N-1)}\xi_s\right\} \frac{(W-E)}{\langle (H^{\nu=1})^2\rangle^m + \langle (H^{\nu=2})^2\rangle^m} \qquad (6\text{-}86)$$

where ξ_s is the renormalized single-particle energy as given by (6-84). Note that the ν=2 part of the Hamiltonian contributes only via the spectral norm of the total H which we can usually regard as well known. Thus a measurement of the occupancy determines the single-particle renormalization and hence a specific Hartree-Fock-like parameter of the Hamiltonian. Such measurements have been used for this purpose and in particular to distinguish between various model interactions [23].

For m-particle or configuration traces of operators which are more complicated than H^2, with H a (0+1+2)-body operator, directly converting the operators into normal or antinormal form is usually out of the question and the method of unitary decomposition, though still more or less essential, by no means solves the problem. Formal methods for doing this, and explicit results for H^3 and H^4, have been given by LI [24], and by GINOCCHIO and his collaborators [25] whose derivations of elegant contraction formulae have been somewhat simplified recently [26] by a direct use of Wick's theorem. A considerable amount of work is in progress towards extending these results, finding good approximations for them, and writing better programs for them. We mention in passing that one encounters the same summations here as one does in the conventional theory of effective interactions which itself is now being studied [27] by these spectral methods.

In order to deal with partitioning by other groups or chains of groups the essentially combinatorial procedure used above may not be at all adequate. For that purpose, and for others as well, it is good to introduce a somewhat more general procedure (which may be regarded as combinatorial also but in a more modern sense), involving trace operators and group invariants. In fact the operator $\binom{n}{k}$ used above is our first example.

If we write $\Psi_\alpha(m)=\psi_\alpha(m)|0\rangle$, where the state operator $\psi_\alpha(m)$ is a linear combination of k-fold products of state operators, we have

$$\rho(m) \equiv \sum \psi_\alpha(m)\psi_\alpha^+(m) = \binom{n}{m} \quad ; \quad \tilde{\rho}(m) = \sum \psi_\alpha^+(m)\psi_\alpha(m) = \binom{N-m}{m} \qquad (6\text{-}87)$$

the first following from the fact that $\rho(m)$ is an m-body operator which gives unity on every m-particle state, and the second from the first by h↔p conjugation. Using now the obvious result that $\langle\langle G\rangle\rangle^m = \langle\langle \tilde{G}\rangle\rangle^{N-m}$, for an arbitrary operator G with h↔p adjoint $\tilde{G}$, we can directly evaluate $\langle\langle F(k)\rangle\rangle^m$; we find the propagation equation,

$$\langle\langle F(k)\rangle\rangle^m = \langle\langle \tilde{\rho}(N-m)F(k)\rangle\rangle^k = \langle\langle \rho(N-m)\tilde{F}(k)\rangle\rangle^{N-k} \qquad (6\text{-}88)$$

which then reproduces our elementary result (6-74). The important thing however is that (6-88) extends to any m-particle subspace Γ if we introduce $\rho_\Gamma(m)$ by simply restricting the α-sum in (6-87) to run over a basis of Γ; note that a similar restriction to Γ_c is imposed thereby on the *complementary* (N-m)-particle space. We have then

$$\langle\langle F(k)\rangle\rangle^{m\Gamma} = \langle\langle \tilde{\rho}_{\Gamma_c}(N-m)F(k)\rangle\rangle^k = \sum_{\Gamma'} \langle\langle \tilde{\rho}_{\Gamma_c}(N-m)F(k)\rangle\rangle^{k\Gamma'}$$

$$\xrightarrow{(s)} \sum_{\Gamma'} \langle\langle \tilde{\rho}_{\Gamma_c}(N-m)\rangle\rangle^{k\Gamma'} \langle F(k)\rangle^{k\Gamma'} = \sum_{\Gamma'} \langle\langle \rho_{\Gamma'}(k)\rangle\rangle^{m\Gamma} \langle F(k)\rangle^{k\Gamma'} \qquad (6\text{-}89)$$

the last two forms of which are valid when $\tilde{\rho}_{\Gamma_c}(N-m)$ behaves as a multiple of unity in each subspace, in which case we have simple (s) propagation, from the space in which F is defined, to the more complicated spaces of interest to us. These forms display beautifully how the trace information then propagates throughout all the subspaces, the propagation coefficient being dependent only on the pair of representations involved, in fact on the weight with which one representation space is found in the other.

It is more or less obvious that only when Γ is defined in terms of the irreducible representations of a group G will the requirement which leads to simple propagation be satisfied by the ρ_Γ operators. But then by construction the $\rho_\Gamma(k)$ are k-body G scalars which satisfy $\rho_\Gamma(k)\psi_\alpha^{\Gamma'}(k) = \delta_{\Gamma\Gamma'}\psi_\alpha^\Gamma$. These properties completely define the operators, since they specify all its k-particle matrix elements, and show that the set of the $\rho_\Gamma(k)$ for fixed k forms a Green's Function which "picks up" the input information in the m=k subspace and propagates it to where it is needed. Just as (6-81) is an extension of (6-74) which deals with operators of mixed particle rank, so there are corresponding extensions of (6-89).

One is accustomed to thinking of group scalars in terms of Casimir operators and indeed a test [20] for simple propagation is to count the number of invariants of the proper maximum particle rank k which one can construct from powers and products of 1,n, the Casimir operators of G, and those of any intermediate group whose irrep is specified. If that number equals the number of irreps with particle numbers up to k the Casimir Green's Function exists, we have simple propagation, and the propagating coefficients are given directly in terms of the Casimir eigenvalues. For example, with averaging over m-particle states with fixed isospin we find 9 operators with maximum particle rank 4, *viz* n^p (p=0-4), $T^2,(T^2)^2,nT^2,n^2T^2$ and 9 irreps with m$\leq$4, *viz* mT=00,1 1/2,20,21,3 1/2,3 3/2,40,41,42. Thus we have simple propagation for operators with particle-rank $\leq$4, and indeed the result is general. The explicit construction of the ρ_Γ operators is straightforward.

This pleasant situation obtains in several important cases [20, 28,29], but in others the operators ρ_Γ are not Casimir representable, other procedures must be adopted to construct them, and their use in (6-89) is difficult. There are other ways to deal with the problems, some of them with very interesting physical motivation, but we shall not discuss them here. See [12,30] for brief discussions.

6. Fluctuations

We have been dealing with smoothed distributions. We now consider the fluctuations, deviations from smooth behavior, for spectra, expectation values and strengths. There is a large body of data. For energy-level fluctuations in particular, which are described by the $I_{f\ell}(E)$ term in (6-1), many long runs of levels of the same exact symmetries (J,π) are observed as resonances in the scattering of slow neutrons on heavy even targets [5,31]. They occur at high excitations, ~6-8 MeV, in the compound nucleus with spacings ~10 e.v., smaller by 10^4 than the spacings near the ground state. A run may contain up to a few hundred levels. Runs are found also in proton scattering on intermediate nuclei [6,32]. Besides that there is some experimental evidence that the kind of fluctuations displayed by these runs extends right to the ground-state domain [33] even though no runs of levels are observable there, nor in the no-man's land which extends from the resonance domain down almost to the ground state; the same thing shows up in shell-model and Monte-Carlo matrix calculations. These results are compatible with the notion that, when proper allowance is made for the rapid change of level spacing with excitation energy, a spectrum forms an analog of a *stationary* discrete random process. It is very reassuring that the theory used to describe the fluctuations predicts this *global* feature of spectra, as well as the *local* fluctuation behavior. We have stressed the behavior of subspectra corresponding to a definite set of exact symmetries. In a sense these form the basic spectra, the fluctuations for mixtures of such spectra being of much less intrinsic interest. We remark also that the stationarity at low excitation energy will often be disturbed near the ground state by the action of collectivities and inexact symmetries.

The questions to be considered are:

1. What is the general nature of the fluctuations?
2. How can one define and calculate observable properties of the fluctuations, and how well do the calculated measures agree with experiment.
3. What information about nuclear structure do the fluctuations carry?
4. What limits do they impose on the accuracy of quantities (e.g. low-lying energies) calculated from smoothed distributions? Do these irreducible errors imply any real loss of significant information?

We have commented above that spectral fluctuations are directly calculable only in terms of high order *spectral* moments. There are no methods available for calculating such quantities and so we introduce an appropriate ensemble [7], and calculate *ensemble* moments, relying on an ergodic property of the ensemble which we must of course demonstrate. Our results will then be valid for "almost all" H's to be found in the ensemble.

What "results" do we calculate? Certainly not the detailed fluctuations, but rather *measures* for them. Natural measures for the energy-level fluctuations are the variances $\Sigma^2(k)$ of the number of levels in an interval of length kD, where D is the locally averaged level spacing; the centroid of the number distribution is of course k, and k may take any positive value, not necessarily integral. Since the number of levels in an interval (x,x+L) is $d\times\int_x^{x+L}\rho(x)dx=d\times\{F(x+L)-F(x)\}$, it follows that $\Sigma^2(k)$ is strictly a *two-point function*, being expressible in terms of a quadratic product of density or distribution functions. Another natural measure is the minimum value of the mean-squared deviations of the observed levels from the positions they would have if the spectrum were uniform, viz

$$\Delta^*(n) = \frac{1}{n} \min_{A,B} \sum_{s=1}^{n} (x_s - A_s - B)^2 \qquad (6\text{-}90)$$

where x_s is the s'th energy value.

Instead of $\Sigma^2(k)$ the experimentalists use $\sigma^2(k)$, the variance of the k'th nearest-neighbor spacings, defined for integral k=0,1,2,... (an unfortunate convention gives k=0 instead of k=1 for nearest neighbors). $\sigma^2(k)$ is not two-point though it is clear that $\sigma^2(k)$ and $\Sigma^2(k)$ are simply related. However it turns out [34] that to high precision, and for almost all ensembles, $\sigma^2(k) = \Sigma^2(k+1) - 1/6$, and thus for all practical purposes the $\sigma^2(k)$ are two-point measures also. Instead of Δ^* the experimentalists use Δ_3, as originally defined by DYSON and MEHTA [35], which roughly speaking assigns weights proportional to the spacings, while Δ^* assigns equal weights to each level. The two Δ's are related in the same way as σ^2 and Σ^2; very closely $\Delta_3(n) = \Delta^*(n) + 1/12$.

There are many other measures, used for various purposes, but essentially all of the useful ones are two-point measures expressible as integrals over $\Sigma^2(x)$ or one of the other two-point forms described at the end of section 3; for example [34], $\bar{\Delta}_3(m) = 2m^{-4} \int_0^m (m^3 - 2m^2 r + r^3) \times \Sigma^2(r)dr$. It is not an accident that useful measures are two-point; more complex fluctuations would require, for any reasonably accurate evaluation, longer runs of levels than the runs ~ 100 which are experimentally found. A first general result is then that the measureable fluctuations in spectra are defined by a two-point correlation function, without significant contributions from k-point with k=3,4,...; in one sense therefore the fluctuations seem to be of the simplest possible structure, described formally by a function which is as close as possible to the density itself. On the other hand, in our usual thinking (6-1) about the separation between secular behavior and fluctuations, we take for granted that these phenomena are as "far apart" as possible, one characterized by very long wave-lengths, $\lambda \sim$ spectrum span, the other by short wavelength, $\sim$level spacing.

There isn't any large paradox here. The fluctuations themselves are immensely complicated, more so (because they are *eigenvalue* fluctuations) than one encounters in many classical systems, but, since only limited runs are available, only relatively simple properties are deducible.

Consider a run of levels of the same exact symmetries, as found in experiment or by matrix diagonalization. A first guess, that the levels are randomly spaced (as are, in time, the pulses from a radioactive target) is wrong; random spacing would give $e^{-x/D}$ as the distribution of the spacings between nearest neighbors; but the distribution which is found, $\sim xe^{-x^2}$, is markedly different from this. It may be regarded as displaying a *repulsion* between levels which makes small spacings quite infrequent, in contrast to the "random" or Poisson distribution. It is plausible in fact to regard the fluctuations quite generally as a manifestation of the WIGNER-von-NEUMANN level repulsion [36].

A heuristic derivation [37] of the nearest-neighbor spacing distribution P(s) introduces r(S), the probability density that, given a level at energy 0, the interval (0,S) will be empty while (S,S+dS) will contain the next level. Then easily

$$P(s)=r(s)\int_{S}^{\infty} P(x)dx \rightarrow P(s)=c\,\exp\left\{-\int^{S}[r(x)-r'(x)/r(x)]dx\right\} \qquad (6\text{-}91)$$

With the assumption that the levels are uncorrelated ($r(x)=1/D$ where D is the mean local spacing) we find the Poisson law, $P(s)=D^{-1}\exp\{-S/D\}$; an assumption of linear repulsion, $r(x)=\alpha x$ gives the Wigner law, $P(s) =\{\pi s/2D^2\}\exp\{-\pi s^2/4D^2\}$, in excellent agreement with experiment, not only at small spacings where linear repulsion is "reasonable" but also at large.

Strangely enough the same distribution follows from a GOE of two-dimensional[18] matrices. The eigenvalue spacing is $S=\{(H_{11}-H_{22})^2 + (2H_{12})^2\}^{1/2}$, the square root of the sum of squares of two identically-distributed, zero-centered, independent Gaussian variables with variances $4v^2$. This is a Rayleigh distribution,its form being exactly that given by the linear-repulsion argument. The repulsion and its linear behavior are made plausible here by the form of S, whose vanishing requires the vanishing of two independent variables.

It is essential however to recognize that a different weighting for the same set of matrices may lead to radically different results. This is obviously true for a singular weighting which would select only a single matrix, but it is so also for different "analytic" weightings. If we represent the ensemble by $H=\alpha 1+\beta\vec{\sigma}\cdot\vec{n}$ and take α,β to be independent Gaussians we will find that S is distributed as the positive half of a Gaussian, with small spacings predominant.

The significant lesson which follows from such considerations is the importance of the measure used to weight the Hamiltonians of the

---

[18] The result is very slightly different [38] for large-dimensional matrices.

ensemble. *"It will follow that neither the nature, nor the origin, nor the consequences of level repulsion can be understood by a probability argument which pays no attention to the specific features of the Hamiltonian"* [39].

6.1 Coin-Tossing Experiments

We describe a series of coin-tossing experiments which are meant to give some feeling for the kind of information which may emerge from fluctuation studies.

Experiment #1: Construct a large d-dimensional real-symmetric matrix by tossing a coin[19] $\frac{1}{2}d(d+1)$ times, assigning +1 to head and -1 to tail, and using the numbers to fill in, one row at a time, the diagonal and upper half of the matrix. Complete the matrix by Hermiticity so that the resultant matrix is real symmetric. Diagonalize and examine the nearest-neighbor-spacing distributions (each spacing being expressed in terms of the locally averaged spacing).

Results: The smoothed eigenvalue density will be semicircular with radius $d^{1/2}$. The nearest-neighbor spacing distribution will be similar to that for experimental results with slow neutrons on an even target. Other measures will not distinguish them either. Observe now that while the matrices are constructed with a random (coin-tossing) input the spectrum generated is by no means *obviously* random. What we are seeing here is the generation of an apparently non-random sequence, not easily distinguishable (if at all) from experimental results, by the non-linear action of matrix diagonalization on a random one. Does it follow from this that the experimental results carry no information?

Experiment #2: Same thing but construct a complex Hermitian matrix, two tosses for each matrix element, one each for the real and imaginary part.

Results: Once again a semicircular density, but with different fluctuations, the nearest-neighbor spacing distribution having a zero slope at the orgin instead of a finite value. Values of $\Sigma^2(r)$ will also be

[19] you thereby construct the random-sign ensemble [7] whose low-order statistical properties are identical with those of the GOE, and indeed of a wider class of real symmetric matrices including those whose elements are independently selected from essentially any zero-centered distribution. Experiment #2 will generate an ensemble similarly equivalent to the Gaussian unitary ensemble, GUE. For reasonable results you should take $d \gtrsim \sim 100$ or else use an ensemble of matrices. If you do things by computer you could equally easily construct the Gaussian ensembles instead of the random sign ones.

smaller, essentially by a factor 2. The fluctuations will not agree[20] with the experiment of #1.

Experiment #3: Same as #1 but construct a block matrix with two disconnected blocks of comparable dimensionality.

Results: A semicircular density again, but the spacing distribution will not have a zero at the origin and, in general terms, the level-repulsion effects will be moderated. The fluctuations will closely resemble those observed with slow neutrons on an odd target and, except for very small target-J, will be indistinguishable from them.

Experiment #4: Same as #3 but construct a multiblock matrix.

Results: The fluctuations will be indistinguishable from those of Poisson, the simplest random process.

The first remark to be made about these experiments is that the results which we quote are really predictions which need not necessarily come true; in #1 for example our process can produce *any* real symmetric matrix of dimensionality d and hence any spectrum; in the unlikely event that our coin tossing produced only heads, we would find the singular spectrum $\{d,0^{d-1}\}$. But our confidence will grow as d increases. And, if we carry out the experiment many times, our predictions will come true (to within a calculable "error") for "almost all" H matrices. It is in this sense that the spectral fluctuations of a large matrix so constructed, or the ensemble fluctuations of an ensemble of such matrices, could serve as a model for the observed fluctuations.

One's immediate reaction to the results of experiment #1, and their agreement with experiment, might be that the slow-neutron fluctuations carry no information at all. But this cannot be the complete story since exeriment #2 gives different results than #1. In fact what is at issue is something about the underlying fundamental symmetries, angular-momentum conservation and time-reversal invariance, for the physical systems being considered.

---

[20] They will in fact agree with a subspectrum realized by choosing alternate eigenvalues from slow neutrons on an odd target with relatively large angular momentum (the latter is the result of a remarkable theorem [18,40] to which we shall not return). We might remark also that there is a third "canonical" ensemble, the Gaussian symplectic ensemble, whose fluctuations we could also reproduce by a separate coin tossing experiment (or by selecting alternate eigenvalues in our spectrum of experiment #1), and which are different again.

The GOE, for which the real random-sign ensemble is a good approximation, satisfies the conditions that: the real-symmetric property is preserved under orthogonal transformations, but under no larger subgroup of the unitary transformations: it supplies a *uniform* measure for the ensemble members, in accord with the standard notion of statistical mechanics that we should have equal *a priori* probabilities, so that all allowed Hamiltonians are treated equally. The latter result follows from the fact that the ensemble is describable in terms of a uniform covering of a sphere in the $d(d+1)/2$-dimensional matrix-element space.

A Hamiltonian which is invariant under time reversal, and for which the angular momentum is preserved or is integral-valued, may be represented by a real (and therefore real symmetric) matrix, and thus the model introduced in experiment #1 is for systems which satisfy these conditions. Different ensembles result if the conditions which give rise to the orthogonal ensemble are not satisfied. In particular, if the Hamiltonian is not time-reversal invariant, then, irrespective of its behavior under rotations, the Hamiltonian matrices are complex and the canonical group which generates the uniform measure for the ensemble is the unitary group. If we have time-reversal invariance, but not rotation invariance, then, for systems with half-integral angular momentum, the H matrices are "quaternion real" and the ensemble is generated by symplectic transformations.

These general results are due originally to DYSON [41]. In an elegant but difficult later paper [42] he establishes connections between the mathematics of the three ensembles (which, in a sense which he defines, are the fundamental "irreducible" ones; hence his "three-fold way") and other mathematical structures.

The different results from experiments #1, 2 (and for the quaternion real matrices, should we construct them) give then a partial answer to our question about the information content of the fluctuations. They do supply a test for two underlying fundamental symmetries, behavior under rotations and time reversal. But unfortunately this test is not sufficiently sensitive, with the data available, to detect violations of these symmetries.

It should be noted that the orthogonal, unitary and symplectic invariances, which have been imposed here, do not derive from any trivial demand that the result of a calculation should be independent of the basis used for it, but rather from the natural desire, in the absence of further information, not to favor any particular subset of the Hamiltonians which are permitted by the fundamental symmetries considered; an extremely nice discussion of these things, in information-theory terms has been given by BALIAN [43]. On the other hand, if we believe it essential to preserve certain other features of H, besides the behavior under rotations and time reversal, (such as e.g. that H should contain a large Q•Q term) we are by no means restricted to the three canonical ensembles. The EGOE mentioned above derives from a requirement that the interaction should have a fixed particle rank, in particular that it should be two-body, in contrast to the canonical ensembles which, because of the statistical independence of the distinct matrix elements, involve only multibody interactions. It is a fact that embedding an ensemble has little or no effect on its fluctuations; this is by no means well understood, but it is fortunate since the more realistic embedded ensembles are technically very much more difficult to deal with than the ordinary ones, such as GOE, which we can continue to use as a realistic model for the fluctuations.

We learn something further about symmetries from our experiments #3, 4, but this time about model-space symmetries which would be the usual source of a matrix block structure which is maintained for an ensemble of Hamiltonians. We see that fluctuations are affected by the "interweaving" of exact symmetries, this of course because this moderates the level repulsion. This would supply an excellent test for hidden symmetries but, at high excitations where long runs of levels are found, one doesn't expect to find any[21]. But it follows also that the fluctuations are affected by weakly broken symmetries which we are liable to find in the ground-state region. Indeed this can probably be regarded as the source of the behavior in the ground-state domain which, from the statistical point of view, would be classed as "non-stationary". But technical difficulties make it hard to give a proper treatment of these problems.

6.2 Two-Point Function and Some Elementary Results

The classical method for dealing with the fluctuations [7,30,38] for the three canonical ensembles makes a direct use of the invariance properties to transform from the matrix element-space (d(d+1)/2 - dimensional for GOE) to the set of d-dimensional spectra, producing thereby the joint-probability distribution for the eigenvalues,

$$P_\beta(E_1 E_2,\dots E_d) = C_\beta \prod_{i<j} |E_i - E_j|^\beta \exp\left\{-\frac{1}{4v_\beta^2}\sum_s E_s^2\right\} \tag{6-92}$$

---

[21] However a simple way to test whether a given model H preserves a non-obvious symmetry is to calculate a shell-model spectrum using it and examine the nearest-neighbor spacing distribution.

with $\beta=1,2,4$ respectively for the orthogonal, unitary and symplectic ensembles, and C_β as a normalization constant. One now has the problem of integrating over all but k energy variables in order to produce the k-point functions, a very difficult problem whose treatment is well described in the references.

Instead of proceeding in this way we use an approximate, though remarkably accurate, method of binary-correlation expansions [14,34] which is relatively simple and is applicable to a wider class of ensembles. For the moments of the two-point density function, (6-44) with $G\to\rho$, we have, always for large d, that

$$\int S^\rho(x,y)x^p y^q dxdy = \Sigma^2_{p,q} = \overline{\langle H^p\rangle\langle H^q\rangle} = \sum_{\zeta>0} \overline{\underbrace{\langle H^p\rangle\langle H^q\rangle}_{\zeta}}$$

$$= \sum_{\zeta>0} \mu^p_\zeta \mu^q_\zeta \overline{\underbrace{\langle H^\zeta\rangle\langle H^\zeta\rangle}_{\zeta}} = \frac{2}{\beta d^2} \sum_{\zeta>0} \zeta \mu^p_\zeta \mu^q_\zeta \Longrightarrow S^\rho(x,y)$$

$$= \frac{2}{\beta d^2} \frac{\partial}{\partial x} \frac{\partial}{\partial y} \left\{\rho_o(x)\rho_o(y) \times \sum_{\zeta>0} \zeta^{-1} v_{\zeta-1}(x) v_{\zeta-1}(y)\right\}$$

$$= -\frac{2\sin^2(\frac{\beta\pi r}{2})}{\beta(\pi r)^2} + \frac{\sin\beta\pi r}{\pi r} \longrightarrow -\frac{2\sin^2(\frac{\beta\pi r}{2})}{\beta(\pi r)^2} + \frac{\delta(x-y)}{d\times\bar{\rho}(x)} \qquad (6\text{-}93)$$

Here the superscript bar denotes ensemble averaging. The identification of the two-point moment in terms of traces follows from (6-25). The next step, the ζ expansion, recognizes that, by the same argument used above for the density, and leading to (6-19), we count only those patterns in which the (p+q) H's are *pairwise* correlated. Such correlations can involve two H's in the same trace ("internally correlated") or one in each trace ("cross-correlated"); ζ is simply the number of pairs of the second type. The $\zeta=0$ term does not enter since it is eliminated by the second term of S^ρ, (6-44). For fixed ζ we could have generated the cross-correlated product by starting with $p=q=\zeta$. For GOE this product is simple to calculate and gives $2\zeta/d^2$, while for the canonical ensembles in general we find $2\zeta/\beta d^2$. We have now to insert into the first factor $(p-\zeta)/2$ correlated pairs in such a way that each pair contracts only over internally correlated pairs. For $\zeta=0$, which we do not encounter here, the number of ways of doing this would be the Catalan number $t_{p-\zeta}$ which we have met in (6-20); for $\zeta\neq0$, by a rather tricky combinatorial argument, we find its value to be $\mu^p_\zeta=\binom{p}{(p-\zeta)/2}$; we have a similar factor for the other trace, and thus a ζ-decomposition of the two-point correlation moments. This moment expansion is inverted by using a result of Table 1 in section 3, and gives $S^{\rho}(x,y)$ in terms of a series of correlated Chebyshev-polynomial excitations of the densities at the two points. That the excitation should be Chebyshev is natural since these are the polynomials with the average density as weight function. There are some uncertainties about the short-wave-length ($\zeta \sim d$) part of this expansion because, for one thing, the discrete nature of the spectrum becomes of consequence here, and this is incompatible with the use of polynomials generated by a continuous (smoothed) density as weight function. We resolve this in a plausible way by requiring that the results give back the known result [8] that the 2-point function is stationary when expressed in terms of the number of levels in the (x,y) interval, $r=(x-y)\bar{\rho}(x)\times d$. Then on using the trigonometrical form for the Chebyshev polynomials, the ζ summation becomes elementary and leads to the penultimate form of (6-93). The final form, the first term of which is the negative of the Y_2 function (6-46), follows by the usual identification of the second term as a δ function (which arises, as in 6-46, from the self-correlation of levels).

The final form is exact for the unitary ensemble and, while not so for the other two, gives measures which are remarkably close to the exact ones, deviating from them in a simple way attributable to a different behavior at very short distances, which can for most purposes be eliminated. All of the two-point properties of spectra, and hence essentially all of the measureable[22] fluctuations, now follow from (6-93), or its exact version; we have only to recall that $d \times \{F(x+L)-F(x)\}$ is the number of levels in the indicated interval. We skip the details [35,34,39].

An essential feature of spectra which will emerge from the two-point function is that spectra are highly rigid "essentially crystalline" [35] structures whose levels move very little as we go across an ensemble; in fact $\delta x_i/D \sim \frac{1}{\pi}\sqrt{\ell n\, d}$ for GOE (and essentially the same for other ensembles) so that a given numbered level (#16,437, say, where the counting begins at the ground state) will be found for various members of the ensemble to move, on an RMS average, only about 1.2 spacings if $d=10^6$ and only 1.7 if $d=10^{12}$. This of course is not

[22]The form of the nearest-neighbor distribution is however not so determined and indeed its slope at the origin depends explicitly on the 3-point function.

a directly measureable quantity, but for an ergodic ensemble it represents equally well the deviations to be found between the levels of a single characteristic spectrum and those derived from its smoothed density via (6-24); there are many Monte-Carlo calculations which verify this behavior, even for the embedded ensembles. Not only are the motions small but they are very highly correlated, the correlation coefficient (6-43) between the motions of levels which are r spacings apart being $1-\ell n(\pi r/2)/\ell n(2d)$, and the variance of the number of levels in a fixed interval $r\bar{D}$ being $\sim \frac{2}{\pi^2} \ell n\, \pi r$.

Correlations between *spacings* behave differently. Adjacent nearest-neighbor spacings are fairly strongly anticorrelated, with correlation coefficient $\zeta=-0.271$ (do not confuse this ζ with the ζ of 6-93). Adjacent spacings of higher order are more strongly anticorrelated; $\zeta \to -1/2$ in the limit. But all correlations between spacings fall off sharply, as r^{-2}, as the separation between them $(r\times D)$ increases.

These general properties agree with what we know about real spectra, and indeed the quantitative comparison between experiment and prediction, via the more complex two-point measures, works out well also [39] with essentially no exceptions or mysteries remaining.

What now is the significance of the fluctuations? With the important exceptions noted above, and with some uncertainties about the effects of symmetries and collectivities, we assert that the fluctuations indeed carry very little information in the sense of parameters which can be deduced from them. They are of course interesting in their own right, and it would be good if a compact and more meaningful description could be given of them beyond the statement that they derive from a two-point function of the general form (6-93). Apart from that it is certainly true, that via the action of the CLT in complicated model spaces, almost all of the microscopic information about the system is really lost, not to be recovered by any kind of shell-model analyzing or other such operations. Most of the information which is not lost is contained in the low-order moments and correlation coefficients which form the subject of the greater part of these lectures.

The fact that almost all the information is filtered away is not surprising at all; it merely tells us that we are really dealing with a subfield of statistical mechanics, for which this general behavior is one of the features taken for granted. Of course the problems which lie at the foundations of statistical mechanics have their counterpart in our immensely simpler domain, in particular the question which asks how it comes about that the completely determined evolution in time can be described by a random process [44]. In our case it is the spectral evolution which concerns us; and, just as it does in conventional statistical mechanics, where the stronger notion of "mixing" [44] is essential, the ergodic behavior of our ensembles goes some distance[23] toward resolving this question.

For ergodicity consider a quantity $f(E;\tau)$, defined for H^{τ}, a member of the ensemble. We define a spectral average over a domain which spans p levels, e.g. by

$$\langle f(E,\tau)\rangle_p = \frac{1}{p}\int_{-p/2}^{p/2} f(E+r\times D;\tau)\,dr \qquad (6\text{-}94)$$

and then ask whether its ensemble variance,

$$\mathrm{Var}_{(e)}\langle f(E;\tau)\rangle_p = \overline{\langle f(E,\tau)\rangle_p^2} - \left\{\overline{\langle f(E;\tau)\rangle_p}\right\}^2 = \frac{2}{p^2}\int_0^p (p-r)\,S^f(r)\,dr \qquad (6\text{-}95)$$

is sharply peaked (about its ensemble average) for large p, in particular whether $\mathrm{Var}_e \xrightarrow{p\to\infty} 0$, and, if so, how fast. The third form here is valid as long as $S^f(E_1,E_2)$ depends only on the energy difference $r\times D$, which we can take to be the case over the energy domains and for most functions of interest. We can use (6-95) for an explicit evaluation of the variance, which relates the ergodic behavior directly to the fluctuations; this is quite feasible for the ergodicity of the one-point functions (since S^f is then known and simple). As an example, one finds [8] for the density itself, ignoring corrections for small p, that

$$\mathrm{Var}_{(e)}\langle \rho(E)\rangle_p = (\bar{\rho}(E))^2\,\Sigma^2(p)/p^2 \simeq \frac{2}{\beta\pi^2}(\bar{\rho}(E))^2\,\frac{\ln(\pi p)}{p^2} \qquad (6\text{-}96)$$

from which we see that the ergodic behavior is very strong, averaging over a small number of levels being adequate for a small variance;

---

[23]But not all the way by any means; it still leaves open the question whether ensemble predictions, which are proved to be valid for "most" members of the ensemble, actually apply to a given physical system. The question here has been recently stated by MAHAUX and WEIDENMÜLLER [45] as follows:

> *Perhaps the most fundamental problem concerns the foundations of both CN* (compound nucleus) *theory and the statistical theory of nuclear spectra: Which properties of the nucleon-nucleon interaction justify the use of random-matrix models for the nuclear Hamiltonian?*

For the smoothed features of spectra, but not for the fluctuations, an answer is at hand, as is implied above.

for the Poisson case we would have $\Sigma^2(p)=p$, for which a much larger averaging domain would be needed. For more complicated functions $f(E)$, and in particular for the $(k>1)$-point correlation functions themselves, the explicit evaluation might well be out of the question. But the conditions for ergodicity can be expressed in terms of the asymptotic behavior of S^f, a sufficient condition for example being given [46,8] by Slutski's theorem that the ensemble variance vanishes if $p^{-1}\int_0^p S^f(r)dr \xrightarrow{p\to\infty} 0$. We shall not pursue these matters.

We have so far discussed only the eigenvalue fluctuations. Further information is carried by the strength fluctuations for a transition operator T. The basic result here, due to PORTER and THOMAS [47], is that the transition amplitude T_{ij}, between Hamiltonian eigenstates in a complex system, when renormalized by its local RMS value, has a Gaussian distribution[24].

This result, which we would come to also in our coin-tossing experiment #1, follows easily if, relying on ergodicity, we introduce a GOE and consider instead the ensemble distribution of the amplitude connecting a pair of numbered states. In the standard treatment one takes for granted that one of these states, Ψ_i, resides in the "statistical" space in which the GOE operates, while the other, Ψ_0, lies outside the space. Then the amplitude T_{i0}, which varies as H moves through the ensemble, is simply the overlap of Ψ_i with the projection, $PT\Psi_0$, of the giant resonance state, $T\Psi_0$, into the statistical space. The distribution, when renormalized is obviously the same for all operators T for which the projection is non-zero; it follows from the result of d-dimensional geometry that, as a unit vector $|i\rangle$ moves uniformly over the sphere, its overlap with a fixed vector $|\lambda\rangle$ has a distribution

$$\rho_{\langle i|\lambda\rangle}(x) = \pi^{-\frac{1}{2}}\,\Gamma(d/2)\,\Gamma\left(\frac{d-1}{2}\right)^{-1}(1-x^2)^{(d-3)/2} \qquad (6\text{-}97)$$

which becomes Gaussian[25] with zero centroid and variance d^{-1} for large d. Different amplitudes involving the same fixed state are essentially uncorrelated, $\zeta \sim -d^{-1}$, and indeed asymptotically independent; on the

[24] The strengths then are distributed as $\chi^2(1)$, the "Porter-Thomas distribution".

[25] Because of the simple nature of the GOE no local strength renormalization is necessary. For ensembles such as EGOE, and with experimental spectra, there is a secular variation of the strength which must be corrected for. To see where this comes from consider for example a particle-hole excitation involving a specific pair of orbits.

other hand, since the projections of two orthogonal states, Ψ_0, Ψ_1, which are outside the statistical space, are not necessarily orthogonal, and may even be parallel, amplitudes T_{i0} and T_{i1} are not necessarily uncorrelated.

This version of the strength distribution would seem appropriate if the states of a transition pair have very different structures, as for example with a scattering resonance in which the resonant state involves many hole-particle excitations of the ground state. The strength then would be greatly fragmented and it is natural to expect general statistical laws to apply. There is evidence however that, just as with the level spectra, the Porter-Thomas distribution is valid more generally, in particular for low-lying transition pairs, both states of which should then be treated statistically [48,39]. Then, with $x_{i\lambda} \equiv \langle i|\lambda\rangle$,

$$T_{ij} = \sum T_{\lambda\mu} x_{i\lambda} x_{j\mu} \longrightarrow \sum t_\lambda x_{i\lambda} x_{j\lambda} \tag{6-98}$$

where the first expansion is with respect to a general basis $|\lambda\rangle, |\mu\rangle$, ..., and the second a basis in which T is diagonal[26] with eigenvalues t_λ. We encounter now the product of two similar zero-centered and asymptotically independent Gaussian random variables; by Table 1 of section 3, or by reference [10], this product, when "standardized" as in (6-29), has a density function $\pi^{-1}K_0(|x|)$ with K_0 the modified Bessel function of the second kind, which then replaces Gaussian as the basic amplitude. The T_{ij} distribution now does depend on the nature of the transition operator, specifically, for an orthogonally invariant ensemble, on its spectrum. If, however, the spectrum is sufficiently "rich" we have (always for large d) the sum of many independent K_0 variables and then again we recover Gaussian via the CLT. The precise criterion here is the one which we have encountered earlier; a non-singular spectrum leads to a Gaussian amplitude and the Porter-Thomas strength distribution, a singular spectrum to something quite different and dependent more specifically on the spectrum.

The ergodicity of the strength distribution for both versions of the theory follows easily from (6-95) by using the asymptotic independence of the basic Gaussian or K_0 amplitudes.

---

[26]More often than not it is more appropriate to introduce separate ensembles for the two members of a transition pair. Then the operator T which induces "one-way" transitions between them is not Hermitian and not diagonalizable. Everything can be expressed however in terms of the Hermitian square T^+T; the results are essentially the same, including the distinction ahead which arises between operators with singular and non-singular spectra.

7. Comments about Applications

We have stressed the principles and methods of statistical spectroscopy. We can make only a few comments about applications. See the Proceedings of the Ames Conference on Moment Methods and Spectral Distributions [49] for further reviews, reports on work in progress and references to many published papers.

7.1 Level Densities, Partition Functions, Low-lying Spectra and Nuclear Masses

Calculations of level densities, and partition functions which carry the same information, are straightforward in principle and feasible in practice, even for huge model spaces. One partitions via configurations and isospin, or equivalently by separate proton and neutron configurations. The common assumption that configuration densities are Gaussian is not really adequate for the low-lying configurations. Third and fourth-moment corrections are desirable for these since, on the one hand, these densities will be most strongly deformed by interactions with higher ones, and, on the other, they largely fix the ground-state energy for which good accuracy is essential. Of particular interest are "no-core" calculations for light nuclei, in which all the nucleons are taken to be active, thus generating huge model spaces [50]. For non-interacting particles spectral calculations are faster than the combinatorial ones commonly used.

There are many calculations of low-lying spectra. For the most part these give decent agreement with shell-model calculations when these are feasible, disagreements however sometimes arising in even nuclei, especially for the 0^+-2^+ spacing. The real significance of the agreements and of the disagreements is not easy to assess. Any calculations derived from a smoothed density, $\langle I(E) \rangle$ of (6-1), has irreducible errors due to the neglect of the fluctuations, and so we encounter the unsolved problem of the information content of the fluctuations in the ground-state region. On the other hand a failure to reproduce a rotational spectrum, for example, would be a significant one. If the "completely statistical" theory which derives from $\langle I(E) \rangle$ is inadequate for spectra it should be possible to treat the lowest states by a combination of statistical and matrix methods, somewhat along the line recently used [27] in studying the effective interaction. We mention too that there has been good success in dealing with absolute binding energies and with Garvey-Kelson and more general mass relationships [51].

7.2 Expectation Values, Sum Rules and Strength Distributions

Sum rules of course determine expectation values which are perhaps the easiest things of all to calculate with satisfactory precision by statistical methods. We have seen in (6-86) how the measurement of occupancies determines renormalized single-particle energies. Recent work [52] on non-energy-weighted and linear-energy-weighted sum rules for Gamow-Teller β decay, as well as on the strength distribution itself, gives a foundation for major extensions of an early intuitive "gross" theory [53] of the β-decay strength. Similar studies [54] of the electromagnetic excitations do the same for the Kurath sum rule for isovector magnetic dipole transitions, and indeed generate usable sum rules for all the multipoles of interest, bounds for strengths in given model spaces, and other related quantities.

7.3 Symmetries

There have been extended studies of the distributions of symplectic [55] and spin-isospin SU(4) [56] symmetries, and also of the representation of Hamiltonians in terms of sums of Casimir operators [57]. Take for example the old "pairing plus quadrupole" model, for which curiously enough, the new calculations have *not* been done. We have $\hat{H}=\alpha P+\beta Q$, with specified forms for P,Q, which is supposed to represent a given detailed microscopic H. We can use (6-50) to calculate α,β *as functions of particle number and isospin*. The norm of the resultant $\hat{H}$, or equivalently its correlation coefficient with H, now gives a quantitative measure of the adequacy of $\hat{H}$, which, if adequate, we might choose to renormalize to the same magnitude as H itself. As a separate aspect of the study of symmetries we mention the problem of finding good measures for symmetry breaking [58].

7.4 Fluctuations

A recent review [39] concluded that agreement with theory is everywhere good for the energy-level fluctuations but that there are some interesting open questions about the strengths. There are however, other ways of thinking about fluctuations and of using random matrix ensembles. In our P+Q model for example, and similarly for other schematic models, we could, instead of discarding that part of H which is orthogonal to the best $\hat{H}$, represent it by a Hamiltonian ensemble of the proper norm, which in large model spaces is always orthogonal to any fixed Hamiltonian vector. Collective excitations characteristic of a schematic model are often unnaturally sharp; by the procedure suggested they would be broadened and in cases where $\hat{H}$ gives a poor representation of H they might be quenched completely. Once certain technical problems of moment evaluations are solved this should give a good method for studying collective modes. At present the only solved problem which makes any use of this procedure (for the level spectrum, not the excitations) is that of the GOE deformation of a pairing interaction [59].

A different application [19] deals with the fluctuations, as we vary the energy W, in the non-energy-weighted sum of the strengths originating in |W>. There are two ingredients, the first being the simple notion that the fluctuations are small when the strength is fragmented and large when it is not; this is taken account of by a local use of the Porter-Thomas distribution, coupled with a spectral calculation for the smoothed energy variation of the sum. The result is an expression, as a ratio of integrals over the calculable strength, R(Z,W) of (6-73), for the effective number of components into which the strength fragments, a good indicator of collectivity. This has been tested by comparison with shell-model calculations but no real applications to the prediction of collective behavior have yet been made.

References

1. N. Bohr: Nature 137, 344 (1936)
2. H. A. Bethe: Phys. Rev. 50, 332 (1936)
3. V. F. Weisskopf, D. H. Ewing: Phys. Rev. 57, 472 (1940)
4. H. H. Barschall: Phys. Rev. 86, 431 (1952)
5. H. Camarda, H. I. Liou, F. Rahn, G. Hacken, M. Slagowitz, W. W. Havens Jr., J. Rainwater: In *Statistical Properties of Nuclei*, ed. by J. B. Garg (Plenum Press, N.Y. 1972)
6. E. G. Bilpuch, A. M. Lane, G. E. Mitchell, J. D. Moses: Phys. Reports 28, 145 (1976)
7. E. P. Wigner: Ann. Math. 62, 548 (1955); SIAM Review 9, 1 (1967)
8. A. Pandey: Ann. Phys. (N.Y.) 119, 170 (1979)
9. C. van Lier, G. E. Uhlenbeck: Physica 4, 531 (1937)
10. M. G. Kendall, A. Stuart: *The Advanced Theory of Statistics; Vol. 1* (Hafner, N.Y. 1969)
11. F. C. Williams: private communication
12. J. B. French, J. P. Draayer: In *Group Theoretical Methods in Physics*, ed. by W. Beiglböck, A. Böhm, E. Takasugi (Springer, Berlin 1979)
13. F. S. Chang: private communication
14. K. K. Mon, J. B. French: Ann. Phys. (N.Y.) 95, 90 (1975)
15. J. B. French, S. S. M. Wong: Phys. Lett. 35B, 5 (1971)
16. H. Cramér: *Mathematical Methods of Statistics* (Princeton Univ. Press, Princeton, N.J. 1946)
17. W. Feller: *An Introduction to Probability Theory and its Applications; Vol. 2* (Wiley, N.Y. 1966)
18. F. J. Dyson: Jour. Math. Phys. 3, 166 (1962)
19. J. P. Draayer, J. B. French, S. S. M. Wong: Ann. Phys. (N.Y.) 106, 472 (1977); F.S.Chang, J.B.French, Phys. Lett. 44B, 131 (1973)

20. J. B. French: In *Nuclear Structure*, ed. by A. Hossain, Harun-ar-Raschid, M. Islam (North-Holland, Amsterdam 1967)
21. J. B. French: In *Dynamical Structure of Nuclear States*, ed. by D. J. Rowe, T. W. Donnelly, L. E. H. Trainor, S. S. M. Wong (Univ. Toronto Press, Toronto 1972)
22. C. M. Vincent: Phys. Rev. 163, 1044 (1967); F. S. Chang, J. B. French, T. H. Thio: Ann. Phys. (N.Y.) 66, 137 (1971)
23. F. S. Chang, J. B. French: Phys. Lett. 44B, 135 (1973); V. Potbhare, S. P. Pandya, Nucl. Phys. A256, 253 (1976); V. K. B. Kota, V. Potbhare: Nucl. Phys. A331, 93 (1979)
24. S. Y. Li: unpublished manuscript
25. J. N. Ginocchio: Phys. Rev. C8, 135 (1973); S. Ayik, J.Ginocchio: Nucl. Phys. A221, 285 (1974)
26. B. D. Chang, S. S. M. Wong: Nucl. Phys. A294, 19 (1978)
27. B. D. Chang: Nucl. Phys. A304, 127 (1978)
28. C. Quesne: Jour. Math. Phys. 16, 2427 (1975)
29. J. C. Parikh: Ann. Phys. (N.Y.) 76, 202 (1973)
30. C. Quesne: Invited Talk at the Ames Conference [49]
31. C. E. Porter: *Statistical Theories of Spectra: Fluctuations* (Academic Press, N.Y. 1965)
32. G. E. Mitchell: Invited Talk at the Ames Conference [49]
33. J. Flores, P. A. Mello: Rev. Mex. Fis. 22, 185 (1973); T. A. Brody, E. Cota, J. Flores, P. A. Mello: Nucl. Phys. A259, 87 (1976)
34. J. B. French, P. A. Mello, A. Pandey: Ann. Phys. (N.Y.) 113, 277 (1978)
35. F. J. Dyson, M. L. Mehta: Jour. Math. Phys. 4, 701 (1963)
36. J. von Neumann, E. Wigner: Phys. Zeitschr. 30, 467 (1929)
37. E. P. Wigner: *Gatlinberg Conf. on Neutron Physics*, ORNL Report #2309 (1957); L. D. Landau and Ya. Smorodinski: *Lectures on the Theory of the Atomic Nucleus* (Consultants Bureau, N.Y. 1958). See also SIAM Review [7]
38. M. L. Mehta: *Random Matrices and the Statistical Theory of Energy Levels* (Academic Press, N.Y. 1967)
39. T. A. Brody, J. Flores, J. B. French, P. A. Mello, A. Pandey, S. S. M. Wong: to be published
40. J. Gunson: Jour. Math. Phys. 3, 752 (1962)
41. F. J. Dyson: Jour. Math. Phys. 3, 140 (1962)
42. F. J. Dyson: Jour. Math. Phys. 3, 1199 (1962)
43. R. Balian: Nuovo Cimento 57B, 183 (1968)
44. N. S. Krylov: *Works on the Foundations of Statistical Physics* (Princeton Univ. Press, Princeton, N.Y. 1979)

45. C. Mahaux, H. A. Weidenmüller: In *Annual Review of Nuclear and Particle Science, Vol. 29* (Annual Reviews Inc., Palo Alto 1979)

46. A. M. Yaglom: *An Introduction to the Theory of Stationary Random Functions* (Prentice-Hall, Englewood Cliffs, N.Y. 1962)

47. C. E. Porter, R. G. Thomas: Phys. Rev. 104, 483 (1956)

48. J. B. French, A. Pandey: to be published

49. Proceedings of Int. Conf. held at Ames, Iowa (Sept. 1979) *Theory and Applications of Moment Methods in Many-Fermion Systems - Spectral Distribution Methods*, ed. by B. J. Dalton, J. P. Vary, S. Williams (to be published by Plenum Press)

50. J. P. Vary: Invited Talk at the Ames Conference [49]

51. J. C. Parikh: Pramana 10, 47 (1978); A. Manare, unpublished

52. K. Kar: Ph.D. Thesis, Univ. of Rochester (1979)

53. K. Takahashi, M. Yamada: Prog. Theor. Phys. 41, 1470 (1969)

54. T. R. Halemane: Ph.D. Thesis, Univ. of Rochester (1979)

55. C. Quesne, S. Spitz: Ann. Phys. (N.Y.) 85, 115 (1974); 112, 304 (1978)

56. R. U. Haq, J. C. Parikh: Nucl. Phys. A220, 349 (1974); K. T. Hecht, J. P. Draayer: Nucl. Phys. A223, 285 (1974)

57. T. R. Halemane, K. Kar, J. P. Draayer: Nucl. Phys. A311, 301 (1978)

58. K. T. Hecht: In *Annual Reviews of Nuclear Science; Vol. 23* (Annual Review, Inc., Palo Alto 1973); C. Quesne, S. Spitz: Jour. Phys. Lett. 38, L237 (1977); J. C. Parikh: private communication (1979)

59. S. F. Edwards, R. C. Jones: J. Phys. A9, 1595 (1976); R. C. Jones, J. M. Kosterlitz, D. J. Thouless: J. Phys. A 11, L45 (1978); A. Pandey, J. B. French: J. Phys. A 12, L83 (1979)

Appendix

NUCLEAR STRUCTURE PUZZLES

G. Bertsch
Michigan State University
East Lansing, Michigan

and

L. Zamick and A. Mekjian
Department of Physics
Rutgers University
New Brunswick, New Jersey

Starting from a nuclear Hamiltonian based on realistic interactions, we can in principle calculate all of the properties of nuclei. But sometimes the theoretical expectation is contradicted by the empirical evidence. We collect below a number of such puzzles. Several of these have already survived a decade of scrutiny and theoretical effort to understand them.

7.1 The Coulomb Energy Problem

In 1969 NOLEN and SCHIFFER [1,2] surprised us by noting that there was about an 8% discrepancy in the calculation of the mass difference of mirror nuclei, e.g. ^{41}Ca - ^{41}Sc, or of the analog-parent differences, e.g. ^{208}Bi - ^{208}Pb. The experimental mass difference of ^{41}Sc and ^{41}Ca is 7.28 MeV. A careful Hartree-Fock calculation by NEGELE [3] yielded only 6.70 MeV, leaving a .58 MeV discrepancy. This theoretical value included direct Coulomb (6.97 keV), exchange Coulomb, finite proton size, spin orbit and vacuum polarization. Since then there has been considerable controversy about the origin of this anomaly, with numerous mechanisms advocated for the resolution of the problem. These include:

1) Smaller Valence Orbits. NOLEN and SCHIFFER [1,2] suggested that the radii of the valence orbits (or neutron excess) should be smaller than the values obtained with the usual Wood-Saxon or Hartree-Fock one-body potentials. Recent elastic magnetic scattering experiments by SICK and others [5] seem to support this conclusion. However, the inclusion of exchange currents in the magnetic scattering calculations gives the same effect on the cross sections as using smaller valence orbits [6]. On the theoretical side, attempts have been made to justify smaller orbits using the velocity- and energy-dependence of the one-body potential, but these have not been successful.

2) Core Polarization. AUERBACH, KAHANA and WENESER [4] suggested that the valence neutron polarize the core, increasing the proton core radius relative to the core neutron radius. This would have exactly the same effect on the Coulomb energy difference as a smaller valence orbit radius. Hartree-Fock theories based on density-dependent inter-

actions give an effect of the correct sign. The magnitude of the effect depends on the interaction that is used. Using the usual range of Skyrme interactions, one can account for 0-250 keV of the 580 keV discrepancy in $^{41}Ca - ^{41}Sc$.

3) Charge-Symmetry Violating Interaction. It has been proposed that there is a short range charge symmetry breaking interaction [7,8]. This is supported by the fact that the mass difference of $^3He - ^3H$ cannot be accounted for by the Coulomb interaction, indeed the discrepancy is about 100 keV. This suggestion is probably correct and would help to resolve the discrepancy in heavy nuclei. However, the microscopic origin of this interaction has not yet been found, within conventional meson-exchange physics. Indeed, the origin of the interaction may be buried in whatever hadronic processes that give the neutron-proton mass difference.

It would help resolve this discussion to measure the difference in proton and neutron radii. For the nucleus ^{48}Ca, the charge radius has been compared with the matter radius deduced from 1 GeV elastic proton scattering [9-11]. The errors are large enough to make a firm conclusion impossible, but it seems that the Hartree-Fock theory gives a reasonable account of the radius difference between neutrons and protons. We should also mention here a pion scattering experiment [12], which indicates a smaller radius difference than Hartree-Fock theory predicts.

A related question to the Nolen-Schiffer anomaly is the isospin mixing between nearby nuclear states [13-17], for example the two 1^+ states in ^{12}C at 12.71 MeV and 15.11 MeV. The observed isospin mixing is too large to be explained with a Coulomb interaction within the p^n shell configurations. However, the levels are nearly unbound and the Coulomb distortion of the single-particle wavefunctions can produce enough isospin mixing to explain the data [18-20].

References

1. J.A. Nolen and J.P. Schiffer, Phys. Lett. 29B (1969) 396.
2. J.A. Nolen and J.P. Schiffer, Ann. Rev. Nucl. Sci. 19 (1969) 471.
3. J.W. Negele, Phys. Rev. C1 (1970) 1260.
4. E.H. Auerbach, S. Kahana and J. Weneser, Phys. Rev. Lett. 23 (1969) 1253.
5. I. Sick, et al., Phys. Rev. Lett. 38 (1977) 1259.
6. A. Arima, et al., Phys. Rev. Lett. 40 (1978) 1001; J. Dubach, Phys. Lett. 81B (1979) 124.
7. J.W. Negele, Nucl. Phys. A165 (1971) 305.
8. H. Sato, Nucl. Phys. A269 (1976) 378.
9. J. Negele, G.K. Varma and L. Zamick, Comments in Nuclear and Particle Physics, to be published.

References (continued)

10. G.K. Varma and L. Zamick, Nucl. Phys. A306 (1978) 343.
11. S. Shlomo and R. Schaeffer, Phys. Lett. 83B (1979) 5.
12. M.J. Jakobson, et al., Phys. Rev. Lett. 38 (1977) 1201.
13. J.M. Lind, et al., Nucl. Phys. A276 (1977) 25.
14. E.G. Adelberger, et al., Phys. Lett. 62B (1976) 29; Phys. Rev. C15 (1977) 484.
15. F.D. Reisman, et al., Nucl. Phys. A153 (1970) 244.
16. W.J. Braithwaite, et al., Phys. Rev. Lett. 29 (1972) 376.
17. C.L. Morris, et al., Phys. Lett. 86B (1979) 31.
18. H. Sato and L. Zamick, Phys. Lett. 70B (1977) 285.
19. F.C. Barker, Aust. J. Phys. 31 (1978) 27.
20. H.C. Wagner and S. Shlomo, Zeit. der Physik, A285 (1978) 283.

7.2 Spin-Orbit Interaction

The problem of accounting for the one-body spin-orbit interaction in nuclei has been a long and outstanding one. The importance of the spin-orbit interaction was realized in 1949 by MAYER [1] and JENSEN [2] who showed that it played a key role in the shell structure of nuclei. The spin-orbit interaction was also incorporated with the optical potential description of nucleon-nucleus scattering where it gives rise to polarization phenomena. Despite its successes in phenomenological descriptions of nuclei, the origin of this interaction based on the underlying two nucleon potential embedded in a many-body framework still remains a puzzle. Some recent attempts at understanding its origin will now be given.

First, the average one-body spin-orbit interaction has a simple phenomenological form,

$$V_{LS}(r) = V_{LS}(\vec{\ell}\cdot\vec{s})\ \frac{1}{r}\frac{d}{dr}\ \rho(r)\,. \qquad (1)$$

The $\rho(r)$ is the nuclear density. The average behavior of Eq. (1) is to give rise to splitting between two different j states of a given ℓ which is $\Delta\varepsilon_{\ell s} = -20(\vec{\ell}\cdot\vec{s})\ \mathrm{MeV}^{2/3}$, with the higher j-state, $j_+ = \ell + 1/2$, lying lower in energy than the lower j-state, $j_- = \ell - 1/2$. In the following discussions, calculations will be considered for two types of closed shell nuclei. One type is called spin-saturated and these nuclei have both j_+, j_- levels fully occupied or fully unoccupied. Examples of spin-saturated nuclei are ^{16}O and ^{40}Ca. On the other hand, a spin-unsaturated nucleus is a system with one level, j_+, fully occupied and the other, j_-, empty. Examples of spin-unsaturated nuclei are ^{48}Ca and ^{208}Pb.

The first calculations of the spin-orbit interaction based on the Brueckner theory of the effective interaction were made by WONG [3]. His spin-orbit splitting was too weak for spin-saturated nuclei, but even worse, it had the wrong sign for spin-unsaturated nuclei. More recently this has been studied by SCHEERBAUM [4]. For spin-saturated nuclei, Scheerbaum finds that both the two-body spin-orbit force and the two-body tensor force give substantial contributions, and the total agrees with experiment in mass 15. This conclusion also follows from simpler considerations based on the nucleon-nucleon scattering phase shifts directly [5]. Thus the spin-saturated nuclei seem to be well understood, except for one relatively minor point. This is the relative splitting of particle and hole orbits at the closed shell. Theoretically, the particle orbit has the larger splitting, due to the higher ℓ. Experimentally, it is the other way around near mass ^{16}O: the $p_{3/2} - p_{1/2}$ splitting is 6.3 MeV, while the $d_{5/2} - d_{3/2}$ splitting is 5.4 MeV.

In spin-unsaturated nuclei, the very serious problem persists with the spin-orbit splitting predicted to be negative [6]. The exchange parts of the central and tensor interactions give rise to this effect in Scheerbaum's calculation. This fallacious prediction of nuclear theory has been confirmed by GOODMAN and BORYSOWICZ [7], who calculate the spin-orbit splitting of the $h_{11/2} - h_{9/2}$ orbits as a function of mass in nuclei near ^{208}Pb. The filling of spin-saturated shells increases this splitting, while the filling of the spin-unsaturated $i_{13/2}$ decreses the splitting. GOODMAN [8] finds support in the experimental data for the predicted dependence on mass, but not of course for the overall magnitude of the splitting.

No obvious suggestions for resolving this discrepancy come to mind. It would be interesting to see what conclusions would follow from more modern interactions, which have weak tensor force, such as the Paris potential [9].

References

1. M.G. Mayer, Phys. Rev. 75 (1949) 1969.
2. O. Haxel, J.H.D. Jensen and H.E. Suess, Phys. Rev. 75 (1949) 1766.
3. C.W. Wong, Nucl. Phys. A108 (1968) 481.
4. R.R. Scheerbaum, Phys. Lett. 61B (1976) 151.
5. J.P. Elliott, et al., Phys. Lett. 24B (1967) 358.
6. R.R. Scheerbaum, Phys. Lett. 63B (1976) 381.
7. A.L. Goodman and J. Borysowicz, Nucl. Phys. A295 (1978) 333.
8. A.L. Goodman, Nucl. Phys. A287 (1977) 1.
9. R. Vinh Mau, "Mesons in Nuclei", ed. M. Rho and D. Wilkinson, Vol. I, (1979) 151.

7.3 Coriolis Interaction

The structure of deformed nuclei is understood in terms of independent particle motion in a rotating frame. The description of a particle(s) plus rotor requires in the Hamiltonian the Coriolis interaction,

$$H_c = -\frac{h^2}{2\Theta}(j_+I_- + j_-I_+)$$

where Θ is the moment of inertia of the even-even core. This interaction makes its presence felt in the spectra, by the decoupling of the $K=\frac{1}{2}$ bands, and as was seen in Faessler's lectures, by the backbending phenomenon. Other aspects of the structure are not so easy to reproduce with the above H_c. As BOHR and MOTTELSON [1] discuss, H_c couples bands with $\Delta K = 1$, introducing to the quadrupole transition amplitudes a term proportional to the quadrupole moment of the core. However, the empirical quadrupole matrix elements are not so large; agreement can only be obtained if H_c is attenuated by a factor of two [2].

There have been several suggestions for resolving this discrepancy. The pairing correlations will reduce the K admixture to the wavefunction. However, a careful analysis by HAMAMOTO [3] indicates that these correlations are not strong enough to reduce H_c to the empirical value.

A likely resolution of this problem was given by RING, et al. [4] who note that these difficulties with the structure do not appear in the Hartree-Fock-Bogoliubov cranking model, which treats all particles on the same footing. Good spectral fits are obtained in situations which require the attenuated H_c in the particle-rotor model [5]. The apparent fault of H_c in the usual rotor model is that the contribution of the odd particle to the angular momentum is neglected. Apparently much of the total angular momentum is due to the last particle. KREINER [6] has formulated a correction associated with this effect; the attenuation factor in H_c is shown to be

$$\left[1 - \frac{\langle j_x^2 \rangle}{\langle j_x I_x \rangle}\right].$$

The contribution of the particle to the angular momentum, $\langle j_x^2 \rangle$, turns out to be large when evaluated numerically in the cranking model.

References

1. A. Bohr and B. Mottelson, "Nucl. Structure", Vol. 2 (Benjamin, New York, 1975) p. 145 ff.
2. S. Hjorth and W. Klamra, Z. Phys. A283 (1977) 287.
3. I. Hamamoto, International School of Nuclear Physics, "Enrico Fermi", Course 64, (Societa Italiana di Fisica, 1977) p. 234.
4. P. Ring and H. Mang, Phys. Rev. Lett. 33 (1974) 1174.
5. P. Ring, et al., Nucl. Phys. A225 (1974) 141.
6. A. Kreiner, Phys. Rev. Lett. 42 (1979) 829.

7.4 Missing M1 Strength in Heavy Nuclei

The M1 strength is a fundamental property of nuclei that shows, among other things, how the spin symmetry is broken in the ground state. It is not difficult to make a plausible theoretical model for the M1 strength in a nucleus such as ^{208}Pb. Using an unperturbed shell model ground state of ^{208}Pb, the giant M1 strength would reside in a linear combination of two states $[h^{\pi}_{9/2}h^{\pi}_{11/2}{}^{-1}]^{1+}$ and $[i^{\nu}_{11/2}i^{\nu}_{13/2}{}^{-1}]^{1+}$. The particle-hole splitting of these two states, as taken from experiment, are respectively 5.36 and 5.86 MeV. The residual interaction should push both states up in energy, and mix them so that the isovector combination lies higher. Most of the M1 strength would reside in the higher state because the magnetic moment of the neutron and proton are of opposite sign. The magnitude of the residual interaction is <1.5 MeV, as calculated in the Brueckner theory. Thus, all of the M1 strength should lie below 8 MeV excitation.

Experimentally, the search for the M1 strength has been a confusing process of elimination, with supposed M1 states turning out to be E1. By 1978, the only strong state left was one at 7.99 MeV, having 25% of the shell-model strength [1]. Now even the M1 character of this state is doubted [2], leaving no important M1 strength below 9 MeV [1-4]. The state would not be observed if it lies above this energy and is highly fragmented.

That the M1 state should be strongly fragmented is no surprise. It has been known for over a decade that several 2 particle-2 hole 1^+ states can come below the 1^+ isovector state in a shell model calculation [5-7]. Refs. [5] and [6] consider a related problem, the hindrance of Gamow-Teller β decays. For example the hindrance of the decay of the ground state of ^{56}Ni to the 1^+ state in ^{56}Co at 1.72 MeV is due to the fact that this lowest 1^+ $T=1$ state in ^{56}Co has very little amplitude (0.016) of the configuration $f^{\nu}_{f/2}f^{\pi}_{7/2}{}^{-1}$, but is predominantly a 2 particle-2 hole state. The work of LEE and PITTEL [7] is directly concerned with ^{208}Pb.

The two links in the theoretical argument that the strength lies below 8 MeV are the single-particle energies and the residual interaction. The residual interaction is probably not at fault; the usual ideas on the nuclear force are adequate for describing the M1 strength in ^{12}C [8], and for the related topic of the $\sigma\tau_-$ strength in ^{90}Zr [9]. The treatment of single-particle energies is called into question by BROWN and SPETH [10]. The mistake, in their view, is that the calculations were made using experimental single-particle energies, rather than Hartree-Fock energies. Since the Brueckner-Hartree-Fock theory gives an effective mass of about 0.7, and the empirical single particle energies are described with an effective mass of about 1.0, Brown and Speth argue that the single particle energies should be raised by 1/0.7. However, for the M1 states, a larger single-particle splitting would require a stronger spin-orbit potential. We have already seen that theory would like to reduce this quantity. It would be interesting in pursuing this problem, to compare the strengths of the spin-orbit potential as deduced from elastic scattering, and as deduced from the single-particle splittings. Another difficulty with this viewpoint is the necessity of reconciling all of the giant vibration data on the same footing. While the isovector excitations call for a reduced effective mass [11,12], the giant quadrupole energy is perfectly consistent with an effective mass of 1.

References

1. R.J. Hold and H.E. Jackson, Phys. Rev. Lett. 36 (1976) 244.
2. W. Knüpfer, et al., Phys. Lett. 77B (1978) 367.
3. S. Raman, M. Mizumoto and R.L. Macklin, Phys. Rev. Lett. 39 (1977) 598.
4. R.M. Laszewski, R.J. Hold and H.E. Jackson, Phys. Rev. Lett. 38 (1977) 813.
5. H. Ejiri, K. Ikeda and J.I. Fujita, Phys. Rev. 176 (1968) 1277.
6. P. Goode and L. Zamick, Phys. Rev. Lett. 22 (1969) 958.
7. T.S.H. Lee and S. Pittel, Phys. Rev. C11, (1975) 607.
8. D. Kurath, Phys. Rev. 134 (1964) B1025.
9. R. Doering, et al., Phys. Rev. Lett. 35 (1975) 1691.
10. G.E. Brown and J. Speth, Distribution of Radiative Strength with Excitation Energy: The E1 and M1 Giant Resonances, to be published.
11. T.T.S. Kuo, J. Blomqvist and G.E. Brown, Phys. Lett. 31B (1970) 93.
12. H.C. Lee, Phys. Rev. Lett. 27, (1971) 200.

7.5 Nuclear Level Densities

Beyond a few MeV excitation, we cannot hope to understand the structure of individual states, unless some quantum number, such as isospin or angular momentum singles them out from their neighbors. In fact, we have seen in the lectures of French that we should not expect such states to show individuality. Thus, the only property left to describe is their numbers. The theory of the level density of a Fermi gas gives the following formula [1,2]

$$\rho(N,Z,E) = \frac{6^{\frac{1}{4}}}{12} \frac{g_0}{(g_0E)^{5/4}} \exp[2(aE)^{\frac{1}{2}}]$$

Here E is the excitation energy, and the single-particle level density is

$$g_0 \approx \frac{2}{\pi^2} mk_F V \approx \frac{3}{2} \frac{A}{\varepsilon_F}$$

This is related to the conventional level density parameter a by $a = \frac{\pi^2}{6} g_0$. The theoretical a is

$$a \approx \frac{3}{2} \frac{A}{\varepsilon_F} \frac{\pi^2}{6} \approx \frac{A}{15}.$$

Experimentally, the level density is higher. The neutron resonance data is best fit with

$$a \approx \frac{A}{8} ,$$

twice as large as theory. This factor of two discrepancy is the problem to be resolved.

BOHR and MOTTELSON [3] suggest that the harmonic oscillator level density be used for g_0. Since the valence particles occupy a relatively larger volume in the harmonic oscillator model, this would increase a. However, the concomitant prediction of the oscillator model, that core nucleons occupy a smaller volume, is easily seen to be incorrect by examining density distributions.

Other possible effects which could influence a are:

--effective mass corrections to the single-particle spectrum. This is certainly important in the description of another Fermi system, liquid ^{3}He. But as we noted previously, the empirical effective mass for nucleons is close to 1 at the Fermi surface.

--the contribution of collective states to level densities. SOLOVIEV, et al. [4] have computed level densities in a model assuming that particle-hole phonons make up the elementary excitations. The result is an increase in level density in certain regions. In particular, the presence of 2^+ and 3^- phonon states can increase the level density by a factor of 10. One difficulty of this description is that it does not respect the Pauli Principle, since the Fermion degrees of freedom are replaced by bosons. We therefore expect that it would overestimate the density of levels.

The problem of level densities appears in a particularly acute form at closed shell nuclei. Let us examine that favorite example, ^{208}Pb. Experimentally, the density of 1^- states at 7-8 MeV excitation is 10/MeV [5,6]. The empirical particle-hole energy gap in ^{208}Pb is 4.2 MeV for protons and 3.4 MeV for neutrons. Thus, in the independent particle description, states at 7-8 MeV excitation will be no more complicated than 2 particle-2 hole, and in fact none of them will have quantum numbers 1^-. To include the residual interactions via the phonon description, we note that the lowest phonon in ^{208}Pb is at 2.7 MeV. It is just possible to make odd parity states in the 7-8 MeV range with two or three phonons. However, the number predicted is an order of magnitude lower than observed.

This problem is also seen at lower excitations, with unanticipated intruder states mingling with the expected shell model states. These are typically strongly deformed states, and it is possible that the increased level density at higher energy is associated with deformation degrees of freedom. The presence of rotational bands in nuclei whose ground state is deformed of course will increase the level density [7]. More interesting, the level density of spherical nuclei can be substantially increased by the presence of deformed states at high excitation [8].

References

1. C. Bloch, Phys. Rev. 93 (1954) 1094.
2. J.M.B. Lang and K. LeCouteur, Proc. Phys. Soc. 67A (1954) 586.
3. A. Bohr and B. Mottelson, "Nuclear Structure", Vol. 1, W.A. Benjamin, Inc. New York and Amsterdam (1969), p. 188.
4. V.G. Soloviev, Ch. Stayanov and I. Vdovin, Nucl. Phys. A224 (1974) 441.
5. H. Baba, Nucl. Phys. A159 (1970) 625.
6. S. Raman, et al., Phys. Rev. Lett. 39 (1977) 813.
7. T. Døssing and A. Jensen, Nucl. Phys. A222 (1974) 493.
8. B.C. Smith, et al., Phys. Rev. C17 (1978) 318.

7.6 Spin Modes

It is only recently that the spin degrees of freedom of the nucleus could be studied as carefully as the charge and density. This has been brought about by better electron scattering experiments, and the availability of intermediate energy protons and pions as nuclear probes.

On the theoretical side, our ignorance is considerable, as we saw in the discussion of the M1 strength. In the words of MOTTELSON [1],

> "At the present time we are only beginning to gain some understanding of the collective modes associated with spin-dependent fields. We are in fact overwhelmed by questions which we cannot properly answer; such as, do the pair of modes $[(\sigma Y_{\lambda-1})^{\lambda}$ and $(\sigma Y_{\lambda+1})^{\lambda}]$ in fact mix strongly? What are the strengths (even the signs) of the collective potentials associated with deformations having these spin dependent structures? Are there incipient instabilities associated with some of these models? (e.g. pion condensates should appear in an analysis of $\tau = 0$, $\kappa = 1$, $\lambda = 0$ modes) etc."

The shell model has particular difficulties with spin densities at high momentum transfer. One interesting example is the elastic scattering of electrons from ^{17}O, measuring the magnetic density [2]. In the shell model this nucleus consists of a $d_{5/2}$ nucleon plus an ^{16}O core. The magnetic moment of ^{17}O is very close to the Schmidt moment and in the early days, this was used as evidence for the validity of the shell model. However, the spin density deviates significantly from the shell model for momentum transfers in the range $q = 1 - 3\ fm^{-1}$. At about $q = 1$ there is a big dip in the experimental cross section, which is called M3 suppression. This can probably be explained by core polarization [3]. Roughly speaking, if we consider the operator $Q_z \sigma_z$ (part of the M3 operator at low momentum transfer), then its expectation value in the ^{16}O core is $(Q_\uparrow - Q_\downarrow)$ where $Q_\uparrow$ is the quadrupole moment of the spin up particles etc. Thus to explain the suppression it is necessary that spin up nucleons have a different quadrupole moment than spin down particles. The M5 region is quite confusing; on the low momentum transfer side the experimental form factor is smaller than the shell model value, on the high momentum transfer side it is larger.

Shell model calculations in the $(Op)^8$ basis predict a smaller cross section at this momentum transfer by an order of magnitude [7]. This is quite surprising, since the low momentum transfer, which relates to the magnetic moment, is quite well fit by such calculations [6]. DUBACH and HAXTON [7] show that a transition density can be constructed within the $(Op)^8$ basis that fits the data. Whether this transition density can be derived from a reasonable Hamiltonian remains to be seen.

There are two 'explanations' of the second peaks which may not be as different as they sound. One is a straight core polarization calculation by SUZUKI, et al. [8], in which it is shown that it is essential to use a tensor interaction. In their calculation, exchange currents are included but do not seem to play an important role.

A more dramatic explanation is afforded by M. ERICSON [9]. She cites the experiment as possible evidence for a precursor to pion condensation. The summed strength of all 1^+ excitations in ^{12}C is equal to the ground state expectation value of the square of the M1(q) operator, and hence is a measure of the fluctuation of this operator. The ^{12}C nucleus in this picture fluctuates between the normal state and the pion condensate state in which the nucleus spins are aligned as in an antiferromagnetic, presumably with a characteristic distance d separating the spin up and spin down nucleus. The peak in the inelastic scattering at $q = 2\ fm^{-1}$ may mean that d is of the order of $1/q$. Ericson sees a strong analogy between the secondary peak and critical opalescence.

It has been suggested by TOKI and WEISE [10] and FAYANS, et al. [11] that perhaps proton inelastic scattering is the more relevant probe of critical opalescence rather than electron scattering. The Fourier transform of the tensor interaction is proportional to $\sigma_1 \cdot q\ \sigma_2 \cdot q$. This interaction may be responsible for spin coherence. The electromagnetic scattering goes as $\vec{\sigma} \times \vec{q} \cdot \varepsilon$. It should be noted that a backward peak is seen in proton inelastic scattering to both the $T = 1, 1^+$ state at 15.11 MeV and the $T = 0, 1^+$ state at 12.7 MeV [12].

References

1. B.R. Mottelson, International School of Nuclear Physics, "Enrico Fermi", Course 64, (Societa Italiana di Fisica, 1977), p. 46.
2. M.V. Hynes, et al., Phys. Rev. Lett. 42 (1979) 1444.
3. A. Arima, Y. Horikawa, H. Hyuga and T. Suzuki, Phys. Rev. Lett. 40 (1978) 1001; L. Zamick, Phys. Rev. Lett. 40 (1978) 381.
4. I. Sick, et al., Phys. Rev. Lett. 38 (1977) 1259.
5. A. Yamaguchi, et al., Phys. Rev. C3 (1971) 1750.
6. S. Cohen and D. Kurath, Nucl. Phys. 73 (1965) 1.
7. J. Dubach and W.C. Haxton, Phys. Rev. Lett. 41 (1978) 1453.
8. H. Sagawa, T. Suzuki, H. Hyuga and A. Arima, Nucl. Phys. A322 (1979) 361.
9. M. Ericson and J. Delorme, Phys. Lett. 76B (1978) 241; M. Ericson, Proceedings 1979 Vancouver Meeting on High Energy Physics and Nuclear Structure, to be published.
10. H. Toki and W. Weise, Phys. Rev. Letters 42 (1979) 1034.
11. S.A. Fayans, E.E. Saperstein and S.V. Tolokonnikov, Nucl. Phys. A326 (1979) 463.
12. J. Comfort, to be published.

Selected Issues from

Lecture Notes in Mathematics

Lecture Notes in Physics

Vol. 91: Computing Methods in Applied Sciences and Engineering, 1977, II. Proceedings, 1977. Edited by R. Glowinski and J. L. Lions. VI, 359 pages. 1979.

Vol. 92: Nuclear Interactions. Proceedings, 1978. Edited by B. A. Robson. XXIV, 507 pages. 1979.

Vol. 93: Stochastic Behavior in Classical and Quantum Hamiltonian Systems. Proceedings, 1977. Edited by G. Casati and J. Ford. VI, 375 pages. 1979.

Vol. 94: Group Theoretical Methods in Physics. Proceedings, 1978. Edited by W. Beiglböck, A. Böhm and E. Takasugi. XIII, 540 pages. 1979.

Vol. 95: Quasi One-Dimensional Conductors I. Proceedings, 1978. Edited by S. Barišić, A. Bjeliš, J. R. Cooper and B. Leontič. X, 371 pages. 1979.

Vol. 96: Quasi One-Dimensional Conductors II. Proceedings 1978. Edited by S. Barišić, A. Bjeliš, J. R. Cooper and B. Leontić. XII, 461 pages. 1979.

Vol. 97: Hughston, Twistors and Particles. VIII, 153 pages. 1979.

Vol. 98: Nonlinear Problems in Theoretical Physics. Proceedings, 1978. Edited by A. F. Rañada. X, 216 pages. 1979.

Vol. 99: M. Drieschner, Voraussage – Wahrscheinlichkeit – Objekt. XI, 308 Seiten. 1979.

Vol. 100: Einstein Symposion Berlin. Proceedings 1979. Edited by H. Nelkowski et al. VIII, 550 pages. 1979.

Vol. 101: A. Martin-Löf, Statistical Mechanics and the Foundations of Thermodynamics. V, 120 pages. 1979.

Vol. 102: H. Hora, Nonlinear Plasma Dynamics at Laser Irradiation. VIII, 242 pages. 1979.

Vol. 103: P. A. Martin, Modèles en Mécanique Statistique des Processus Irréversibles. IV, 134 pages. 1979.

Vol. 104: Dynamical Critical Phenomena and Related Topics. Proceedings, 1979. Edited by Ch. P. Enz. XII, 390 pages. 1979.

Vol. 105: Dynamics and Instability of Fluid Interfaces. Proceedings, 1978. Edited by T. S. Sørensen. V, 315 pages. 1979.

Vol. 106: Feynman Path Integrals, Proceedings, 1978. Edited by S. Albeverio et al. XI, 451 pages. 1979.

Vol. 107: J. Kijowski, W. M. Tulczyjew, A Symplectic Framework for Field Theories. IV, 257 pages. 1979.

Vol. 108: Nuclear Physics with Electromagnetic Interactions. Proceedings, 1979. Edited by H. Arenhövel and D. Drechsel. IX, 509 pages. 1979.

Vol. 109: Physics of the Expanding Universe. Proceedings, 1978. Edited by M. Demiański. V, 210 pages. 1979.

Vol. 110: D. A. Park, Classical Dynamics and Its Quantum Analogues. VIII, 339 pages. 1979.

Vol. 111: H.-J. Schmidt, Axiomatic Characterization of Physical Geometry. V, 163 pages. 1979.

Vol. 112: Imaging Processes and Coherence in Physics. Proceedings, 1979. Edited by M. Schlenker et al. XIX, 577 pages. 1980.

Vol. 113: Recent Advances in the Quantum Theory of Polymers. Proceedings 1979. Edited by J.-M. André et al. V, 306 pages. 1980.

Vol. 114: Stellar Turbulence. Proceedings, 1979. Edited by D. F. Gray and J. L. Linsky. IX, 308 pages. 1980.

Vol. 115: Modern Trends in the Theory of Condensed Matter. Proceedings, 1979. Edited by A. Pekalski and J. A. Przystawa. IX, 597 pages. 1980.

Vol. 116: Mathematical Problems in Theoretical Physics. Proceedings, 1979. Edited by K. Osterwalder. VIII, 412 pages. 1980.

Vol. 117: Deep-Inelastic and Fusion Reactions with Heavy Ions. Proceedings, 1979. Edited by W. von Oertzen. XIII, 394 pages. 1980.

Vol. 118: Quantum Chromodynamics. Proceedings, 1979. Edited by J. L. Alonso and R. Tarrach. IX, 424 pages. 1980.

Vol. 119: Nuclear Spectroscopy. Proceedings, 1979. Edited by G. F. Bertsch and D. Kurath. VII, 250 pages. 1980.